D0829099

Contributors

Kenneth E. Boulding, *Department of Economics, University of Colorado*

John Cobb, *School of Theology at Claremont, California*

Herman E. Daly, *The World Bank*

Anne H. Ehrlich, *Department of Biological Sciences, Stanford University*

Paul R. Ehrlich, *Department of Biological Sciences, Stanford University*

Nicholas Georgescu-Roegen, *Department of Economics, Vanderbilt University*

Garrett Hardin, *Department of Biology, University of California, Santa Barbara*

John P. Holdren, *Energy and Resources Group, University of California, Berkeley*

M. King Hubbert (1903–1989), *formerly of the United States Geological Survey*

C. S. Lewis (1898–1963), *former Professor of Medieval and Renaissance Literature at Cambridge University*

E. F. Schumacher (1911–1977), *former economist, Surrey, England*

Gerald Alonzo Smith, *Department of Economics, Mankato State University*

T. H. Tietenberg, *Department of Economics, Colby College*

Kenneth N. Townsend, *Department of Economics, Hampden-Sydney College*

Valuing the Earth

Economics, Ecology, Ethics

edited by Herman E. Daly
and Kenneth N. Townsend

The MIT Press
Cambridge, Massachusetts
London, England

Fourth Printing, 1993

This book was set in 10/13 Palatino by DEKR Corp. and was printed and bound in the United States of America.

Library of Congress Cataloging-in-Publication Data

Valuing the earth : economics, ecology, ethics / edited by Herman E. Daly and Kenneth N. Townsend.
p. cm.
Includes bibliographical references and index.
ISBN 0-262-04133-2 (hc). — ISBN 0-262-54068-1 (pb)
1. Economic development—Environmental aspects. 2. Environmental policy—Moral and ethical aspects. 3. Environmental protection—Moral and ethical aspects. 4. Human ecology—Moral and ethical aspects. I. Daly, Herman E. II. Townsend, Kenneth N. (Kenneth Neal), 1955–
HD75.6.V36 1993
333.7'2—dc20 92-22098
CIP

Contents

Preface

This book originated as a signal of an emerging paradigm shift in economics. Appearing in 1973 under the title *Toward a Steady-State Economy,* it explored changing interpretations of the usefulness of the prescription of continued economic growth as a panacea for problems originating in underdevelopment and maldistribution of wealth. In its original form this book attempted to demonstrate two things: that economic growth is constrained, ultimately, by a general, inescapable scarcity of both source and sink environmental resources, as well as by thermodynamic limits to investments in technical efficiencies of resource utilization; and that these constraints to economic growth matter to people who daily choose patterns of resource consumption/wealth accumulation/pollution. The *modus operandi* of the original effort was the examination of economics, in light of inescapable general scarcity, from the vantage points of ecological science, ethics, and economics proper. The result was a collection of essays, drawn from many and varied points of view, that pointed to the desirability of rehabilitating the notion, popular with classical economists from Malthus to Mill, that an economic *stationary state* is not only inevitable but, given the grim consequences of a slavish adherence to the twentieth-century dictum that healthy economies are growing economies, desirable as well.

Toward a Steady-State Economy received promising albeit considerably mixed attention from economists and others interested in economic growth, generating controversy. A few years later, what began as an effort to revise the work produced sufficient change in the collection of essays to merit a new title. In 1980 *Economics, Ecology, Ethics: Essays toward a Steady-State Economy* offered additional contributions that served to update the evolution of the limits-to-growth controversy, as well as evidence of the acknowledgment of the emerging paradigm in economics literature.

The current text continues the work of *Economics, Ecology, Ethics.* Many of the essays from the two previous collections have been retained, in view of their seminal significance for the development of the steady-state economy argument. At the margin, essays have been added that track the continued evolution of the steady-state paradigm. In particular, the shift in focus from attempts to economize at the margin of source utilization to efforts to economize on exploitation of sinks has necessitated both the inclusion of new material and, regrettably, the omission of several fine essays. What remains is the message that, beyond some point—perhaps soon for developed countries, and ultimately for all countries—economic growth is both physically and economically unsustainable, as well as morally undesirable.

Herman E. Daly
The World Bank
Washington, D.C.

Kenneth N. Townsend
Hampden-Sydney College
Hampden-Sydney, Virginia

Valuing the Earth

Introduction

Herman E. Daly and
Kenneth N. Townsend

The basic rule that for every independent policy goal we must have an independent policy instrument has been emphasized by Professor Jan Tinbergen (*On the Theory of Economic Policy,* Amsterdam: North Holland Press, 1952) but seems to have been forgotten in recent discussion. Yet we all recognize that "you can't kill two birds with one stone," at least not if the birds are flying independently. If they are flying in tandem or sitting on the same fence, then one might manage to do it. In economic theory today we are trying to kill three birds with two stones. This book argues, among other things, that we need a third stone because the birds are flying independently. The "birds" are the three goals of allocation, distribution, and scale. The first two have a long history in economic theory and have their two specific independent policy instruments. The third, scale, has not yet been formally recognized and has no corresponding policy instrument.

Practice, however, responding to real problems, has moved ahead of theory by implicitly recognizing scale, as well as allocation and distribution, and forcing a clear distinction among the three. The practical context in which this has happened is the policy of tradable pollution permits (discussed in part III).

Allocation refers to the relative division of the resource flow among alternative product uses—how much goes to production of cars, to shoes, to plows, to teapots, etc. A good allocation is one that is efficient, i.e., that allocates resources among product uses in conformity with individual preferences as weighted by the ability of the individual to pay. The policy instrument that brings about an efficient

Adapted in part from "Towards an Environmental Macroeconomics," *Land Economics,* 67, no. 2 (May 1991), 255–259, by permission of the publisher.

allocation is relative prices determined by supply and demand in competitive markets. Allocative issues are dealt with in part III of this collection.

Distribution refers to the relative division of the resource flow, as embodied in final goods and services, among alternative people: how much goes to you, to me, to others, to future generations, and how much is reserved for other species with whom we share the planet. A good distribution is one that is just or fair—or at least one in which the degree of inequality is limited within some acceptable range. The policy instrument for bringing about a more just distribution is transfers—taxes and welfare payments. Issues related to distribution are mainly found in part II but also in part III.

Scale refers to the physical volume of the flow of matter-energy from the environment as low-entropy raw materials and back to the environment as high-entropy wastes. It may be thought of as the product of population times per capita resource use. It is measured in absolute physical units, but its significance is relative to the natural capacities of the ecosystem to regenerate the inputs and absorb the waste outputs on a sustainable basis. The economy is viewed as a subsystem of the larger, but finite and nongrowing, ecosystem, and its scale is significant relative to the fixed size of the ecosystem. A good scale is one that is at least sustainable, that does not erode environmental carrying capacity over time. An optimal scale is one that is at least sustainable, but beyond that it is one at which we have not yet sacrificed ecosystem services that are at present worth more at the margin than the production benefits derived from the growth in the scale of resource use. Scale issues are the focus of part I but appear also in part III.

Economic theory has abstracted from scale in two rather opposite ways. First, it has assumed that environmental sources and sinks are infinite relative to the scale of the economic subsystem. Second, by assuming that scale is total rather than infinitesimal, i.e., that nature is just one more sector like agriculture or industry, and that each micro allocative decision for each resource includes the *in natura* use among the set of alternative uses. Consequently, in this view there is no separate macro issue of scale, and no policy instrument for the control of scale is needed.

To the extent that our policy instruments do affect scale, e.g., through growth-stimulating macro policies, the consequence is nearly always to expand scale, which of course creates no problem if sources and sinks are infinite. But scale has become important because the economic subsystem has grown to the point that its physical demands on the ecosystem are far from trivial. We have moved from a relatively empty world to a relatively full world from the point of view of human beings. Since scale can no longer be considered infinitesimal, its dismissal now rests on the view that it is total, and that the ecosystem is not the containing natural matrix of the economy but just one more sector within the all-inclusive economy waiting for its due allocation of resources according to individual willingness to pay for its service or product. To put it starkly, in the neoclassical view the economy contains the ecosystem; in the steady-state view the ecosystem contains the economy, to which it supplies a throughput of matter-energy taken from *in natura* uses according to some rule of sustainable yield rather than according to individual willingness to pay. This difference in view is rather like the difference between Ptolemy and Copernicus—is the earth or the sun the center of the universe? If the earth is the center we have to deal with too many epicycles to "save the appearance" of uniform circular orbits. If the economy is to contain everything then we have to internalize too many externalities to save the appearance of individualistic willingness to pay as our "uniform circular orbit."

The latter approach is capable of yielding many insights, although in our view it is ultimately unsatisfactory. Many insights have, nevertheless, come from the field of environmental economics. This field has evolved quickly from its origins in the 1960s to its present state of development. Along the way it has become essentially a branch of applied microeconomics, motivated by A. C. Pigou's development of the principle of externalities, i.e., that rationally self-interested agents will frequently act so as to "externalize" all or part of the costs associated with an activity, passing to society the brunt of the expenses and thereby producing a misallocation of resources.[1] A further challenge came from Ronald Coase's pioneering work on the claim that such externalities do not present *prima facie* evidence of the need for regulation of private activity.[2] Economists sought to determine the optimal disposition of property rights between polluter and "pollutee" through private bargaining.

The suboptimal treatment of common environmental resources has been a major theme in environmental economics. Suboptimization has two sources: destruction of a joint resource, shared by all, known as *environmental quality,* resulting from man's impact on the environment; and the public provision of services to deal with man's impact on the environment, which may or may not be directly connected with the externalities of environmental destruction. Consequently, much of the theoretical and empirical research in environmental economics has been conducted from the point of view of the economist as discoverer, and agent of restoration, of market failures. This focus can be seen quite clearly in the excellent book by economists William J. Baumol and Wallace E. Oates, which is worth quoting at length:

Man's influence on the quality of the environment depends on two things: the damage he does and the effort devoted to undoing that damage. This statement is hardly more than a tautology. Yet, most discussions in the theoretical literature treat only the first of these activities. Focusing upon the problem of externalities, they examine systematically why man's activities are likely to have deleterious consequences for the quality of life, and why, moreover, they are likely to go beyond anything that can be defended on grounds of economic efficiency. But they typically have not considered the otner side of the matter: the resources devoted to the public services that are designed to improve the quality of life at the same time that external costs continue to eat away at it.[3]

This focus has produced excellent insight into the *microanalytical* nature of environmental problems. Outstanding work abounds in a literature distributed across several excellent journals and many monographs that clearly demonstrates the economist's contribution to a better understanding of how to optimally allocate scarce environmental resources. Formal recognition of the need to treat pollution as an inevitable by-product of economic production, and consequently of the need for a "cradle-to-grave" approach to the production/pollution phenomenon, is clearly seen in the seminal work by Ayers and Kneese outlining what has come to be known as the "materials balance" methodology in environmental economics.[4]

For all the good that has come of the research effort in environmental economics to deal with the externalities phenomenon, there remains a gaping hole in the methodology. This void is reflected in the fact that, whereas microeconomics considers openly the phenomenon of environmental destruction, the field of macroeconomics seems to ignore the environment, even the branch of macroeconomics known as growth theory, which posits the goal of continuous and limitless future economic growth as its leitmotif.[5]

Preanalytic Visions of the World

As a discipline, economics has come under fire for its analytical opportunism. This tendency is exacerbated because economists seek and generally find a market for their ideas in the public forum. That economics has not evolved into either the unified body of received theory envisioned by classical and neoclassical economists, or the positive science of empirical induction called for by nineteenth-century historicists, has been a source of public controversy in an intellectual discipline that, like it or not, impinges upon critical economic and political policy. Some of the controversy originates in the legitimate disagreement among economists of different stripes over methodology. Much of the disagreement can also be traced to what economist Milton Friedman has described as different predictions about the economic consequences of "taking action," rather than from "differences in values, differences about which men can ultimately only fight."[6]

While we strongly disagree with Friedman's view that value differences are ultimately immune to reasoned persuasion and must end in fighting (as evident from part II of this book), we do recognize that many disagreements are not about values but about how the world works. Substantial disagreement between persons considering the desirability of taking action stem from what Joseph Schumpeter termed different "preanalytic visions." As much as anything else, the determinants of the characteristics of a Lakatosian research program or a Kuhnian paradigm in economics are the researcher's well-formed and sometimes *proto*-analytically derived visions of what classical economists dubbed "the nature of things." People bring to an intellectual endeavor all sorts of preconceptions. Schumpeter recognized this intellectual baggage as the result of preanalytic cognitive acts:

> In practice we all start our own research from the work of our predecessors, that is, we hardly ever start from scratch. But suppose we did start from scratch, what are the steps we should have to take? Obviously, in order to be able to posit to ourselves any problems at all, we should first have to visualize a distinct set of coherent phenomena as a worthwhile object of our analytic effort. In other words, analytic effort is of necessity preceded by preanalytic cognitive act that supplies the raw material for the analytic effort. In this book, the preanalytic cognitive act will be called Vision.[7]

In economics the dominant precognitive conception of nature is of a stern physical force that exacts careful, rational behavior from *homo economicus,* owing to nature's imposition of particular scarcities of desirable resources. Orthodox economics abounds with well-reasoned theories that account for the way in which all manner of resources may be rationed among competing claims for them, given particular resource scarcity. Price theory may be understood as the body of received theory accounting for the human response to particular scarcities, which arises from the basic awareness that beyond Eden nature yields resources stingily. In macroeconomics, the common preanalytic vision is the circular flow, in which the economy is viewed as a system isolated from its surrounding environment. According to such a vision, problems associated with resource exhaustion, overpopulation, and overpollution cannot exist—the economy simply does not depend upon a physical environment but rather upon the continuous flow of money income, which is, in theory, boundless. Lacking in the preanalytic visions of many macroeconomists is an awareness of the presence of general resource scarcity, as well as its consequences for growth economies. Many other economists would acknowledge the principle of general scarcity but conceive of the world as being sufficiently large, relative to the scale of the human endeavor, as to render immediate consideration of the consequences of economic growth in a finite environment impractical. Finally, some economists cling to the notion that human ingenuity and technical efficiency provide mankind with an "ultimate resource" that provides a *modus vivendi* for a finite world.[8]

The lesson of much of physical science has been that nature imposes both particular and general scarcities upon resource availability and pollution sink space. A species is always at risk of losing ready access to certain critical resources. Additionally, as populations grow, species risk exceeding the effective carrying capacity of the immediate environment. To be sure, humankind has the ability to substitute new resources for dwindling critical resources, to improve the service yield from reduced supplies of resources generally, and to keep pollution associated with bioeconomic activity to a minimum. One seemingly natural conclusion, which appears to inform a popular vision of "the nature of things," is to imagine that the terrestrial dowry of resources and environmental sink space is adequately large to provide sustenance for even larger populations and economies than now exist. Also, many people seem to bring to discussions of growth economics the preconception that technological advance can occur at rates that exceed the rates at which resources and sinks are exhausted. In either case, the general conclusion seems to be that the world's economy can continue to grow, without disaster, for the foreseeable future.

In contrast with this vision is the preanalytic conception, popular among classical economists from Malthus to Mill, that the processes of growth, both demographic and economic, are bounded by inviolable resource scarcities. Malthus's *An Essay on the Principle of Population* was one of those rare arguments that are accepted without criticism. His claim that population growth proceeds geometrically toward subsistence at carrying capacity was generally regarded as self-evident, leaving its author with little obligation to defend it.[9] When Mill articulated the concept of an economic stationary state, in which the prospect of future economic growth was absent, he based his vision upon the Malthusian view. According to Mill, the human economy would reach a stationary state beyond which economic growth was impossible, leaving human ingenuity the task of improving the quality of lives through arts, culture, and improved distribution of incomes.[10]

An Impossibility Theorem

Contemporaneous with the rise of macroeconomics in the twentieth century has been a preoccupation, both in the capitalist West and the socialist East, with economic growth. Belief in the efficacy of exponential growth of both the human population and human economy doubtless is grounded in the preanalytic vision of nature as a generous benefactor.

Marx viewed nature as providing spontaneous gifts of land, minerals, and energy resources. Belief in the socialist world that such spontaneous gifts are properly the birthright of all resulted in the treatment of the environment in the East as a common resource long after such an arrangement proved ruinous. In the West, overestimation of the magnitude of ecological sources and sinks resulted in overly optimistic projections of sustainable growth in the economy. Whether the normative prescription of economic growth results from a belief in infinitely large sources and sinks or an infinitely expandable technological acumen operating on finite sources and sinks, the growth prescription is ubiquitous.

As authors Paul and Anne Ehrlich, John Holdren, Nicholas Georgescu-Roegen, and M. King Hubbert demonstrate in essays included in part I of the book, however, a finite environment—even one within a seemingly infinite universe—cannot sustain growth indefinitely. Regardless of future technologies that may be applied to transforming energy and matter, no perfect recycling is permitted in our ecosystem. Consequently, it becomes desirable and even necessary to begin to take steps to identify the optimal scale for both the human population and its economy.

The Macroeconomics of Optimal Scale

The term *scale* is shorthand for "the physical scale or size of the human presence in the ecosystem, as measured by population times per capita resource use." Optimal *allocation* of a given scale of resource flow within the economy is one thing (a microeconomic problem). Optimal *scale* of the whole economy relative to the ecosystem is an entirely different problem (a macro-macro problem). The micro allocation problem is analogous to allocating optimally a given amount of weight in a boat. But once the best relative location of weight has been determined, there is still the question of the absolute amount of weight the boat should carry, even when optimally allocated. This absolute optimal scale of load is recognized in the maritime institution of the Plimsoll line. When the watermark hits the Plimsoll line the boat is full, it has reached its safe *carrying capacity.* Of course if the weight is badly allocated the waterline will touch the Plimsoll mark sooner. But eventually, as the absolute load is increased, the watermark will reach the Plimsoll line even for a boat whose load is optimally allocated. Optimally loaded boats will still sink under too much weight—even though they may sink optimally! The major task of environmental macroeconomics is to design an economic institution analogous to the Plimsoll mark—to keep the weight, the absolute scale, of the economy from sinking our biospheric ark.

Microeconomics has not discovered in the price system any built-in tendency to grow only up to the scale of aggregate resource use that is optimal (or even merely sustainable) in its demands on the biosphere. *Optimal scale, like distributive justice, full employment, or price level stability, is a macroeconomic goal.* And it is a goal that is likely to conflict with other macroeconomic goals. The traditional solution to unemployment is growth in production, which means a larger scale. Frequently the solution to inflation is also thought to be growth in real output and a larger scale. And most of all the issue of distributive justice is "finessed" by the claim that aggregate growth will do more for the poor than redistributive measures. Macroeconomic goals tend to conflict, and certainly optimal scale conflicts with any goal that requires further growth, once the optimum has been reached.

Optimal Scale and the Steady State

One measure of the ubiquity of the growthmanic preanalytic vision is humankind's preoccupation with the discovery of boundless, cheap energy. After the conclusion of World War II, the United States government heralded the development of fission-generated electricity as the innovation that would make electricity too cheap and plentiful to meter. More recently, the discovery of superconductivity shows promise of enabling the development of extremely low-loss transmission of electric energy, resulting in fantastic energy efficiencies. Finally, the global attention given the mistaken "discovery" of cold fusion in a bottle reflects our obsession with the ultimate Promethean exercise of power: energy to fuel the human economy for millennia. Yet as the twentieth century draws to a close it becomes clearer and clearer that neither infinite resource availability in the absence of an infinite ecological sink, nor an infinite sink in the face of resource scarcity, permits economic growth to be sustained. Thus exercises in relaxing energy constraints, given nature's finite absorptive capacity for thermal and material waste, become problems rather than solutions to problems. Even the advent of commercially viable fusion energy, which would stretch terrestrial availability of energy resources considerably, not only would fail to eliminate pollution constraints to economic growth, it would actually hasten humankind's progress toward the brink of ecological disaster. Ultimately, only a steady-state economy, one in which population and the stock of exosomatic artifacts are maintained at a constant level with minimal throughput of materials and energy, can be sustained, and even a steady state cannot be sustained forever. Under the discipline of a steady-state economy free energy would be a blessing; within the context of a growth economy it would be a curse.

The purpose of this collection of essays is threefold: to demonstrate the biophysical impossibility of sustainable growth of people and their things, even in an efficient economy; to convince the reader of the moral reasons for voluntarily getting humankind off the growth treadmill; and to provide examples of economic policies associated with a steady-state economy that promise to lengthen potential human history. A steady state is as critical for sustainable development within a socialist state as it is in a capitalist regime.

Ultimately, people must begin to ask themselves difficult questions like: "What is a becoming existence for humankind?" or "Assuming an optimal, efficient microeconomy, how big should the macroeconomy be?" Finally, what is the goal to be served by growth, if it were sustainable? This book takes a look at the moral, biophysical, and economic aspects of macroenvironmental economics and attempts to develop partial answers to the question of how to lead a good life on a finite earth.

Notes

1. A. C. Pigou, *The Economics of Welfare,* 2d ed. (London: Macmillan, 1950).

2. R. H. Coase, "The Problem of Social Cost," *Journal of Law and Economics* 3 (October 1960), 1–44.

3. William J. Baumol and Wallace E. Oates, *The Theory of Environmental Policy: Externalities, Public Outlays, and the Quality of Life* (Englewood Cliffs, N.J.: Prentice-Hall, 1975), p. 1.

4. R. Ayers and A. Kneese, "Production, Consumption, and Externalities," *American Economic Review* 59 (1969), 282–297.

5. The dearth of environmental issues in macroeconomics is explored more fully in Herman E. Daly, "Towards an Environmental Macroeconomics," *Land Economics* 67 (May 1991), 255–259.

6. Milton Friedman, "The Methodology of Positive Economics," *Essays on Positive Economics* (Chicago: University of Chicago Press, 1953), p. 5.

7. Joseph Schumpeter, *History of Economic Analysis* (New York: Oxford University Press, 1954), p. 41.

8. Julian Simon, *The Ultimate Resource* (Princeton, N.J.: Princeton University Press, 1981).

9. T. R. Malthus, *An Essay on the Principle of Population* (London: Pelican Classics, 1970).

10. J. S. Mill, *Principles of Political Economy,* vol. 2 (London: John W. Parker, 1857), pp. 320–326.

Introduction to *Essays toward a Steady-State Economy*

Herman E. Daly

Paradigms in Political Economy

This book is a part of an emerging paradigm shift in political economy. The terms paradigm and *paradigm shift* come from Thomas Kuhn's insightful book *The Structure of Scientific Revolutions,*[1] in which Kuhn explores the ways that entire patterns of thought—a kind of gestalt for which he uses the word *paradigm*—are established and changed. Kuhn contends that paradigm shifts—occasional discontinuous, revolutionary changes in tacitly shared points of view and preconceptions of science—are an integral part of scientific thought. They form the necessary complement to *normal science,* which is what Kuhn calls the day-to-day cumulative building on the past, the puzzle solving, and the refining of models that fit within the paradigm shared by all the scientists of a particular discipline. Indeed, science students are taught to accept the prevailing paradigm so their work will adhere to the same designs, rules, and standards, thus assuring the *cumulative* building of knowledge.

Just as we are unconscious of the lenses in our own eyeglasses until we have trouble seeing clearly, so we are unconscious of paradigms until the clarity of scientific thought becomes blurred by anomaly. Even under the stress of facts that do not seem to fit, paradigms are not easily abandoned. If they were, the cohesion and coherence necessary to form a scientific *community* would be lacking. Most anomalies, after all, do become resolved within the paradigm; they must, if the paradigm is to command the loyalty of scientists. To

Published in part in *The Patient Earth,* ed. John Harte and Robert H. Socolow (New York: Holt, Rinehart and Winston, Inc., 1971). Adapted and reprinted by permission of the editors.

abandon one paradigm in favor of another is to change the entire basis of intellectual community among the scientists within a discipline, which is why Kuhn calls such changes scientific revolutions. Discontinuous with the preceding paradigm, a new paradigm must at first rely on its own criteria for justification, for many of the questions that can be asked and many of the answers that can be found are likely to be absent from the previous paradigm. Indeed, even logical debate between adherents of different paradigms is often very limited, for proponents of two paradigms may not agree on what is a problem and what is a solution.

The history of science contains numerous examples of anomalies that brought crisis to old paradigms and were answered with new ones. Shall we take the earth or the sun as the center of our cosmos? Does a stone swinging on a string represent constrained fall or pendulum motion? Are species fixed or slowly evolving? And problems arise in political economy that may require more than normal puzzle solving. Shall we conceive of economic growth as a permanent normal process of a healthy economy or as a temporary passage from one steady state to another? Shall we take the flow of income or the stock of wealth as the magnitude most directly responsible for the satisfaction of human wants? Shall we conceive of land, labor, and capital as each being productive, and think in terms of three sources of value, or shall we conceive of labor as the only productive factor, the only source of value, and find that land and capital enhance the productivity of labor?

In a way, it all depends on how we want to look at it. And yet, there is far more to it than that. Which point of view is simpler or more appealing aesthetically? Which removes the intellectually or socially most vexing anomalies? Which is likely to suggest the most interesting and fruitful problems for future research? These kinds of criteria are not reducible to logical or factual differences. They involve a gestalt, an element of faith, personal commitment, and values.

That revolutionary paradigm shifts, both large ones and small ones, are historically and logically descriptive of the physical sciences has been admirably shown by Kuhn in his book and by Arthur Koestler in *The Sleepwalkers*.[2] Michael Polanyi takes a related viewpoint in his admirable book *Personal Knowledge*.[3] The focus of all three writers is physical science, and Koestler focuses especially on astronomy. But scientific revolutions characterize all of science, including political economy. Since values are a larger part of social science and also influence the acceptance or rejection of paradigms, such shifts may be even more characteristic of the social sciences.

The history of economic thought brings several such shifts to mind. In the mercantilist paradigm of the Renaissance period, wealth meant precious metal, treasure easily convertible into armies and national power. The way to attain wealth was from mines or from a favorable balance of international trade. The implication of this paradigm was that the way to riches was to devote a nation's labor power to digging up metal that had no other use than as coinage, or to making goods to be given to foreigners in exchange for such minimally useful metal. Moreover, maintaining a surplus balance of trade required low prices on goods exported for sale in competitive markets, which meant low wages to home workers inasmuch as labor was the major cost of production. Making sure that the supply of laborers was large was one means of keeping wages low. The anomalous outcome was that, for a mercantilist nation to be "wealthy," it needed a large number of poor laborers.

The physiocrats of mid-eighteenth-century France—the first economic theorists—tried to explain economics in accordance with natural law and saw agriculture and Mother Earth as the source of all net value. Reproduction of plants and animals provided the paradigm by which all other increase in wealth was understood. Money was sterile. The concept that it "reproduced" through interest was rejected, because it did not fit the paradigm. But the anomaly of interest did not disappear, and the process of tracing all net value back to land became very complex.

The classical economists, witnesses to the problems of mercantilism as well as the beginnings of the Industrial Revolution, saw labor as the source of wealth and division of labor and improvement in the "state of the arts" as the source of productivity. Their main concern was how the product of labor got distributed among the social classes that cooperated to produce it. Adam Smith believed that an "invisible hand"—competition—would control the economy and that a certain natural order would keep atomistic individuals from exploiting each other, thereby harnessing individual self-interest to the social good. Classical economists thought that, over the long run, population growth and diminishing returns would unavoidably channel the entire economic surplus into rent, thus reducing profit to zero and terminating economic growth. What was anomalous about classical economics was not its long-run implications, however, but the then-existing misery of the working class, misery that gave the lie to the belief that the invisible hand could effectively prevent exploitation.

Karl Marx was largely a classical economist, to the extent that he saw labor as the source of net economic product. But in place of atomistic individuals acting in natural harmony and short-run cooperation among three classes—landlords, laborers, and capitalists—Marx saw two classes in direct day-to-day conflict: the owners of the means of production and the nonowners. The owners kept the net product of labor, paying the workers only what their replacement would cost. Atomistic competition would continue to exist *within* each class; but the essential idea of Marxist economics is the exploitative relation *between* classes, which Marx believed would lead to revolution. The earlier classical economists recognized the likelihood of long-run class conflicts, but Marx emphasized this as a central economic factor. This emphasis constituted a paradigm shift.

The neoclassical economists shifted the paradigm back to atomism, though adding an analysis of imperfect competition as they did so. Their big change, however, was to conceive of net value as the result of psychic want satisfaction rather than the product of labor. The origin of value was subjective, not objective. The focus was not on distribution among classes but on efficiency of allocation—how could a society get the maximum amount of want satisfaction from scarce resources, *given* a certain distribution of wealth and income among individuals and social classes? Pure competition provided the optimal allocation.

John Maynard Keynes, observing the economic problems of the 1930s, could not accept the anomaly presented by the wide disuse of resources that were supposed to be optimally allocated. He was less concerned that resources be "optimally" allocated in some refined sense than that they should not lie unused. Classical and neoclassical economics, with Say's Law among their premises, required that unemployment be viewed as an aberration. Social reality, however, insisted that unemployment was central. Keynes changed the theoretical viewpoint accordingly.

The present-day Keynesian-neoclassical synthesis seeks full macroeconomic employment and optimal microeconomic allocation of resources. The *summum bonum* to be maximized is no longer psychic want satisfaction, which is unmeasurable, but annual aggregate real output, GNP—Gross National Product—a value index of the quantity flow of annual production. Distribution recedes into the background;

the goal becomes to make the total pie bigger, thereby enabling everyone to get absolutely more without changing the relative size of parts. Both full employment and efficient allocation serve to increase the growth of real GNP. Conversely, and perhaps more importantly, growth of GNP is necessary to maintain full employment. In one of the first important contributions to growth theory, Evesy Domar stated the issue very well:

> The economy finds itself in a serious dilemma: if sufficient investment is not forthcoming today, unemployment will be here today. But if enough is invested today, still more will be needed tomorrow.
>
> It is a remarkable characteristic of a capitalistic economy that while, on the whole, unemployment is a function of the difference between its actual income and its productive capacity, most of the measures (i.e., investment) directed toward raising national income also enlarge productive capacity. It is very likely that the increase in national income will be greater than that of capacity, but the whole problem is that the increase in income is temporary and presently peters out (the usual multiplier effect), while capacity has been increased for good. So far as unemployment is concerned, investment is at the same time a cure for the disease and the cause of even greater ills in the future.[4]

Thus, continual growth in both capacity (stock) and income (flow) is a central part of the neoclassical growth paradigm. But in a finite world continual growth is impossible.[5] Given finite stomachs, finite lifetimes, and the kind of man who does not live by bread alone, growth becomes undesirable long before it becomes impossible. But the tacit, and sometimes explicit, assumption of the Keynesian-neoclassical growthmania synthesis is that aggregate wants are infinite and should be served by trying to make aggregate production infinite, and that technology is an omnipotent *deus ex machina* who will get us out of any growth-induced problems.

To call the ideas and resultant changes hastily sketched above *paradigm shifts* is to use Kuhn's term with a bit of poetic license. In the physical sciences, to which Kuhn applied the term, reality does not change except on an evolutionary time scale. The *same* things are perceived in different ways. But social reality changes more rapidly. This, however, can be viewed as an additional reason for the periodic necessity, in the social sciences, of regrinding our lenses to a new prescription.

Ideology, ethical apology, and ethical criticism are also sources of paradigm shifts in the social sciences. As Marx said, the goal is not just to interpret the world but to change it. And he was right. Even if we wish to be neutral or "value-free," we cannot, because the paradigm by which people try to understand their society is itself one of the key determining features of the social system. No one denies that the distinction between *is* and *ought* is an elementary rule of clear thinking. To say *is* when we should say *ought* is wishful thinking. To say *ought* when we should say *is* (or never to say *ought* at all) is apology for the status quo. But these distinctions belong in the mind of the individual thinker. They are not proper lines for division of labor between individuals, much less between professions. Attempts to divide thought in this way contribute heavily to the schizophrenia of the modern age.

Kuhn notes that paradigm shifts are usually brought about by the young or by people new to a discipline, those relatively free of the established preconceptions. Accordingly, we find that thought on a steady-state economy has been more eagerly received by physical scientists and biologists than economists and by the relatively young among economists. The interests of the physical and life sciences in the issue of growth versus steady state is evident from the program of the American Association for the Advancement of Science (AAAS) 1971 meetings. Consider the following report:

> Another way of interpreting the content of the AAAS meeting is to describe major themes that keep recurring. . . . Three topics appear this year in a variety of forms and contexts. They seek answers to:
> How to live on *a finite* earth?
> How to live a *good life* on a finite earth?
> How to live a good life on a finite earth *at peace and without destructive mismatches?*[6]

The many sessions in which these themes appear are then listed, including the presidential address.

Simultaneously with the AAAS meetings in Philadelphia, the American Economic Association (AEA) held meetings in New Orleans, where, judging from the detailed program, not one of these questions was even on the agenda. Yet the question "How to live a *good* life on a finite earth?" would seem to be of more direct concern to economists than to physicists and biologists. Why this striking discrepancy? Do economists have more important questions on their minds? I think not. It is simply that economists must undergo a revolutionary paradigm shift and sacrifice large intellectual (and material?) vested interests in the perpetual growth theories and policies of the last thirty years before they can really come to grips with these questions. The advantage of the physical scientists is that, unlike economists, they are viscerally convinced that the world is a finite, open system at balance in a near steady state, and they have not all invested time and energy in economic growth models. As Kuhn points out,

> Scientific revolutions . . . need seem revolutionary only to those whose paradigms are affected by them . . . astronomers, for example, could accept X-rays as a mere addition to knowledge, for their paradigms were unaffected by the existence of the new radiation. But for men like Kelvin, Crookes, and Roentgen, whose research dealt with radiation theory or with cathode ray tubes, the emergence of X-rays necessarily violated one paradigm as it created another. That is why these rays could be discovered only by something's first going wrong with normal research.[7]

A steady-state economy fits easily into the paradigm of physical science and biology—the earth approximates a steady-state open system, as do organisms. Why not our economy also, at least in its physical dimensions of bodies and artifacts? Economists forgot about physical dimensions long ago and centered their attention on value. But the fact that wealth is measured in value units does not annihilate its physical dimensions. Economists may continue to maximize value, and value could conceivably grow forever, but the physical mass in which value inheres must conform to a steady state, and the constraints of physical constancy on value growth will be severe and must be respected.

Perhaps this explains why many of the essays in this volume on political economy were written by physicists and biologists. But lest I be unfair to my own profession, I must observe that some leading economists, particularly Kenneth Boulding and Nicholas Georgescu-Roegen, have made enormous contributions toward reorienting economic thought along lines more congruent with a finite physical world. It is time for the profession to follow their lead.[8]

Ends, Means, and Economics

Chemistry has outgrown alchemy, and astronomy has emerged from the chrysalis of astrology, but the moral science of political economy has degenerated into the amoral game of politic economics. Political economy was concerned with scarcity and the resolution of the social conflicts engendered by scarcity. Politic economics tries to buy off social conflict by abolishing scarcity—by promising more things for more people, with less for no one, for ever and ever—all vouchsafed by the amazing grace of compound interest. It is not politic to remember, with John Ruskin,

> the great, palpable, inevitable fact—the root and rule of all economy—that what one person has, another cannot have; and that every atom of substance, of whatever kind, used or consumed, is so much human life spent; which if it issue in the saving present life or the gaining more, is well spent, but if not is either so much life prevented, or so much slain.[9]

Or, as Ruskin more succinctly put it in the same discussion, "there is no wealth but life."

Nor is it considered politic economics to take seriously the much more compelling demonstration of the same insight by Georgescu-Roegen, who has made us aware that

> the maximum of life quantity requires the minimum rate of natural resources depletion. By using these resources too quickly, man throws away that part of solar energy that will still be reaching the earth for a long time after he has departed. And everything that man has done in the last two hundred years or so puts him in the position of a fantastic spendthrift. There can be no doubt about it: any use of natural resources for the satisfaction of nonvital needs means a smaller quantity of life in the future. If we understand well the problem, the best use of our iron resources is to produce plows or harrows as they are needed, not Rolls Royces, not even agricultural tractors.[10]

Significantly, the masterful contribution of Georgescu-Roegen is not so much as mentioned in the Journal of Economic Literature's 1976 survey of the literature on environmental economics. The first sentence of that survey beautifully illustrates the environmental hubris of growth economics: "Man has probably always worried about his environment because he *was once* totally dependent on it" (emphasis added).[11] Contrary to the implication, our dependence on the environment is still total, and it is overwhelmingly likely to remain so. Nevertheless, Robert Solow suggests that, thanks to the substitutability of other factors for natural resources, it is not only conceivable but likely that "the world can, in effect, get along without natural resources."[12] In view of such statements, it is evidently impossible to insist too strongly that, in Frederick Soddy's words,

> life derives the whole of its physical energy or power, not from anything self-contained in living matter, and still less from an external deity, but solely

> from the inanimate world. It is dependent for all the necessities of its physical continuance primarily upon the principles of the steam-engine. The principles and ethics of human convention must not run counter to those of thermodynamics.[13]

Lack of respect for the principles of the steam engine also underlies the basic message of the very influential book *Scarcity and Growth,* by Harold Barnett and Chandler Morse. We are told that "nature imposes particular scarcities, not an inescapable general scarcity," and we are asked to believe that

> advances in fundamental science have made it possible to take advantage of the uniformity of matter/energy—a uniformity that makes it feasible, without preassignable limit, to escape the quantitative constraints imposed by the character of the earth's crust. . . . Science, by making the resource base more homogeneous, erases the restrictions once thought to reside in the lack of homogeneity. In a neo-Ricardian world, it seems, the particular resources with which one starts increasingly become a matter of indifference. The reservation of particular resources for later use, therefore, may contribute little to the welfare of future generations.[14]

Unfortunately for the politic economics of growth, it is not the uniformity of matter-energy that makes for usefulness but precisely the opposite. If all materials and all energy were uniformly distributed in thermodynamic equilibrium, the resulting "homogeneous resource base" would be no resource at all. It is nonuniformity—differences in concentration and temperature—that makes for usefulness. The mere fact that all matter-energy may ultimately consist of the same basic building blocks is of little significance if it is the *potential for ordering those blocks* that is ultimately scarce, as the entropy law tells us is the case. Only a Maxwell's Sorting Demon[15] could turn a lukewarm soup of electrons, protons, neutrons, quarks, and whatnot into a resource. And the entropy law tells us that Maxwell's demon does not exist. In other words, nature really *does* impose "an inescapable general scarcity," and it is a serious delusion to believe otherwise.

The differences in viewpoint cited above could hardly be more fundamental. It seems necessary, therefore, to start at the very beginning if we are to root out the faddish politic economics of growth and replant the traditional political economy of scarcity. Standard textbooks have long defined economics as the study of the allocation of scarce means among competing ends; thus a reconsideration of ends and means will provide our starting point. Modern economics' excessive devotion to growth will be explained in terms of an incom-

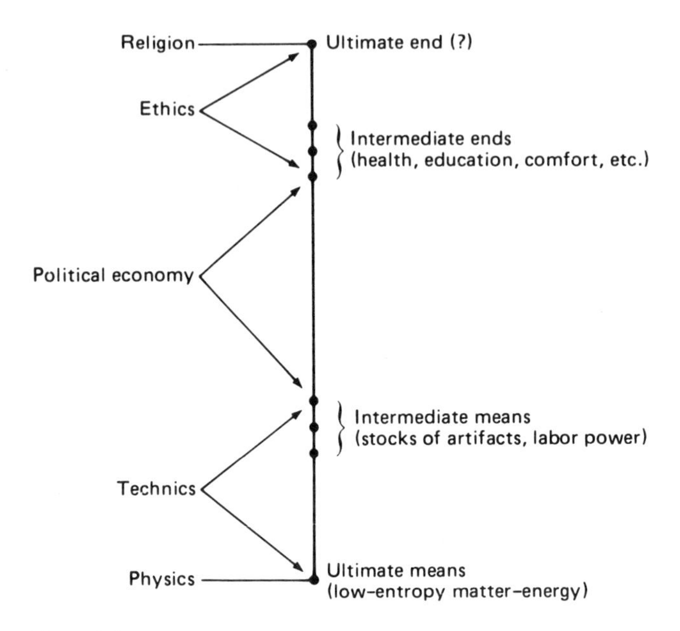

Figure I.1
The ends-means spectrum.

plete view of the total ends-means spectrum. The arguments of the two main traditions—the "scarce means arguments" and the "competing higher ends arguments"—provide the basic organizing principle for this volume.

In the largest sense, humanity's ultimate economic problem is to use ultimate means wisely in the service of the Ultimate End. It is thus not hard to understand our tendency to divide up the single, overwhelming problem into a number of smaller subproblems, as illustrated in figure I.1. This is a good procedure as long as we do not forget about other parts of the spectrum in our zeal to solve the problem of one segment.

At the top of the spectrum is the Ultimate End—that which is intrinsically good and does not derive its goodness from any instrumental relation to some higher good. At the bottom is ultimate means, the useful stuff of the world, low-entropy matter-energy, which we can only use up and cannot create or replenish, and whose net production, therefore, cannot possibly be the end of any human activity. Each intermediate category on the spectrum is an end with respect to lower categories and a means with respect to higher categories. Below the Ultimate End we have a hierarchy of intermediate ends, which are in a sense means in the service of the Ultimate End. Intermediate ends are ranked with reference to the Ultimate End. The mere fact that we speak of priorities among our goals presumes a first place, an ordering principle, an Ultimate End. We may not be able to define it very well, but logically we are forced to recognize its existence. Above ultimate means are intermediate means (physical stocks), which can be viewed as ends directly served by the use of ultimate means (the entropic flow of matter-energy, the *throughput).*

On the left of the spectrum line are listed the traditional disciplines of study that correspond to each segment of the spectrum. The central, intermediate position of economics is highly significant. In looking only at the middle range, economics has naturally not dealt with ultimates or absolutes, which are found only at the extremes, and has falsely assumed that the middle-range pluralities, relativities, and substitutabilities among competing ends and scarce means were representative of the whole spectrum. Absolute limits are absent from the economists' paradigm because absolutes are encountered only in confrontation with the ultimate poles of the spectrum, which have been excluded from the focus of our attention. Even ethics and technics exist for the economist only at the very periphery of professional awareness.

In terms of this diagram, economic growth implies the creation of ever more intermediate means (stocks) for the purpose of satisfying ever more intermediate ends. Orthodox growth economics, as we have seen, recognizes that particular resources are limited but does not recognize any general scarcity of all resources together. The orthodox dogma is that technology can always substitute new resources for old, without limit. Ultimate means are not considered scarce. Intermediate means are scarce, it is argued, only because our capacity to transform ultimate means has not yet evolved very far toward its unlimited potential. Growth economists also recognize that any single intermediate end or want can be satisfied for any given individual. But new wants keep emerging (and new people as well), so the aggregate of all intermediate ends is held to be insatiable, or infinite in number if not in intensity. The growth economists' vision is one of continuous growth in intermediate means (unconstrained by any scarcity of ultimate means) in order to satisfy ever more intermediate ends (unconstrained by any impositions from the

Ultimate End). Infinite means plus infinite ends equals growth forever.

A consideration of the ultimate poles of the spectrum, however, gives us a very different perspective, forcing us to ask two questions: (1) What, precisely, are our ultimate means, and are they limited in ways that cannot be overcome by technology? (2) What is the nature of the Ultimate End, and is it such that, beyond a certain point, further accumulation of intermediate means (bodies and artifacts) not only fails to serve the Ultimate End but actually renders a disservice? It will be argued in this volume that the answer to both sets of questions is *yes*. The absolute scarcity of ultimate means limits the *possibility* of growth (part I). The competition from other ends, which contribute more heavily at the margin toward the Ultimate End, limits the *desirability* of growth (part II). Moreover, the interaction of desirability and possibility provides the *economic* limit to growth, which is the most stringent, and should be the governing, limit (part III).

Paradoxically, growth economics has been both too materialistic and not materialistic enough. In ignoring the ultimate means and the laws of thermodynamics, it has been insufficiently materialistic. In ignoring the Ultimate End and ethics, it has been too materialistic.

Critics of growth can be classified into ends-based (moral) and means-based (biophysical). Many writers are, to some extent, in both traditions. This is to be expected, because the two traditions are not as logically independent as may at first appear. For example, many moral issues regarding distributive justice and intergenerational equity hardly arise if one believes that continual economic growth is biophysically possible. Likewise, if one's arena of moral concern excludes the poor, future generations, and subhuman life, then many biophysical constraints are no longer of interest. To crack the nut of growthmania, it is not enough to hammer from above with moral arguments, because there is sufficient "give" underneath for optimistic biophysical assumptions to cushion the blow (space colonies, green revolutions, breeder reactors, etc.) Hammering only from below with biophysical arguments leaves too much room for elastic morality to absorb the blow. (The interest rate automatically looks after the future; growth itself is the Ultimate End, or as close as we can come to it; our manifest destiny is to colonize space as the earth is a mere dandelion gone to seed; etc.) Growth chestnuts have to be placed on the unyielding anvil of biophysical realities and then crushed with the hammer of moral argument. The entropy law and ecology provide the biophysical anvil. Concern for future generations and subhuman life and inequities in current wealth distribution provide the moral hammer.

Human beings are both material creatures in absolute dependence upon their physical environment and rational beings who have purposes and strive to become better. These two aspects must be consistent with each other. Improvement presupposes survival, and survival in an entropic and evolving world is impossible without continual striving for improvement. Biophysically based conclusions about economic growth, or any other subject, should be in accord with morally based conclusions. A discrepancy indicates a flawed understanding of the natural world or a warped set of values. That ends-based and means-based arguments should converge in their rejection of growthmania is both comforting and not unexpected.

The overall problem is how to use ultimate means to serve best the Ultimate End. We might call this ultimate political economy, or *stewardship*. To state the problem in this way is to emphasize at once both its wholeness and the necessity of breaking it into more manageable subproblems, for the overall problem must be tackled one step at a time. Yet one step is valueless without the others, and one correct step is worse than valueless if the steps it takes for granted were false steps. If our concept of the Ultimate End is evil rather than good, then an inverted ethics is better for us than a consistent ethics. If our ethical priorities are upside down, then an inverted or incorrect imputation of value to intermediate means is better than a correct imputation. If our intermediate means are incorrectly valued, then a technology that efficiently and powerfully converts ultimate means into the most valuable intermediate means is worse than a weak technology. And an erroneous physics that will cause technology to stumble rather than advance an evil end efficiently is better than a correct physics.

The parts of the total economic problem are related not only from the top down but also from the bottom up. Our customary ethical ordering of intermediate ends conditions our perception of the Ultimate End. We tend to take our conventional priorities as given and then deduce the nature of the Ultimate End as that which legitimates the conventional priorities. We tend also to order our intermediate ends in such a way that we can effectively serve them with the existing evaluation of intermediate means. Further, there is a tendency to value the intermediate means according to the technical and physical possibilities for producing them. If it is possible, we must do it.

I do not mean to say that working only in one direction is always proper and in the other always improper. The point is that the parts of the problem are highly interrelated and cannot be dealt with in isolation, and, even though ideally our starting point should be the Ultimate End, we can only see that end dimly and may find clues to its nature in our experience with ethical, economic, and even technical problems encountered on the way.

The total problem of relating the five subproblems—theology, ethics, political economy, technology, physics—is more delicate than any of the subproblems themselves, but not for that reason any less imperative. Surely we must have a vision of the total problem, otherwise we do not understand what our specialties are. It is hoped that the collection of articles in this book will help to fill out such a total vision. Clearly, each stage can be dealt with only in a partial and incomplete manner. But the premise on which this volume rests is that it is better to deal incompletely with the whole than to deal wholly with the incomplete.

Let us now turn to an overview of the particular paradigm this collection seeks to develop, one that will lead to a steady-state economy. The terms *steady state* and *stationary state* are used synonymously. The former is common in physical sciences, the latter common in economics and demography.

The Steady-State Economy

Any discussion of the relative merits of the steady, stationary, or no-growth economy, and its opposite, the economy in which wealth and population are growing, must recognize some important quantitative and qualitative differences between rich and poor countries and between rich and poor classes within countries. To see why this is so, consider the familiar ratio of Gross National Product (GNP) to total population (P). This ratio, per capita annual product (GNP/P), is the measure usually employed to distinguish rich from poor countries, and, in spite of its many shortcomings, it does have the virtue of reflecting in one ratio the two fundamental life processes of production and reproduction. Let us ask two questions of both numerator and denominator for both rich and poor countries—namely, what is the quantitative rate of growth; and, qualitatively, exactly what is it that is growing?

1. The rate of growth in the denominator, P, is much higher in poor countries than in rich countries. Although mortality is tending to equality at low levels throughout the world, fertility[16] in poor nations remains roughly *twice* that of rich nations. The average Gross Reproduction Rate (GRR)[17] for rich countries is around 1.5, and that for poor countries is around 3.0 (that is, on the assumption that all survive to the end of reproductive life, each mother would be replaced by 1.5 daughters in rich countries and 3 in poor countries). Moreover, all poor countries have a GRR greater than 2.0, and all rich countries have a GRR less than 2.0, with practically no countries falling in the area of the 2.0 dividing point. No other social or economic index divides the world so clearly and consistently into "developed" and "underdeveloped" as does fertility.[18]

2. Qualitatively, the incremental population in poor countries consists largely of hungry illiterates; in rich countries it consists largely of well-fed members of the middle class. The incremental person in poor countries contributes negligibly to production but makes few demands on world resources—although from the point of view of his poor country, these few demands of many new people can easily dissipate any surplus that might otherwise be used to raise productivity.[19] The incremental person in the rich country contributes to his country's GNP, and to feed his high standard of living contributes greatly to depletion of the world's resources and pollution of its spaces.

3. The numerator, GNP, has grown at roughly the same rate in rich and poor countries, around 4 or 5 percent annually, with the poor countries probably growing slightly faster. Nevertheless, because of the poor countries' more rapid population growth, their per capita income has grown more slowly than that of rich countries. Consequently, the gap between rich and poor has widened.[20]

4. The incremental GNP of rich and poor nations has an altogether different qualitative significance. This follows from the two most basic laws of economics: (a) the law of diminishing marginal utility, which really says nothing more than that people satisfy their most pressing wants *first*—thus each additional dollar of income or unit of resource is used to satisfy a less pressing want than the previous dollar or unit; and (b) the law of increasing marginal cost, which says that producers first use the best qualities of factors (most fertile land,

most experienced worker, and so on) and the best combination of factors known to them. They use the less efficient (more costly) qualities and combinations only when they run out of the better ones, or when one factor, such as land, becomes fixed (nonaugmentable). Also, in a world of scarcity, as more resources are devoted to one use, fewer are available for other uses. The least important alternative uses are sacrificed first, so that as more of any good is produced, progressively more important alternatives must be sacrificed; that is, a progressively higher price (opportunity cost) must be paid. Applied to GNP, the first law means that the marginal (incremental) benefits from equal increments of output are decreasing, and the second law means that the marginal cost of equal increments in output is increasing.

At some point, perhaps already passed in the United States, an extra unit of GNP costs more than it is worth. Technological advances can put off this point, but not forever. Indeed, they may bring it to pass sooner because more powerful technologies tend to provoke more powerful ecological backlashes and to be more disruptive of habits and emotions. To put things more concretely, growth in GNP in a poor country means more food, clothing, shelter, basic education, and security, whereas for the rich country it means more electric toothbrushes, yet another brand of cigarettes, more tension and insecurity, and more force-feeding through more advertising. In sum, extra GNP in a poor country, assuming it does not go mainly to the richest class of that country, represents satisfaction of relatively basic wants, whereas extra GNP in a rich country, assuming it does not go mainly to the poorest class of that country, represents satisfaction of relatively trivial wants.

For our purposes, the upshot of these differences is that, for the poor, growth in GNP is still a good thing, but for the rich it is probably a bad thing. Growth in population, however, is a bad thing for both: For the rich, population growth is bad because it makes growth in GNP (a bad thing) less avoidable. For the poor, population growth is bad because it makes growth in GNP, and especially in per capita GNP (a good thing), more difficult to attain. We shall be concerned in this book mainly with a rich, affluent-effluent economy such as that of the United States. Our purposes will be to define more clearly the concept of steady state, to see why it is necessary, to consider its economic and social implications, and finally to comment on an emerging political economy of finite wants and nongrowth.

The Nature and Necessity of the Stationary State

The term *stationary state* (steady state) is used here in its classical sense.[21] Over a century ago, John Stuart Mill, the great synthesizer of classical economics, spoke of the stationary state in words that could hardly be more relevant today, and they will serve as the starting point in our discussion.

> But in contemplating any progressive movement, not in its nature unlimited, the mind is not satisfied with merely tracing the laws of its movement; it cannot but ask the further question, to what goal? . . .
>
> It must always have been seen, more or less distinctly, by political economists, that the increase in wealth is not boundless: that at the end of what they term the progressive state lies the stationary state, that all progress in wealth is but a postponement of this, and that each step in advance is an approach to it . . . if we have not reached it long ago, it is because the goal itself flies before us [as a result of technical progress].
>
> I cannot . . . regard the stationary state of capital and wealth with the unaffected aversion so generally manifested towards it by political economists of the old school. I am inclined to believe that it would be, on the whole, a very considerable improvement on our present condition. I confess I am not charmed with the ideal of life held out by those who think that the normal state of human beings is that of struggling to get on; that the trampling, crushing, elbowing, and treading on each other's heels which form the existing type of social life, are the most desirable lot of human kind, or anything but the disagreeable symptoms of one of the phases of industrial progress. The northern and middle states of America are a specimen of this stage of civilization in very favorable circumstances; . . . and all that these advantages seem to have yet done for them (notwithstanding some incipient signs of a better tendency) is that the life of the whole of one sex is devoted to dollar-hunting, and of the other to breeding dollar-hunters.
>
> . . . Those who do not accept the present very early stage of human improvement as its ultimate type may be excused for being comparatively indifferent to the kind of economical progress which excites the congratulations of ordinary politicians; the mere increase of production and accumulation. . . . I know not why it should be a matter of congratulation that persons who are already richer than anyone needs to be, should have doubled their means of consuming things which give little or no pleasure except as representative of wealth. . . . It is only in the backward countries of the world that increased production is still an important object: in those most advanced, what is economically needed is better distribution, of which one indispensable means is a stricter restraint on population.

> There is room in the world, no doubt, and even in old countries, for a great increase in population, supposing the arts of life to go on improving, and capital to increase. But even if innocuous, I confess I see very little reason for desiring it. The density of population necessary to enable mankind to obtain, in the greatest degree, all the advantages both of cooperation and of social intercourse, has, in all the most populous countries, been attained. A population may be too crowded, though all be amply supplied with food and raiment. It is not good for a man to be kept perforce at all times in the presence of his species. . . . Nor is there much satisfaction in contemplating the world with nothing left to the spontaneous activity of nature; with every rood of land brought into cultivation, which is capable of growing food for human beings; every flowery waste or natural pasture plowed up, all quadrupeds or birds which are not domesticated for man's use exterminated as his rivals for food, every hedgerow or superfluous tree rooted out, and scarcely a place left where a wild shrub or flower could grow without being eradicated as a weed in the name of improved agriculture. If the earth must lose that great portion of its pleasantness which it owes to things that the unlimited increase of wealth and population would extirpate from it, for the mere purpose of enabling it to support a larger, but not a happier or a better population, I sincerely hope, for the sake of posterity, that they will be content to be stationary, long before necessity compels them to it.
>
> It is scarcely necessary to remark that a stationary condition of capital and population implies no stationary state of human improvement. There would be as much scope as ever for all kinds of mental culture, and moral and social progress; as much room for improving the Art of Living and much more likelihood of its being improved, when minds cease to be engrossed by the art of getting on. Even the industrial arts might be as earnestly and as successfully cultivated, with this sole difference, that instead of serving no purpose but the increase of wealth, industrial improvements would produce their legitimate effect, that of abridging labor.[22]

The direction in which political economy has evolved in the last hundred years is not along the path suggested by Mill. In fact, most economists are hostile to the classical notion of stationary state and dismiss Mill's discussion as "strongly colored by his social views"[23] (as if the neoclassical theories were not so colored!), and "nothing so much as a prolegomenon to Galbraith's *Affluent Society*" (which also received a hostile reception from the economics profession). While giving full credit to Mill for his many other contributions to economics, most economists consider his discussion of the stationary state as something of a personal aberration. Also his "relentless insistence that every conceivable policy measure must be judged in terms of its effects on the birth rate" is dismissed as "hopelessly dated." The truth is, however, that Mill is even more relevant today than in his own time.

With this historical background, let us now analyze the steady state with a view toward clarifying what Mill somewhat mistakenly thought "must have always been seen more or less distinctly by political economists," namely, "that wealth and population are not boundless."

By *steady state* is meant a constant stock of physical wealth (capital), and a constant stock of people (population).[24] Naturally, these stocks do not remain constant by themselves. People die, and wealth is physically consumed, that is, worn out, depreciated. Therefore, the stocks must be maintained by a rate of inflow (birth, production) equal to the rate of outflow (death, consumption). But this equality may obtain, and stocks remain constant, with a high rate of throughput (equal to both the rate of inflow and the rate of outflow) or with a low rate. Our definition of steady state is not complete until we specify the rates of throughput by which the constant stocks are maintained. For a number of reasons we specify that the rate of throughput should be as low as possible. For an equilibrium stock, the average age at "death" of its members is the reciprocal of the rate of throughput. The faster the water flows through the tank, the less time an average drop spends in the tank. For the population, a low rate of throughput (a low birth rate and an equally low death rate) means a high life expectancy, and it is desirable for that reason alone—at least within limits. For the stock of wealth, a low rate of throughput (low production and equally low consumption) means greater life expectancy or durability of goods and less time sacrificed to production. This means more "leisure" or nonjob time to be divided into consumption time, personal and household maintenance time, culture time, and idleness.[25] This, too, seems socially desirable, at least within limits.

To these reasons for the desirability of a low rate of throughput we must add some reasons for the impracticability of high rates. Since matter and energy cannot be created, production inputs must be taken from the environment, which leads to depletion. Since matter and energy cannot be destroyed, an equal amount of matter and energy in the form of waste must be returned to the environment, leading to pollution. Hence lower rates of throughput lead to less depletion and pollution, higher rates to more. The limits regarding what rates of depletion and pollution are tolerable must be supplied by ecology. A definite limit to the size of maintenance flows of matter and energy is set by ecological thresholds which, if exceeded, cause a breakdown of the system. To keep flows below these limits, we can operate on two variables: the *size* of the stocks and the *durability* of the stocks. As long as we are well below these thresholds, economic cost-benefit calculations of depletion and pollution can be relied on as a guide. But as these thresholds are approached, marginal cost and marginal benefit become meaningless, and Alfred Marshall's erroneous motto that "nature does not make jumps" and most of neoclassical marginalist economics become inapplicable. The "marginal" cost of one more step may be to fall over the precipice.

Of the two variables—size of stocks and durability of stocks—only the second requires further clarification. *Durability* means more than just how long a particular commodity lasts. It also includes the efficiency with which the after-use "corpse" of a commodity can be recycled as an input to be born again as the same or a different commodity. Within certain limits, to be discussed below, durability of stocks ought to be maximized in order that depletion of resources might be minimized.

We might suppose that the best use of resources would imitate the model that nature has furnished: a closed-loop system of material cycles powered by the sun (what A. J. Lotka called the "mill wheel of life" or the "world engine").[26] In such an "economy," durability is maximized, and the resources on earth could presumably last as long as the sun continues to radiate the energy to turn the closed material cycles.

We can set up an economy in imitation of nature in which all waste products are recycled. Instead of the sun, however, we use other sources of energy because of the scale of our industrial activity. Even modern agriculture depends as much on geologic capital (to make fertilizers, machines, and pesticides) as on solar income. This capital (fossil fuels and fission materials), from which we now borrow, may not last more than a couple of centuries, but there is another possible energy source, controlled thermonuclear fusion, which may someday provide a practically inexhaustible supply of energy with little radioactive waste, thereby alleviating problems of resource depletion and radioactive contamination. At least that is the claim of fusion enthusiasts.

Nevertheless, the serious problem of waste heat remains. The second law of thermodynamics tells us that it is impossible to recycle energy and that eventually all energy will be converted into waste heat. Also, it is impossible to recycle materials with one hundred percent completeness. Some material is irrecoverably lost in each cycle. Eventually, all life will cease as entropy or chaos approaches its maximum. But the second law of thermodynamics implies that, even before this very long-run universal thermodynamic heat-death occurs, we will be plagued by thermal pollution, for whenever we use energy, we must produce unusable waste heat. When a localized energy process causes a part of the environment to heat up, thermal pollution can have serious effects on ecosystems, since life processes and climatic phenomena are regulated by temperature.

We have already argued that, given the size of stocks, the throughput should be minimized, since it is really a cost. But the throughput is in two forms, matter and energy, and the ecological cost will vary, depending on how the throughput is apportioned between them. The amount of energy throughput will depend on the rate of material recycling. If we recycle none of our used material goods, then we must expend energy to replace those goods from raw materials, and this energy expenditure is in many instances greater than the energy needed to recycle the product. For example, the estimated energy needed to produce a ton of steel plate from iron ore is 2700 kilowatt-hours, whereas merely 700 kilowatt-hours are needed to produce the same ton by recycling scrap steel.[27] However, this is not the whole story. The mere expenditure of energy is not sufficient to close material cycles, since energy must work through the agency of material implements. To recycle aluminum cans requires more trucks to collect the cans as well as more energy to run the trucks. More trucks require more steel, glass, rubber, and so forth, which require more iron ore and coal, which require still more trucks. This is the familiar web of interindustry interdependence reflected in an input-output table.[28] All of these extra intermediate activities required to recycle the aluminum cans involve some inevitable pollution as well. If we think of each industry as adding recycling to its production process, then this will generate a whole chain of direct and indirect demands on matter and energy resources that must be taken away from final demand uses and devoted to the intermediate activities of recycling. It will take more intermediate products and activities to support the same level of final output.

As we attempt to recycle more and more of our produced goods, we will reach the point of diminishing returns; the energy expenditure alone will give rise to a ruinous amount of waste heat or thermal pollution. On the other hand, if we recycle too small a fraction of our produced goods, then nonthermal pollution and resource depletion become a severe problem.

The introduction of material recycling permits a trade-off; that is, it allows us to choose that combination of material and energy depletion and pollution which is least costly in the light of specific local conditions. *Cost* here means total ecological cost, not just pecuniary costs, and it is extremely difficult to measure.

In addition to the trade-offs involved in minimizing the ecological cost of the throughput for a given stock, we must recognize that the total stock (consisting of wealth and people) is variable both in total size and in composition. Since there is a direct relationship between the size of the stock and the size of the throughput necessary to maintain the stock, we have a trade-off between size of total stock (viewed as benefit) and size of the flow of throughput (viewed as a cost); in other words, an increase in benefit implies an increase in cost. Furthermore, a given throughput can maintain a constant total stock consisting of a large substock of wealth and a small substock of people or a large substock of people and a small substock of wealth. Here we have a trade-off in the form of an inverse relationship between two benefits. This latter trade-off between people and wealth is imposed by the constancy of the total stock and is limited by minimal subsistence per capita wealth at one extreme and by minimal technological requirements for labor to maintain the stock of wealth at the other extreme. Within these limits this trade-off essentially represents the choice of a standard of living. Economics and ecology can at best specify the terms of this trade-off; the actual choice depends on ethical judgments.

In sum, the steady state of wealth and population is maintained by an inflow of low-entropy matter-energy (depletion) and an outflow of an equal quantity of high-entropy matter-energy (pollution). Stocks of wealth and people, like individual organisms, are open systems that feed on low entropy.[29] Many of these relationships are summarized in figure I.2.

The classical economists thought that the steady state would be made necessary by limits on the depletion side (the law of increasing cost or diminishing returns), but in fact the main limits seem to be occurring on the pollution side. In effect, pollution provides another foundation for the law of increasing costs, but it has received little attention in this regard, since pollution costs are social, whereas depletion costs are usually private. On the input side, the environ-

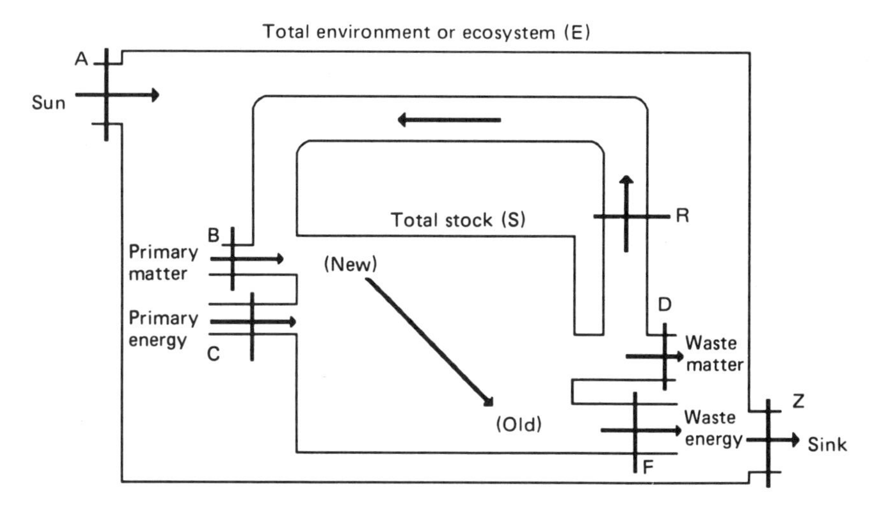

Figure I.2
Rectangle E is the total ecosystem, which contains the total stock (S) of wealth and people as one of its mutually dependent components. The ecosystem imports energy from outer space (sun, A) and exports waste heat to outer space (sink, Z). The stock contains matter in which a considerable amount of available energy is stored (mined coal, oil in tanks, water on high ground, living things, wood products, and the like), as well as matter in which virtually no available energy is stored. Matter and energy in the stock must be separately maintained. The stock is maintained in a steady state when B is equal to D and C is equal to F. In the steady state, throughput equals either input (B plus C) or output (D plus F), since input and output are equal to each other. When input and output are not equal, then the throughput is measured by the smaller of the two.

From the second law of thermodynamics, we know that energy cannot be recycled. Matter may be recycled (R), but only by using more energy (and matter) to do it. In the diagram, energy moves only from left to right, whereas matter moves in both directions.

For a constant S, the lower the rate of throughput the more durable or longer-lived is the total stock. For a given throughput, the lower the rate of recycling (R), the more durable are the individual commodities. The optimum durability of an individual commodity is attained when the marginal production cost of increased durability equals the marginal recycling cost of not having increased durability further. *Cost* is total ecological cost and is extremely difficult to measure.

Both the size of the stock and the rate of throughput must not be so large relative to the total environment that they obstruct the natural ecological processes that form the biophysical foundations of wealth. Otherwise, the total stock and its associated throughput become a cancer that kills the total organism.

ment is partitioned into spheres of private ownership. Depletion of the environment coincides, to some degree, with depletion of the owner's wealth and inspires at least a minimum of stewardship. On the output side, however, the waste absorption capacity of the environment is not subject to partitioning and private ownership. Air and water are used freely by all, and the result is a competitive, profligate exploitation—what biologist Garrett Hardin calls the "commons effect," what welfare economists call "external diseconomies," and what I like to call the "invisible foot." Adam Smith's invisible hand leads private self-interest unwittingly to serve the common good. The invisible foot leads private self-interest to kick the common good to pieces. Private ownership and private use under a competitive market give rise to the invisible hand. Public ownership with unrestrained private use gives rise to the invisible foot. Public ownership with public restraints on use gives rise to the visible hand (and foot) of the planner. Depletion has been partially restrained by the invisible hand, while pollution has been encouraged by the invisible foot. It is therefore not surprising to find limits occurring mainly on the pollution side—which, of course, is not to deny depletion limits.

It is interesting that the first school of economists, the physiocrats, emphasized human beings' dependence on nature. For them only the "natural" activity of agriculture was capable of producing a net product of value. Indeed, the word *physiocracy* meant rule of nature. Something of the physiocrats' basic vision, if not their specific theories, is badly needed in economics today.

Economic and Social Implications of the Steady State

The economic and social implications of the steady state are enormous and revolutionary. The physical flows of production and consumption must be *minimized, not maximized,* subject to some desirable population and standard of living.[30] The central concept must be the stock of wealth, not, as presently, the flow of income and consumption. Furthermore, the stock must not grow. For several reasons, the important issue of the steady state will be distribution, not production. The problem of relative shares can no longer be avoided by appeals to growth. The argument that everyone should be happy as long as his absolute share of the wealth increases, regardless of his relative share, will no longer be available. Absolute and relative shares will move together, and the division of physical wealth will be a zero-sum game. In addition, the arguments justifying inequality in wealth as necessary for savings, investment, and growth will lose their force. With production flows (which are really *costs* of maintaining the stock) kept low, the focus will be on the distribution of the stock of wealth, not on the distribution of the flow of income. Marginal productivity theories and "justifications" pertain only to flow and therefore are not available to explain or "justify" the distribution of stock ownership. Also, even though physical stocks remain constant, increased income in the form of leisure will result from continued technological improvements. How will it be distributed, if not according to some ethical norm of equality? The steady state would make fewer demands on our environmental resources but much greater demands on our moral resources. In the past, a good case could be made that leaning too heavily on scarce moral resources, rather than relying on abundant self-interest, was the road to serfdom. But in an age of rockets, hydrogen bombs, cybernetics, and genetic control, there is simply no substitute for moral resources and no alternative to relying on them, whether they prove sufficient or not.

On the question of maximizing versus minimizing the flow of production, there is an interesting analogy with ecological succession. Young ecosystems (early stages of succession) are characterized by a high production efficiency, and mature ecosystems (late stages of succession) are characterized by a high maintenance efficiency. For a given B (biomass stock), young ecosystems tend to maximize P (production flow), giving a high production efficiency P/B; mature ecosystems, on the other hand, tend to minimize P for a given B, thus attaining a high maintenance efficiency, B/P. According to ecologist Eugene P. Odum, young ecosystems seem to emphasize production, growth, and quantity, whereas mature ecosystems emphasize protection, stability, and quality.[31] For the young system, the flow of production is the quantitative source of growth and is maximized. For the mature, the flow of production is the maintenance cost of protecting the stability and quality of the stock and is minimized. If we conceive of the human economy as an ecosystem moving from an earlier to a later stage of succession (from the "cowboy economy" to the "spaceman economy," as Boulding puts it), then we would expect, by analogy, that production, growth, and quantity would be replaced by protective maintenance, stability, and quality as the major social goals. The cardinal virtues of the past become the cardinal sins of the present.

With constant physical stocks, economic growth must be in nonphysical goods: service and leisure.[32] Taking the benefits of technological progress in the form of increased leisure is a reversal of the historical practice of taking the benefits mainly in the form of goods and has extensive social implications. In the past, economic development has increased the physical output of a day's work while the number of hours in a day has, of course, remained constant, with the result that the opportunity cost of a unit of time in terms of goods has risen. Time is worth more goods, a good is worth less time. As time becomes more expensive in terms of goods, fewer activities are "worth the time." We become goods-rich and time-poor. Consequently, we crowd more activities and more consumption into the same period of time in order to raise the return on nonwork time so as to bring it into equality with the higher returns on work time, thereby maximizing the total returns to total time. This gives rise to what Staffan Linder has called the "harried leisure class." We use not only work time but also personal consumption time more efficiently, and we even try to be efficient in our sleep by attempting subconscious learning. Time-intensive activities (friendships, care of the aged and children, meditation, and reflection) are sacrificed in favor of commodity-intensive activities (consumption). At some point, people will feel rich enough to afford more time-intensive activities, even at the higher price. But advertising, by constantly extolling the value of material-intensive commodities, postpones this point. From an ecological view, of course, this is exactly the reverse of what is called for. What is needed is a low relative price of time in terms of material commodities. Then time-intensive activities will be substituted for material-intensive activities. To become less materialistic in our habits, we must raise the relative price of matter. Keeping physical stocks constant and using technology to increase leisure time will do just that. Thus a policy of nonmaterial growth, or leisure-only growth, in addition to being necessary for keeping physical stocks constant, has the further beneficial effect of encouraging a more generous expenditure of time and a more careful use of physical goods. A higher relative price of material-intensive goods may at first glance be thought to encourage their production. But material goods require material inputs, so costs as well as revenues would increase, thus eliminating profit incentives to expand.

In the 1930s Bertrand Russell proposed a policy of leisure growth rather than commodity growth and viewed the unemployment question in terms of the distribution of leisure. The following words are from his delightful essay "In Praise of Idleness."

> Suppose that, at a given moment, a certain number of people are engaged in the manufacture of pins. They make as many pins as the world needs, working (say) eight hours a day. Someone makes an invention by which the same number of men can make twice as many pins as before. But the world does not need twice as many pins. Pins are already so cheap that hardly any more will be bought at a lower price. In a sensible world,. everybody concerned in the manufacture of pins would take to working four hours instead of eight, and everything else would go on as before. But in the actual world this would be thought demoralizing. The men still work eight hours, there are too many pins, some employers go bankrupt, and half the men previously concerned in making pins are thrown out of work. There is, in the end, just as much leisure as on the other plan, but half the men are totally idle while half are still overworked. In this way it is insured that the unavoidable leisure shall cause misery all round instead of being a universal source of happiness. Can anything more insane be imagined?[33]

In addition to this strategy of leisure-only growth, and the resulting reinforcement of an increased price of material-intensity relative to time-intensity, we can internalize some pollution costs by charging pollution taxes. Economic efficiency requires only that a price be placed on environmental amenities; it does not tell us who should pay the price. The producer may claim that the use of the environment to absorb waste products is a right that all organisms and firms must of necessity enjoy, and whoever wants air and water to be cleaner than it is at any given time should pay for it. Consumers may argue that the use of the environment as a source of clean inputs of air and water takes precedence over its use as a sink, and that whoever makes the environment dirtier than it otherwise would be should be the one to pay. Again the issue becomes basically one of distribution—not what the price should be, but who should pay it. The fact that the price takes the form of a tax automatically decides who will receive it—the government. But this raises more distribution issues; and the "solutions" to these problems are ethical, not technical.

Another possibility of nonmaterial growth is to redistribute wealth from the low-utility uses of the rich to the high-utility uses of the poor, thereby increasing total "social utility." Joan Robinson has noted that this egalitarian implication of the law of diminishing marginal utility was "sterilized . . . mainly by slipping from utility to physical output as the object to be maximized."[34] As we move back from physical output to nonphysical utility, the egalitarian implications become "unsterilized."

Economic growth has kept at bay two closely related problems. First, growth is necessary to maintain full employment. Only if it is possible for nearly everyone to have a job can the income-through-jobs ethic of distribution remain workable. Second, growth takes the edge off of distributional conflicts. If everyone's absolute share of income is increasing, there is a tendency not to fight over relative shares, especially since such fights may interfere with growth and even lead to a lower absolute share for all. But these problems cannot be kept at bay forever, because growth cannot continue indefinitely.

Growth, by allowing full employment, permits the old principles of distribution (income-through-jobs link) to continue in effect. But with no growth in physical stocks, and a policy of using technological progress to increase leisure, full employment and income-through-jobs are no longer workable mechanisms for distribution. Furthermore, we add a new dimension to the distribution problem—how to distribute leisure. The point is that distribution issues must be squarely faced and not left to work themselves out as the by-product of full-employment policies aimed at promoting growth.

A stationary population, with low birth and death rates, would imply a greater percentage of old people than in the present growing population, though hardly a geriatric society as some youth worshipers claim. The average age, assuming that current U.S. mortality holds, would change from twenty-seven to thirty-seven. One hears much nonsense about the conservatism and reactionary character of older populations and the progressive dynamism of younger populations, but a simple comparison of Sweden (old but hardly reactionary) with Brazil (young but hardly progressive) should make us cautious about such facile relationships. It is also noted that the age pyramid of a stationary U.S. population would be essentially rectangular up to about age fifty and then would rapidly taper off, and that the age "pyramid" would no longer be roughly congruent with the pyramid of authority in hierarchical organizations, with the result that the general correlation between increasing age and increasing authority would not hold for very many people. Quite true, but a salutary result could well be that more people will seek their personal fulfillment outside the structure of hierarchical organizations and that fewer people would rise to levels of their incompetence within bureaucracies. Since old people do not work, this further accentuates the distribution problem. However, the percentage of children will diminish, so in effect there will be mainly a change in the direction that payments are transferred. More of the earnings of working adults will be transferred to the old, and less to children.

What institutions will provide the control necessary to keep the stocks of wealth and people constant, with the minimum sacrifice of individual freedom? This, I submit, is the question we should be struggling with. It would be far too simpleminded to blurt out "socialism" as the answer, since socialist states are as badly afflicted with growthmania as capitalist states. The Marxist eschatology of the classless society is based on the premise of complete abundance; consequently, economic growth is exceedingly important in socialist theory and practice. Also, population growth, for the orthodox Marxist, cannot present problems under socialist institutions. This latter tenet has weakened a bit in recent years, but the first continues in full force. However, it is equally simpleminded to believe that the present big capital, big labor, big government, big military type of private profit capitalism is capable of the required foresight and restraint and that the addition of a few pollution and severance taxes here and there will solve the problem. The issues are much deeper and inevitably impinge on the distribution of income and wealth.

All economic systems are subsystems within the big biophysical system of ecological interdependence. The ecosystem provides a set of physical constraints to which all economic systems must conform. The facility with which an economic system can adapt to these constraints is a major, if neglected, criterion for comparing economic systems. This neglect is understandable, because in the past ecological constraints showed no likelihood of becoming effective. But population growth, growth in the physical stock of wealth, and growth in the power of technology all combine to make ecological constraints effective. Perhaps this common set of constraints will be one more factor favoring convergence of economic systems.

Why do people produce junk and cajole other people into buying it? Not out of any innate love for junk or hatred of the environment, but simply in order to earn an income. If, with the prevailing distribution of wealth, income, and power, production governed by the profit motive results in the output of great amounts of noxious junk, then something is wrong with the distribution of wealth and power, the profit motive, or both. We need some principle of income distribution independent of and supplementary to the income-through-jobs link.[35] Perhaps a start in this direction was made by Oskar Lange in his *On the Economic Theory of Socialism*,[36] in which he attempted to combine some socialist principles of distribution with the allocative efficiency advantages of the market system. However, at least as much remains to be done here as remains to be done in designing institutions for stabilizing population. But before much progress can be made on these issues, we must recognize their necessity and blow the whistle on growthmania.

An Emerging Political Economy of Finite Wants and Nongrowth

Although the ideas expressed by Mill have been totally dominated by growthmania, a growing number of economists have frankly expressed their disenchantment with the growth ideology. Arguments stressing ecologically sound limits to wealth and population have been made by Boulding and by Spengler (both past presidents of the American Economic Association).[37] Recently E. J. Mishan, Tibor Scitovsky, and Staffan Linder have made penetrating antigrowth arguments.[38] There is also much in Galbraith that is antigrowth—at least against growth of commodities for which the want must be manufactured along with the product.[39]

In spite of these beginnings, most economists are still hung up on the assumption of infinite wants, or the postulate of nonsatiety, as the mathematical economists call it. Any single want can be satisfied, but all wants in the aggregate cannot be. Wants are infinite in number if not in intensity, and the satisfaction of some wants stimulates other wants. If wants are infinite, growth is always justified—or so it would seem.

Even while accepting the foregoing hypothesis, we could still object to growthmania on the grounds that, given the completely inadequate definition of GNP, "growth" simply means the satisfaction of ever more trivial wants while simultaneously creating ever more powerful externalities that destroy ever more important environmental amenities. To defend ourselves against these externalities, we produce even more, and instead of subtracting the purely defensive expenditures, we add them! For example, the medical bills paid for treatment of cigarette-induced cancer and pollution-induced emphysema are added to GNP when, in a welfare sense, they should clearly be subtracted. This should be labeled *swelling*, not growth. The satisfaction of wants created by brainwashing and "hogwashing" the public through the mass media also represents mostly swelling. A policy of maximizing GNP is practically equivalent to a policy of maximizing depletion and pollution.

We may hesitate to say "maximizing" pollution on the grounds that the production inflow into the stock can be greater than the consumption outflow as long as the stock increases, as it does in our growing economy. To the extent that wealth becomes more durable, the production of waste can be kept low by expanding the stock. But is this in fact what happens? In the present system, if we want to maximize production, we must have a market for it. Increasing the durability of goods reduces the replacement demand. The faster things wear out, the greater can be the flow of production and income. To the extent that consumer apathy and weakening competition permit, there is every incentive to minimize durability. Planned obsolescence and programmed self-destruction and other waste-making practices, so well discussed by Vance Packard, are the logical result of maximizing a marketed physical flow.[40] If we must maximize something, it should be the stock of wealth, not the flow—but with full awareness of the ecological limits that constrain this maximization.

But why this perverse emphasis on flows, this "flow fetishism" of standard economic theory? Again, I believe the underlying issue is distribution. There is no theoretical explanation, much less justification, for the distribution of the stock of wealth. It is a historical datum. But the distribution of the flow of income is at least partly explained by marginal productivity theory, which at times is even misinterpreted as a justification. Everyone gets a part of the flow—call it wages, interest, rent, or profit—and it all looks rather fair. But not everyone owns a piece of the stock, and that does not seem quite so fair. Looking only at the flow helps to avoid disturbing thoughts.

Even the commonsense arguments for infinite wants—that the rich seem to enjoy their high consumption—cannot be generalized without committing the fallacy of composition. If all earned the same high income, a consumption limit occurs sooner than if only a minority had high incomes. The reason is that a large part of the consumption by plutocrats is consumption of personal services rendered by the poor, which would not be available if all were rich. Plutocrats can easily spend large sums on consumption, since all the maintenance work of the household can be done by others. By hiring the poor to maintain and even purchase commodities for them, the rich devote their limited consumption time only to the most pleasurable aspects of consumption. The rich only ride their horses; they do not clean, comb, saddle, and feed them, nor do they clean out the stable. If all did their own maintenance work, consumption would perforce be less. Time sets a limit to consumption.

The big difficulty with the infinite wants assumption, however, is that pointed out by Keynes, who, in spite of the use made of his theories in support of growth, was certainly no advocate of unlimited growth, as can be seen in the following quotation:

> Now it is true that the needs of human beings may seem to be insatiable. But they fall into two classes—those needs which are absolute in the sense that we feel them whatever the situation of our fellow human beings may be, and those which are relative in the sense that we feel them only if their satisfaction lifts us above, makes us feel superior to, our fellows. Needs of the second class, those which satisfy the desire for superiority, may indeed be insatiable; for the higher the general level, the higher still are they. But this is not so true of the absolute needs—a point may soon be reached, much sooner perhaps than we are all of us aware of, when those needs are satisfied in the sense that we prefer to devote our further energies to noneconomic purposes.[41]

For Keynes, real absolute needs are those that can be satisfied and do not require inequality and invidious comparison for their very existence; relative wants are the wants of vanity and are insatiable. Lumping the two categories together and speaking of infinite wants in general can only muddy the waters. The same distinction is implicit in the quotation from Mill, who spoke disparagingly of "consuming things which give little or no pleasure except as representative of wealth."

Some two and a half millennia before Keynes, the prophet Isaiah, in a discourse on idolatry, developed the theme more fully.

> [Man] cuts down cedars; or he chooses a holm tree or an oak and lets it grow strong among the trees of the forest; he plants a cedar and the rain nourishes it. Then it becomes fuel for a man; and he takes a part of it and warms himself, he kindles a fire and bakes bread; also he makes a god and worships it, he makes a graven image and falls down before it. Half of it he burns in the fire; over the half he eats flesh, he roasts meat and is satisfied; also he warms himself and says, "Aha, I am warm, I have seen the fire!"

> And the rest of it he makes into a god, his idol; and he falls down to it and worships it; he prays to it and says; "Deliver me, for thou art my god!"
>
> They know not, nor do they discern; for he has shut their eyes so that they cannot see, and their minds so that they cannot understand. No one considers, nor is there knowledge or discernment to say, "Half of it I burned in the fire, I also baked bread on its coals, I roasted flesh and have eaten; and shall I make the residue of it an abomination? Shall I fall down before a block of wood?" He feeds on ashes, a deluded mind has led him astray, and he cannot deliver himself or say, "Is there not a lie in my right hand?" (Isa. 44:14–20)

The first half of the tree burned for warmth and food, the finite absolute wants of Keynes, the bottom portion of GNP devoted to basic wants—these are all approximately synonymous. The second or surplus half of the tree used to make an idol, Keynes's infinite relative wants or wants of vanity, the top or surplus (growing) portion of GNP used to satisfy marginal wants—these are also synonymous. Furthermore, the surplus half of the tree used to make an idol, an abomination, is symbolic of the use made of the economic surplus throughout history of enslaving and coercing others by gaining control over the economic surplus and obliging people to "fall down before a block of wood." The controllers of the surplus may be a priesthood that controls physical idols made from the surplus and used to extract more surplus in the form of offerings and tribute. Or they may be feudal lords who, through the power given by possession of the land, extract a surplus in the form of rent and the *corvee;* or capitalists (state or private) who use the surplus in the form of capital to gain more surplus in the form of interest and quasi rents. If growth must cease, the surplus becomes less important, and so do those who control it. If the surplus is not to lead to growth, then it must be consumed, and ethical demands for equal participation in the consumption of the surplus could not be countered by arguments that inequality is necessary for accumulation. Accumulation in excess of depreciation, and the privileges attached thereto, would not exist.

We no longer speak of worshiping idols. Instead of idols we have an abomination called GNP, large parts of which, however, bear such revealing names as Apollo, Poseidon, and Zeus. Instead of worshiping the idol, we maximize it. The idol has become rather more abstract and conceptual and rather less concrete and material, while the mode of adoration has become technical rather than personal. But fundamentally, idolatry remains idolatry, and we cry out to the growing surplus, "Deliver me, for thou art my god!" Instead we should pause and ask with Isaiah, "Is there not a lie in my right hand?"

Notes

1. Thomas S. Kuhn, *The Structure of Scientific Revolutions*, 2d ed. (Chicago: University of Chicago Press, 1969).

2. Arthur Koestler, *The Sleepwalkers* (New York: Macmillan, 1968).

3. Michael Polanyi, *Personal Knowledge* (New York: Harper & Row, 1964).

4. Evesy Domar, "Expansion and Employment," *American Economic Review* 37 (March 1947), 34–55.

5. This, of course, is a *physical* axiom. If a "quantity" has no physical dimensions, it is not limited by physical finitude. Thus "psychic income" or welfare may increase forever. But the physical stock that yields want-satisfying services, and the physical flows that maintain that stock, are limited. What about GNP? If we choose to measure GNP in such a way that it reflects total want satisfaction, then presumably it could increase forever. However, this is emphatically *not* the way we measure GNP at present. Prices (exchange values) and quantities are the basis of GNP. Prices bear no relation whatsoever to total utility or want satisfaction. Quantity probably has borne a direct relation to welfare in the past. Whether it still does today in affluent countries is very debatable. But, in any event, quantities are limited by physical considerations. Even quantities of "services rendered" have some irreducible physical dimension. It is always some *thing* that yields a service—for example, a machine or a skilled person.

6. *Science* (November 19, 1971), 847–848.

7. Kuhn, *Structure of Scientific Revolutions*, p. 92.

8. The high professional reputations of Boulding and Georgescu-Roegen are based on their many contributions within the orthodox paradigm. Their work outside that paradigm has probably diminished rather than enhanced their academic prestige.

9. John Ruskin, *Unto This Last: Four Essays on the First Principles of Political Economy*, ed. Lloyd J. Hubenka (Lincoln: University of Nebraska Press, 1967), p. 86 (originally published 1860).

10. Nicholas Georgescu-Roegen, *The Entropy Law and the Economic Process* (Cambridge, Mass.: Harvard University Press, 1971), p. 21.

11. A. C. Fisher and F. M. Peterson, "The Environment in Economics: A Survey," *Journal of Economic Literature* 14 (March 1976), 1.

12. Robert Solow, "The Economics of Resources or the Resources of Economics," *American Economic Review* 64 (May 1974), 11.

13. Frederick Soddy, *Cartesian Economics: The Bearing of Physical Science upon State Stewardship* (London: Hendersons, 1922), p. 9.

14. Harold Barnett and Chandler Morse, *Scarcity and Growth: The Economics of Natural Resource Availability,* Resources for the Future (Baltimore: Johns Hopkins University Press, 1963), p. 11.

15. Imagine a demon who opens and shuts a window in a partition separating two volumes of gas so as to let the fast-moving molecules go from right to left and the slow-moving molecules go from left to right, thereby sorting the two. If the two compartments were originally at equilibrium (equal temperature), then the sorting action of the demon would cause a "spontaneous" increase in temperature of the left compartment and a decrease in the right compartment. This would contradict the entropy law, which forbids spontaneous movement away from equilibrium. The natural or "downhill" direction is toward more mixing of entities; sorting of entities would be like going uphill, which should use energy rather than liberate energy. The temperature differential created by the demon would provide a source of continuous energy, a kind of perpetual motion machine, which is contrary to the laws of physics.

16. *Fertility* refers to actual reproduction, as opposed to fecundity, which refers to reproductive potential or capacity. One measure of fertility is the Gross Reproduction Rate, defined in note 17.

17. GRR is roughly the ratio of one generation to the preceding generation, assuming that all children born survive to the end of their reproductive life. It is usually defined in terms of females only. The length of a generation is the mean age of mothers at childbirth.

18. United Nations, *Population Bulletin of the United Nations,* no. 7, 1963 (New York: United Nations, 1965).

19. Goran Ohlin, *Population Control and Economic Development* (Paris: Development Centre of the Organisation for Economic Co-operation and Development, 1967).

20. According to Robert E. Baldwin: "In the 1957–58 to 1963–64 period, the less developed nations maintained a 4.7 percent annual growth rate in gross national product compared to a 4.4 percent rate in the developed economies. The gap in per capita income widened because population increased at only 1.3 percent annually in the developed countries compared to a 2.4 percent annual rate in the less developed economies." (*Economic Development and Growth* [New York: Wiley, 1966], p. 8.)

21. The term *stationary state* has been burdened with two distinct meanings in economics. The classical meaning is that of an actual state of affairs toward which the real world is supposed to be evolving; that is, a teleological or eschatological concept. The neoclassical sense of the term is entirely mechanistic—an epistemologically useful fiction like an ideal gas or frictionless machine—and describes an economy in which tastes and techniques are

constant. The latter sense is more current in economics today, but the former meaning is the relevant one in this discussion.

22. J. S. Mill, *Principles of Political Economy*, vol. 2 (London: John W. Parker, 1857), pp. 320–326, with omissions.

23. All quotes in this paragraph are from Mark Blaug, *Economy Theory in Retrospect* (Homewood, Ill.: Irwin, 1968), pp. 214–221. Blaug's views are, I think representative of orthodox economists.

24. By *stock* is meant a quantity measured at a point in time; for example, a population census or a balance sheet of assets and liabilities as of a certain date. By *flow* is meant a quantity measured across some actual or conceptual boundary over a period of time; for example, births and deaths per year or an income and loss statement for a given year. The boundary lines separating the stock of wealth from the rest of the physical world may sometimes be fuzzy. But the main criterion is that physical wealth must in some way have been transformed by human beings to increase its usefulness over its previous state as primary matter or energy. For example, coal in the ground is primary matter and energy; coal in the inventory of firms and households is physical wealth; coal after use in the form of carbon dioxide and soot is waste matter. The heat produced by the coal is partly usable and partly unusable. Eventually, all the heat becomes unusable or waste heat, but, while it is usable, it is a part of the physical stock of wealth. For some purposes, we may wish to define proven reserves in mines as part of wealth, but that presents no problems.

25. Staffan B. Linder, *The Harried Leisure Class* (New York: Columbia University Press, 1970).

26. A. J. Lotka, *Elements of Mathematical Biology* (New York: Dover, 1957). Republication. See especially chapter 24.

27. Report of the Committee for Environmental Information before Joint Congressional Committee on Atomic Energy, January 29, 1970. Quoted in "The Space Available Report of the Committee for Environmental Information," *Environment* 12 (March 1970), 7.

28. Herman E. Daly, "On Economics as a Life Science," *Journal of Political Economy* 76 (July 1968), 392–406. Reprinted in this volume.

29. Erwin Schrödinger, *What Is Life?* (New York: Macmillan, 1945).

30. Kenneth E. Boulding, "The Economics of the Coming Spaceship Earth," in Henry Jarrett, ed., *Environmental Quality in a Growing Economy* (Baltimore: Johns Hopkins University Press, 1966). Reprinted in this volume.

31. Eugene P. Odum, "The Strategy of Ecosystem Development," *Science* (April 18, 1969).

32. Services are included in GNP and are not in themselves physical outputs. However, increasing service outputs often require increases in physical inputs to the service sector, so that there is an indirect physical component.

Leisure is not counted in GNP, and more physical inputs are not necessarily required as the amount to leisure is increased.

33. Bertrand Russell, *In Praise of Idleness and Other Essays* (London: Allen and Unwin, 1935), pp. 16–17.

34. Joan Robinson, *Economic Philosophy* (London: Watts, 1962), p. 55.

35. Robert Theobald, *Free Men and Free Markets* (Garden City, N.Y.: Doubleday, 1965).

36. Oskar Lange, *On the Economic Theory of Socialism*, ed. Benjamin E. Lippincott (New York: McGraw-Hill, 1964).

37. Boulding, "Economics of the Coming Spaceship Earth"; J. J. Spengler, public address, Yale Forestry School, Summer 1969.

38. E. J. Mishan, *The Costs of Economic Growth* (New York: Praeger, 1967); Tibor Scitovsky, "What Price Economic Growth," *Papers on Welfare and Growth* (Stanford, Calif.: Stanford University Press, 1964); Linder, *Harried Leisure Class*.

39. J. K. Galbraith, *The Affluent Society* (Boston: Houghton Mifflin, 1958).

40. Vance Packard, *The Waste Makers* (New York: Pocket Books, 1963).

41. J. M. Keynes, "Economic Possibilities for Our Grandchildren," in *Essays in Persuasion* (New York: Norton, 1963; originally published 1931).

I Ecology: Ultimate Means and Biophysical Constraints

Any physical object which by its influence deteriorates its environment commits suicide.

Alfred North Whitehead, *Science and the Modern World,* 1925

Introduction

Herman E. Daly and
Kenneth N. Townsend

"All flesh is grass," said the prophet Isaiah. That is probably the most concise statement ever made of the ecological constraints on human life. But such visions of unity and wholeness have been fragmented by the specialization of modern thought. The economist's abstract world of commodities, with its laws of motion and equilibrium, has very few points of contact left with "grass" and is even in danger of losing touch with "flesh." Seemingly, economics has become detached from its own biophysical foundations. Standard textbooks do little to counteract this trend, representing the economic process—according to economist Nicholas Georgescu-Roegen—with a mechanistic diagram of a circular flow, "a pendulum movement between production and consumption within a completely closed system." In modern economic growth theory, aggregate production functions generally ignore nature and natural resources completely. Physical scientists, such as M. King Hubbert, find this neglect rather perplexing:

> One speaks of the rate of growth of GNP. I haven't the faintest idea what this means when I try to translate it into coal, and oil, and iron, and the other physical quantities which are required to run an industry. So far as I have been able to find out, the quantity GNP is a monetary bookkeeping entity. It obeys the laws of money. It can be expanded or diminished, created or destroyed, but it does not obey the laws of physics.[1]

Even though GNP is an abstract entity, it is a value *index* of an aggregate of *physical* quantities. Value is the product of prices times those quantities. In calculating real GNP and its growth rate, we hold constant both absolute and relative prices for the purpose of isolating and measuring *quantitative* change. Although GNP cannot be expressed in simple physical units, it remains an index of physical quantities and therefore should be very much subject to laws of physics. Economic models that ignore this dependence are grossly deficient and are in large part responsible for our present ecological crisis.

Earlier thinkers have called attention to this deficiency, and some of their words bear repeating. J. A. Hobson, a British economic heretic of the late nineteenth and early twentieth century (familiar to economics students for his theory of underconsumption, which influenced Keynes, and his theory of imperialism, which influenced Lenin), noted that

> all serviceable organic activities consume tissue and expend energy, the biological costs of the services they render. Though this economy may not correspond in close quantitative fashion to a pleasure and pain economy or to any conscious valuation, it must be taken as the groundwork for that conscious valuation. For most economic purposes we are well-advised to prefer the organic test to any other test of welfare, bearing in mind that many organic costs do not register themselves easily or adequately in terms of conscious pain or disutility, while organic gains are not always interpretable in conscious enjoyment.[2]

The mathematical biologist A. J. Lotka, noted for his contributions to demography in the early years of the twentieth century, expressed a similar insight:

> Underlying our economic manifestations are biological phenomena which we share in common with other species; and . . . the laying bare and clearly formulating of the relations thus involved—in other words the analysis of biophysical foundations of economics—is one of the problems coming within the program of physical biology.[3]

Clearly, Hobson and Lotka saw the importance of natural bases for economic thinking.

And so do the authors in part I. The articles here deal with the biophysical foundations of economics and the groundwork for conscious valuation. In "Why Isn't Everyone as Scared as We Are?" biologists Paul and Anne Ehrlich provide an overview of the issue in terms of a historical choice point between the "growthmanic" path of orthodox economists and the "sustainable society" path of the ecologists and bioeconomists. Perhaps the major blind spots of growth economists are their lack of appreciation of the second law of thermodynamics and its manifold implications and their failure to recognize the magnitude and fragility of the life support services provided by the very natural systems whose proper functioning is being disrupted by the ever-larger entropic flow of materials and energy required by economic growth.

An introduction to "Availability, Entropy, and the Laws of Thermodynamics" is provided by Ehrlich, Ehrlich, and Holdren, setting the stage for Nicholas Georgescu-Roegen's discussion of "The Entropy Law and the Economic Problem," which is essentially the introduction to his magistral work *The Entropy Law and the Economic Process,* a book that made a major contribution toward reuniting economics with its biophysical foundations. The implications of this analysis are extended in the excerpts from his "Energy and Economic Myths," and some general policy implications, which he calls a "minimal bioeconomic program," are spelled out. Georgescu-Roegen's reservations about the concept of a steady-state economy are also included (and replied to in the postscript to this volume).

M. K. Hubbert's discussion of "Exponential Growth as a Transient Phenomenon in Human History" gives a geologist's analysis of the biophysical constraints to economic growth, and of the "exponential-growth culture" that renders mankind seemingly incapable of dealing with problems associated with nongrowth.

Renewable resources provide a means of tapping the relatively permanent flow of solar energy, which, if managed on a sustainable-yield basis, can provide a quasi-permanent source of useful low entropy. But renewable resources can easily be rendered nonrenewable by overexploitation. The institutional conditions leading to overexploitation are analyzed by biologist Garrett Hardin in his classic "Tragedy of the Commons," now supplemented by a new section of "Second Thoughts" on the subject, which Dr. Hardin kindly wrote at the editors' invitation.

Notes

1. M. K. Hubbert in F. F. Darling and J. P. Milton, eds., *Future Environments of North America* (Garden City, N.Y.: Natural History Press, 1966), p. 291.

2. J. A. Hobson, *Economics and Ethics* (Boston: Heath, 1929), p. xxi.

3. A. J. Lotka, *Elements of Mathematical Biology* (New York: Dover, 1957).

1 Why Isn't Everyone as Scared as We Are?

Paul R. Ehrlich and
Anne H. Ehrlich

In the early 1930s, when we were born, the world population was just 2 billion; now it is more than two and a half times as large and still growing rapidly.[1] The population of the United States is increasing much more slowly than the world average, but it has more than doubled in only six decades—from 120 million in 1928 to 250 million in 1990.[2] Such a huge population expansion within two or three generations can by itself account for a great many changes in the social and economic institutions of a society. It also is very frightening to those of us who spend our lives trying to keep track of the implications of the population explosion.

A Slow Start

One of the toughest things for a population biologist to reconcile is the contrast between his or her recognition that civilization is in imminent serious jeopardy and the modest level of concern that population issues generate among the public and even among elected officials.

Much of the reason for this discrepancy lies in the slow development of the problem. People aren't scared because they evolved biologically and culturally to respond to short-term "fires" and to tune out long-term "trends" over which they had no control.[3] Only if we do what doesn't come naturally—if we determinedly focus on what seem to be gradual or nearly imperceptible changes—can the outlines of our predicament be perceived clearly enough to be frightening.

Consider the *very* slow-motion origins of our predicament. It seems reasonable to define humanity as having first appeared some four million years ago in the form of australopithecines, small-brained upright creatures like "Lucy."[4] Of course, we don't know the size of this first human population but it's likely that there were never more than 125,000 australopithecines at any given time.

Our own species, *Homo sapiens*,[5] evolved a few hundred thousand years ago. Some ten thousand years ago, when agriculture was invented, probably no more than five million people inhabited Earth—fewer than now live in the San Francisco Bay area. Even at the time of Christ, two thousand years ago, the entire human population was roughly the size of the population of the United States today; by 1650 there were only 500 million people, and in 1850 only a little over a billion. Since there are now well past 5 billion people, the vast majority of the population explosion has taken place in less than a tenth of one percent of the history of *Homo sapiens*.

This is a remarkable change in the abundance of a single species. After an unhurried pace of growth over most of our history, expansion of the population accelerated during the Industrial Revolution and really shot up after 1950. Since midcentury, the human population has been growing at annual rates ranging from about 1.7 to 2.1 percent per year, doubling in forty years or less. Some groups have grown significantly faster; the population of the African nation of Kenya was estimated to be increasing by over 4 percent annually during the 1980s—a rate that if continued would double the nation's population in only seventeen years.[6] That rate did continue for over a decade and only recently has shown slight signs of slowing. Meanwhile, other nations, such as those of northern Europe, have grown much more slowly in recent decades.

But even the highest growth rates are still *slow-motion changes compared to events we easily notice and react to*. A car swerving at us on the highway is avoided by actions taking a few seconds. The Alaskan oil spill caused great public indignation but faded from the media and the consciousness of most people in a few months. America's participation in World War II spanned less than four years. During the last four years, even Kenya's population grew by only about 16 percent—a change hardly perceptible locally, let alone from a distance. In four years, the world population expands only a little more than 7 percent. Who could notice that? Precipitous as the population explosion has been in historical terms, it is occurring at a snail's pace in an individual's perception. It is not an event, it is a trend that must be analyzed in order for its significance to be appreciated.

Exponential Growth

The time it takes a population to double in size is a dramatic way to picture rates of population growth, one that most of us can understand more readily than percentage growth rates. Human populations have often grown in a pattern described as "exponential."[7] Exponential growth occurs in bank accounts when interest is left to accumulate and itself earns interest. Exponential growth occurs in populations because children, the analogue of interest, remain in the population and themselves have children.[8]

A key feature of exponential growth is that it often seems to start slowly and finish fast. A classic example used to illustrate this is the pond weed that doubles each day the amount of pond surface covered and is projected to cover the entire pond in thirty days. The question is, how much of the pond will be covered in twenty-nine days? The answer, of course, is that just half of the pond will be covered in twenty-nine days. The weed will then double once more and cover the entire pond the next day. As this example indicates, exponential growth contains the potential for big surprises.[9]

The limits to human population growth are more difficult to perceive than those restricting the pond weed's growth. Nonetheless, like the pond weed, human populations grow in a pattern that is essentially exponential, so we must be alert to the treacherous properties of that sort of growth. The key point to remember is that *a long history of exponential growth in no way implies a long future of exponential growth.* What begins in slow motion may eventually overwhelm us in a flash.

The last decade or two has seen a slight slackening in the human population growth rate—a slackening that has been prematurely heralded as an "end to the population explosion." The slowdown has been only from a peak annual growth rate of perhaps 2.1 percent in the early 1960s to about 1.8 percent in 1990. To put this change in perspective, the population's doubling time has been extended from thirty-three years to thirty-nine. Indeed, the world population *did* double in the thirty-seven years from 1950 to 1987. But even if birthrates continue to fall, the world population will continue to expand (assuming that death rates don't rise), although at a slowly slackening rate, for about another century. Demographers think that growth will not end before the population has reached 10 billion or more.[10]

So, even though birthrates have declined somewhat, *Homo sapiens* is a long way from ending its population explosion or avoiding its consequences. In fact, the biggest jump, from 5 to 10 billion in well under a century, is still ahead. But this does not mean that growth couldn't be ended sooner, with a much smaller population size, if we—all of the world's nations—made up our minds to do it. The trouble is, many of the world's leaders and perhaps most of the world's people still don't believe that there are compelling reasons to do so. They are even less aware that if humanity fails to act, *nature may end the population explosion for us*—in very unpleasant ways—well before 10 billion is reached.

Those unpleasant ways are beginning to be perceptible. Humanity in the 1990s will be confronted by more and more intransigent environmental problems, global problems dwarfing those that worried us in the late 1960s. Perhaps the most serious is that of global warming, a problem caused in large part by population growth and overpopulation. It is not clear whether the severe drought in North America, the Soviet Union, and China in 1988 was the result of the slowly rising surface temperature of Earth, but it is precisely the kind of event that climatological models predict as more and more likely with continued global warming.[11] In addition to more frequent and more severe crop failures, projected consequences of the warming include coastal flooding, desertification, the creation of as many as 300 million environmental refugees,[12] alteration of patterns of disease, water shortages, general stress on natural ecosystems, and synergistic interactions among all these factors.[13]

Continued population growth and the drive for development in already badly overpopulated poor nations will make it *exceedingly* difficult to slow the greenhouse warming—and impossible to stop or reverse it—in this generation at least. And, even if the warming should miraculously not occur, contrary to accepted projections,[14] human numbers are on a collision course with massive famines anyway.

Making the Population Connection

Global warming, acid rain, depletion of the ozone layer, vulnerability to epidemics, and exhaustion of soils and groundwater are all, as we shall see, related to population size. They are also clear and present dangers to the persistence of civilization. Crop failures due to global warming alone might result in the premature deaths of a billion or more in the next few decades, and the AIDS epidemic could slaughter hundreds of millions. Together these would constitute a harsh "population control" program provided by nature in the face of humanity's refusal to put into place a gentler program of its own.

We shouldn't delude ourselves: the population explosion will come to an end before very long. The only remaining question is whether it will be halted through the humane method of birth control, or by nature wiping out the surplus. We realize that religious and cultural opposition to birth control exists throughout the world; but we believe that people simply don't understand the choice that such opposition implies. Today, anyone opposing birth control is unknowingly voting to have the human population size controlled by a massive increase in early deaths.

Of course, the environmental crisis isn't caused just by expanding human numbers. Burgeoning consumption among the rich and increasing dependence on ecologically unsound technologies to supply that consumption also play major parts. This allows some environmentalists to dodge the population issue by emphasizing the problem of malign technologies. And social commentators can avoid commenting on the problem of too many people by focusing on the serious maldistribution of affluence.

But scientists studying humanity's deepening predicament recognize that a major factor contributing to it is rapidly worsening overpopulation. The Club of Earth, a group whose members all belong to both the U.S. National Academy of Sciences and the American Academy of Arts and Sciences, released a statement in September 1988 that said in part:

> Arresting global population growth should be second in importance only to avoiding nuclear war on humanity's agenda. Overpopulation and rapid population growth are intimately connected with most aspects of the current human predicament, including rapid depletion of nonrenewable resources, deterioration of the environment (including rapid climate change), and increasing international tensions.[15]

When three prestigious scientific organizations cosponsored an international scientific forum, "Global Change," in Washington in 1989, there was general agreement among the speakers that population growth was a substantial contributor toward prospective catastrophe. Newspaper coverage was limited, and while the population component was mentioned in *The New York Times*'s article,[16] the point that population limitation will be essential to resolving the predicament was lost. The coverage of environmental issues in the media has been generally excellent in the last few years, but there is still a long way to go to get adequate coverage of the intimately connected population problem.

Even though the media occasionally give coverage to population issues, some people never get the word. In November 1988, Pope John Paul II reaffirmed the Catholic Church's ban on contraception. The occasion was the twentieth anniversary of Pope Paul VI's anti-birth-control encyclical, *Humanae Vitae.*

Fortunately, the majority of Catholics in the industrial world pay little attention to the encyclical or the Church's official ban on all practical means of birth control. One need only note that Catholic Italy at present has the smallest average completed family size (1.3 children per couple) of any nation. Until contraception and then abortion were legalized there in the 1970s, the Italian birth rate was kept low by an appalling rate of abortion.

The bishops who assembled to celebrate the anniversary defended the encyclical by announcing that "the world's food resources theoretically could feed 40 billion people."[17] In one sense they were right. It's "theoretically possible" to feed 40 billion people—in the same sense that it's theoretically possible for your favorite major league baseball team to win every single game for fifty straight seasons, or for you to play Russian roulette ten thousand times in a row with five out of six chambers loaded without blowing your brains out.

One might also ask whether feeding 40 billion people is a worthwhile goal for humanity, even if it could be reached. Is any purpose served in turning Earth, in essence, into a gigantic human feedlot? Putting aside the near certainty that such a miracle couldn't be sustained, what would happen to the *quality* of life?

We wish to emphasize that the population problem is in no sense a "Catholic problem," as some would claim. Around the world, Catholic reproductive performance is much the same as that of non-Catholics in similar cultures and with similar economic status. Nevertheless, the *political* position of the Vatican, traceable in no small part to the extreme conservatism of Pope John Paul II, is an important barrier to solving the population problem.[18] Non-Catholics should be very careful not to confuse Catholics or Catholicism with the Vatican—most American Catholics don't. Furthermore, the Church's position on contraception is distressing to many millions of Catholics, who feel it morally imperative to follow their own consciences in their personal lives and disregard the Vatican's teachings on this subject.

Nor is unwillingness to face the severity of the population problem limited to the Vatican. It's built into our genes and our culture. That's one reason many otherwise bright and humane people behave like fools when confronted with demographic issues. Thus, an economist specializing in mail-order marketing can sell the thesis that the human population could increase essentially forever because people are the "ultimate resource," and a journalist can urge more population growth in the United States so that we can have a bigger army![19] Even some environmentalists are taken in by the frequent assertion that "there is no population problem, only a problem of distribution." The statement is usually made in a context of a plan for conquering hunger, as if food shortage were the only consequence of overpopulation.

But even in that narrow context, the assertion is wrong. Suppose food *were* distributed equally. If everyone in the world ate as Americans do, less than half the *present* world population could be fed on the record harvests of 1985 and 1986.[20] Of course, everyone doesn't have to eat like Americans. About a third of the world grain harvest—the staples of the human feeding base—is fed to animals to produce eggs, milk, and meat for American-style diets. Wouldn't feeding that grain directly to people solve the problem? If everyone were willing to eat an essentially vegetarian diet, that additional grain would allow perhaps a billion more people to be fed with 1986 production.

Would such radical changes solve the world food problem? Only in the *very* short term. The additional billion people are slated to be with us by the end of the century. Moreover, by the late 1980s, humanity already seemed to be encountering trouble maintaining the production levels of the mid-1980s, let alone keeping up with population growth. The world grain harvest in 1988 was some 10 percent *below* that of 1986. And there is little sign that the rich are about to give up eating animal products.

So there is no reasonable way that the hunger problem can be called "only" one of distribution, even though redistribution of food resources would greatly alleviate hunger today. Unfortunately, an important truth, that maldistribution is a cause of hunger now, has been used as a way to avoid a more important truth—that overpopulation is critical today and may well make the distribution question moot tomorrow.

The food problem, however, attracts little immediate concern among well-fed Americans, who have no reason to be aware of its severity or extent. But other evidence that could make everyone face up to the seriousness of the population dilemma is now all around us, since problems to which overpopulation and population growth make major contributions are worsening at a rapid rate. They often appear on the evening news, although the population connection is almost never made.

Consider the television pictures of barges loaded with garbage wandering like The Flying Dutchman across the seas, and news stories about "no room at the dump."[21] They are showing the results of the interaction between too many affluent people and the environmentally destructive technologies that support that affluence. Growing opportunities to swim in a mixture of sewage and medical debris off American beaches can be traced to the same source. Starving people in sub-Saharan Africa are victims of drought, defective agricultural policies, and an overpopulation of both people and domestic animals—with warfare often dealing the final blow. All of the above are symptoms of humanity's massive and growing negative impact on Earth's life-support systems.

Recognizing the Population Problem

The average person, even the average scientist, seldom makes the connection between such seemingly disparate events and the population problem, and thus remains unworried. To a degree, this failure to put the pieces together is due to a taboo against frank discussion of the population crisis in many quarters, a taboo generated partly by pressures from the Catholic hierarchy and partly by other groups who are afraid that dealing with population issues will produce socially damaging results.

Many people on the political left are concerned that focusing on overpopulation will divert attention from crucial problems of social justice (which certainly need to be addressed *in addition* to the population problem). Often those on the political right fear that dealing with overpopulation will encourage abortion (it need not) or that halting growth will severely damage the economy (it could, if not handled properly). And people of varied political persuasions who are unfamiliar with the magnitude of the population problem believe in a variety of far-fetched technological fixes—such as colonizing outer space—that they think will allow the need for regulating the size of the human population to be avoided forever.[22]

Even the National Academy of Sciences avoided mentioning controlling human numbers in its advice to President Bush on how to deal with global environmental change. Although Academy members who are familiar with the issue are well aware of the critical population component of that change, it was feared that all of the Academy's advice would be ignored if recommendations were included about a subject taboo in the Bush administration. That strategy might have been correct, considering Bush's expressed views on abortion and considering the administration's weak appointments in many environmentally sensitive positions. After all, the Office of Management and Budget even tried to suppress an expert evaluation of the potential seriousness of global warming by altering the congressional testimony of a top NASA scientist, James Hansen, to conform with the administration's less urgent view of the problem.[23]

All of us naturally lean toward the taboo against dealing with population growth. The roots of our aversion to limiting the size of the human population are as deep and pervasive as the roots of human sexual behavior. Through billions of years of evolution, outreproducing other members of your population was the name of the game. It is the very basis of natural selection, the driving force of the evolutionary process.[24] Nonetheless, the taboo must be uprooted and discarded.

Overcoming the Taboo

There is no more time to waste; in fact, there wasn't in 1968 when *The Population Bomb* was published. Human inaction has already condemned hundreds of millions more people to premature deaths from hunger and disease. The population connection must be made in the public mind. Action to end the population explosion *humanely* and start a gradual population *decline* must become a top item on the human agenda: the human birthrate must be lowered to slightly below the human death rate as soon as possible. There still may be time to limit the scope of the impending catastrophe, but not *much* time. Ending the population explosion by controlling births is necessarily a slow process. Only nature's cruel way of solving the problem is likely to be swift.

Of course, if we do wake up and succeed in controlling our population size, that will still leave us with all the other thorny problems to solve. Limiting human numbers will not alone end warfare, environmental deterioration, poverty, racism, religious prejudice, or sexism; it will just buy us the opportunity to do so. As the old saying goes, whatever your cause, it's a lost cause without population control.[25]

America and other rich nations have a clear choice today. They can continue to ignore the population problem and their own massive contributions to it. Then they will be trapped in a downward spiral that may well lead to the end of civilization in a few decades. More frequent droughts, more damaged crops and famines, more dying forests, more smog, more international conflicts, more epidemics, more gridlock, more drugs, more crime, more sewage swimming, and other extreme unpleasantness will mark our course. It is a route already traveled by too many of our less fortunate fellow human beings.

Or we can change our collective minds and take the measures necessary to lower global birthrates dramatically. People can learn to treat growth as the cancerlike disease it is and move toward a sustainable society. The rich can make helping the poor an urgent goal, instead of seeking more wealth and useless military advantage over one another. Then humanity might have a chance to manage all those other seemingly intractable problems. It is a challenging prospect, but at least it will give our species a shot at creating a decent future for itself. More immediately and concretely, taking action now will give our children and their children the possibility of decent lives.

Notes

1. The world population in 1990 is about 5.3 billion. Most demographic information is from *1989 World Population Data Sheet*, issued by the Population Reference Bureau (PRB), 777 Fourteenth St. NW, Suite 800, Washington, D.C. 20005. In some cases, as above, we have made simple extrapolations for the 1990 figures. Besides the fine annual data sheet, PRB produces several very useful publications on population issues.

2. Note that the U.S. population was growing much faster before then, spurred by substantial numbers of immigrants. It *quadrupled* in the 6 decades before 1928, turning a post–Civil War society largely restricted to the eastern half of the nation into a cosmopolitan world power spanning the continent.

3. This evolutionary blind spot is discussed at length in R. Ornstein and P. Ehrlich, *New World/New Mind* (New York: Doubleday, 1988).

4. D. Johanson and M. Edey, *Lucy: The Beginnings of Mankind* (New York: Simon and Schuster, 1981). While there is still controversy over details of human history, there is no dispute that an erect, small-brained hominid something like Lucy was one of our ancestors. This exciting book beautifully presents the view of human origins of one outstanding group of scientists. For more on the controversies and on other discoveries, see R. Lewin's excellent *Bones of Contention* (New York: Simon and Schuster, 1987).

5. Note that we are considering *Homo sapiens* as the latest human species and are applying the term *human* to all hominids since the australopithecines (just as the term *ape* is applied to several species). Some people would restrict the term *human* to *Homo sapiens.*

6. When annual growth rates are under 5 percent, a working estimate of the number of years required to double the population at that rate can be obtained by simply dividing the percentage rate into 70. Thus, with Kenya's growth rate of 4.1 percent, the estimate of doubling time is 70/4.1 = 17.1 years. A recent decline in Kenya's birthrate was reported in J. Perlez, "Birth Control Making Inroads in Populous Kenya," *New York Times*, September 10, 1989, but the population still has a doubling rate of less than 20 years.

7. Exponential growth occurs when the increase in population size in a given period is a *constant* percentage of the size at the beginning of the period. Thus a population growing at 2 percent annually or a bank account growing at 6 percent annually will be growing exponentially. Exponential growth does not have to be fast; it can go on at very low rates or, if the rate is negative, can be exponential shrinkage.

Saying a population is "growing exponentially" has almost come to mean "growing very fast," but that interpretation is erroneous. True exponential growth is rarely seen in human populations today, since the percentage of growth has been changing. In most cases, the growth rate has been gradually declining since the late 1960s. Nevertheless, it is useful to be aware of the exponential model, since it is implied every time we project a population size into the future with qualifying statements such as "if that rate continues."

8. For mathematical details on exponential growth, see P. R. Ehrlich, *Environment* (San Francisco: Freeman, 1977), pp. 100–104. The term *exponential* comes from the presence in the equation for growth of a constant, *e*, the base of natural logarithms, raised to a power (exponent) that is variable (the growth rate multiplied by the time that rate will be in effect).

9. The potential for surprise in repeated doublings can be underlined with another example. Suppose you set up an aquarium with appropriate life support systems to maintain 1,000 guppies, but no more. If that number is exceeded, crowding will make the fishes susceptible to "ich," a parasitic disease that will kill most of the guppies. You then begin the population with a pair of sex-crazed guppies. Suppose that the fishes reproduce fast enough to double their population size every month. For eight months everything is fine, as the population grows 2→4→8→16→32→64→128→256→512. Then within the ninth month the guppy population surges through the fatal 1,000 barrier, the aquarium becomes overcrowded, and most of the fishes perish. In fact, the last 100 guppies appear in less than five days—about 2 percent of the population's history.

10. Note that "doubling times" represent what would happen if the growth rates of the moment continued unchanged into the future. Demographic projections include changes in growth rates, usually caused by reductions in birthrates and/or *declines* in death rates (demographers classically don't consider rises in death rates in their global projections). Projections therefore often show the population taking more, and occasionally less, time to double than was indicated by the "doubling time" of a recent year.

11. For a fine discussion of climate models, see S. H. Schneider, *Global Warming* (San Francisco: Sierra Club Books, 1989).

12. "Eco-Refugees Warning," *New Scientist*, June 10, 1989.

13. Synergisms occur when the joint impact of two (or more) factors is greater than the sum of their separate impacts.

14. See Schneider, *Global Warming*, and extensive references therein.

15. Statement released September 3, 1988, at the Pugwash Conference on Global Problems and Common Security, at Dagomys, near Sochi, USSR. The signatories were Jared Diamond, UCLA; Paul Ehrlich, Stanford; Thomas Eisner, Cornell; G. Evelyn Hutchinson, Yale; Gene E. Likens, Institute of Ecosystem Studies; Ernst Mayr, Harvard; Charles D. Michener, University of Kansas; Harold A. Mooney, Stanford; Ruth Patrick, Academy of Natural Sciences, Philadelphia; Peter H. Raven, Missouri Botanical Garden; and Edward O. Wilson, Harvard.

The National Academy of Sciences and the American Academy of Arts and Sciences are the top honorary organizations for American scientists and scholars, respectively. Hutchinson, Patrick, and Wilson are also laureates of the Tyler Prize, the most distinguished international award in ecology.

16. May 4, 1989, by Philip Shabecoff, a fine environmental reporter. In general, the *Times* coverage of the environment is excellent. But even this best of American newspapers reflects the public's lack of understanding of the urgency of the population situation.

17. *Washington Post*, November 19, 1988, p. C15.

18. Italy is not a freak case. Catholic France has an average completed family size of 1.8 children, the same as Britain and Norway; Catholic Spain, with less than half the per capita GNP of Protestant Denmark, has the same completed family size of 1.8 children. We are equating "completed family size" here with the *total fertility rate,* the average number of children a woman would bear in her lifetime, assuming that current age-specific birth and death rates remained unchanged during her childbearing years—roughly 15–49. In the United States, a Catholic woman is more likely to seek abortion than a non-Catholic woman (probably because she is likelier to use less-effective contraception). By 1980, Catholic and non-Catholic women in the U.S. (except Hispanic women, for whom cultural factors are strong) had virtually identical family sizes. (W. D. Mosher, "Fertility and Family Planning in the United States: Insights from the National Survey of Family Growth," *Family Planning Perspectives* 20, no. 5 [September/October 1988], 202–217) On the role of the Vatican, see, for instance, Stephen D. Mumford, "The Vatican and Population Growth Control: Why an American Confrontation?," *The Humanist* (September/October 1983), and Penny Lernoux, "The Papal Spiderweb," *The Nation,* April 10 and 17, 1989.

19. J. Simon, *The Ultimate Resource* (Princeton, N.J.: Princeton University Press, 1981); B. Wattenberg, *The Birth Dearth* (New York: Pharos Books, 1987).

20. R. W. Kates, R. S. Chen, T. E. Downing, J. X. Kasperson, E. Messer, S. R. Millman, *The Hunger Report: 198* (Providence, R.I.: The Alan Shawn Feinstein World Hunger Program, Brown University, 1988). The data on distribution in this paragraph are from this source.

21. The name of a series of reports on KRON-TV's news programs, San Francisco, the week of May 8, 1989.

22. For an amusing analysis of the "outer space" fairy tale, see Garrett Hardin's classic essay "Interstellar Migration and the Population Problem," *Journal of Heredity* 50 (1959), 68–70, reprinted in G. Hardin, ed., *Stalking the Wild Taboo,* 2d ed. (Los Altos, Calif.: William Kaufmann, 1978). Note that some things have changed; to keep the population of Earth from growing today, we would have to export to space 95 million people annually!

23. This story received broad coverage in both electronic and print media; for instance, *New York Times,* May 8, 1989.

24. For a discussion of natural selection and evolution written for nonspecialists, see P. R. Ehrlich, *The Machinery of Nature* (New York: Simon and Schuster, 1986).

25. "Population control" does not require coercion, only attention to the needs of society.

2 Availability, Entropy, and the Laws of Thermodynamics

Paul R. Ehrlich,
Anne H. Ehrlich, and
John P. Holdren

Many processes in nature and in technology involve the transformation of energy from one form into others. For example, light from the sun is transformed, upon striking a meadow, into thermal energy in the warmed soil, rocks, and plants; into latent heat of vaporization as water evaporates from the soil and through the surface of the plants; and into chemical energy captured in the plants by photosynthesis. Some of the thermal energy, in turn, is transformed into infrared electromagnetic radiation heading skyward. The imposing science of thermodynamics is just the set of principles governing the bookkeeping by which one keeps track of energy as it moves through such transformations. A grasp of these principles of bookkeeping is essential to an understanding of many problems in environmental sciences and energy technology.

The essence of the accounting is embodied in two concepts known as the first and second laws of thermodynamics. No exception to either one has ever been observed. The first law, also known as the law of conservation of energy, says that energy can neither be created nor destroyed. If energy in one form or one place disappears, the same amount must show up in another form or another place. In other words, although transformations can alter the *distribution* of amounts of energy among its different forms, the *total* amount of energy, when all forms are taken into account, remains the same. The term *energy consumption,* therefore, is a misnomer; energy is used, but it is not really consumed. One can speak of fuel consumption, because fuel, as such, does get used up. But when we burn gasoline,

From *Ecoscience,* by Paul R. Ehrlich, Anne H. Ehrlich, and John P. Holdren (San Francisco: W. H. Freeman and Company, 1977); reprinted by permission of the authors.

the amounts of energy that appear as mechanical energy, thermal energy, electromagnetic radiation, and other forms are exactly equal all together to the amount of chemical potential energy that disappears. The accounts must always balance; apparent exceptions have invariably turned out to stem from measurement errors or from overlooking categories. The immediate relevance of the first law for human affairs is often stated succinctly as, "You can't get something for nothing."

Yet, if energy is stored work, it might seem that the first law is also saying, "You can't lose!" (by saying that the total amount of stored work in all forms never changes). If the amount of stored work never diminishes, how can we become worse off? One obvious answer is that we can become worse off if energy flows to places where we can no longer get at it—for example, infrared radiation escaping from Earth into space. Then the stored work is no longer accessible to us, although it still exists. A far more fundamental point, however, is that *different kinds of stored work are not equally convertible into useful, applied work.* We can therefore become worse off if energy is transformed from a more convertible form to a less convertible one, even though no energy is destroyed and even if the energy has not moved into an inaccessible place. The degree of convertibility of energy—stored work into applied work—is often called *availability.*

Energy in forms having high availability (that is, in which a relatively large fraction of the stored work can be converted into applied work) is often called high-grade energy. Correspondingly, energy of which only a small fraction can be converted to applied work is called low-grade energy, and energy that moves from the former category to the latter is said to have been degraded. Electricity and the chemical energy stored in gasoline are examples of high-grade energy; the infrared radiation from a light bulb and the thermal energy in an automobile exhaust are corresponding examples of lower-grade energy. The quantitative measure of the availability of thermal energy is temperature. More specifically, the larger the *temperature difference* between a substance and its environment, the more convertible into applied work is the thermal energy the substance contains; in other words, the greater the temperature difference, the greater the availability. A small pan of water boiling at 100°C in surroundings that are at 20°C represents considerable available energy because of the temperature difference; the water in a swimming pool at the same 20°C temperature as the surroundings contains far more total thermal energy than the water in the pan, but the availability of the thermal energy in the swimming pool is zero, because there is no temperature difference between it and its surroundings.

With this background, one can state succinctly the subtle and overwhelmingly important message of the second law of thermodynamics: *all physical processes, natural and technological, proceed in such a way that the availability of the energy involved decreases.* (Idealized processes can be constructed theoretically in which the availability of the energy involved stays constant, rather than decreasing, but in all real processes there is *some* decrease. The second law says that an *increase* is not possible, even in an ideal process.) As with the first law, apparent violations of the second law often stem from leaving something out of the accounting. In many processes, for example, the availability of energy in some *part* of the affected system increases, but the decrease of availability elsewhere in the system is always large enough to result in a net decrease in availability of energy overall. What is consumed when we use energy, then, is not energy itself but its availability for doing useful work.

The statement of the second law given above is deceptively simple; whole books have been written about equivalent formulations of the law and about its implications. Among the most important of these formulations and implications are the following:

1. In any transformation of energy, some of the energy is degraded.
2. No process is possible whose sole result is the conversion of a given quantity of heat (thermal energy) into an equal amount of useful work.
3. No process is possible whose sole result is the flow of heat from a colder body to a hotter one.
4. The availability of a given quantity of energy can only be used once; that is, the property of convertibility into useful work cannot be "recycled."
5. In spontaneous processes, concentrations (of anything) tend to disperse, structure tends to disappear, order becomes disorder.

That statements 1 through 4 are equivalent to or follow from our original formulation is readily verified. To see that statement 5 is related to the other statements, however, requires establishing a formal connection between order and availability of energy. This connection has been established in thermodynamics through the concept of *entropy*, a well-defined measure of disorder that can be shown to be a measure of unavailability of energy as well. A statement of the second law that contains or is equivalent to all the others is: *all physical processes proceed in such a way that the entropy of the universe increases.* (Not only can't we win—we can't break even, and we can't get out of the game!)

Consider some everyday examples of various aspects of the second law. If a partitioned container is filled with hot water on one side and cold water on the other and is left to itself, the hot water cools and the cold water warms—heat flows from hotter to colder. Note that the opposite process (the hot water getting hotter and the cold getting colder) does not violate the first law, conservation of energy. That it does not occur illustrates the second law. Indeed, many processes can be imagined that satisfy the first law but violate the second and therefore are not expected to occur. As another example, consider adding a drop of dye to a glass of water. Intuition and the second law dictate that the dye will spread, eventually coloring all the water—concentrations disperse, order (the dye/no dye arrangement) disappears. The opposite process, the spontaneous concentration of dispersed dye, is consistent with conservation of energy but not with the second law.

A more complicated situation is that of the refrigerator, a device that certainly causes heat to flow from cold objects (the contents of the refrigerator—say, beer—which are made colder) to a hot one (the room, which the refrigerator makes warmer). But this heat flow is not the *sole* result of the operation of the refrigerator: energy must be supplied to the refrigeration cycle from an external source, and this energy is converted to heat and discharged to the room, along with the heat removed from the interior of the refrigerator. Overall, availability of energy has decreased, and entropy has increased.

One illustration of the power of the laws of thermodynamics is that in many situations they can be used to predict the maximum efficiency that could be achieved by a perfect machine, without specifying any details of the machine! (Efficiency may be defined, in this situation, as the ratio of useful work to total energy flow.) Thus, one can specify, for example, what *minimum* amount of energy is necessary to separate salt from seawater, to separate metals from their ores, and to separate pollutants from auto exhaust without knowing any details about future inventions that might be devised for these purposes. Similarly, if one is told the temperature of a source of thermal energy—say, the hot rock deep in Earth's crust—one can calculate rather easily the maximum efficiency with which this thermal energy can be converted to applied work, regardless of the cleverness of future inventors. In other words, *there are some fixed limits to technological innovation, placed there by fundamental laws of nature.* . . .

More generally, the laws of thermodynamics explain why we need a continual input of energy to maintain ourselves, why we must eat much more than a pound of food in order to gain a pound of weight, and why the total energy flow through plants will always be much greater than that through plant-eaters, which in turn will always be much greater than that through flesh-eaters. They also make it clear that *all* the energy used on the face of the Earth, whether of solar or nuclear origin, will ultimately be degraded to heat. Here the laws catch us both coming and going, for they put limits on the efficiency with which we can manipulate this heat. Hence, they pose the danger . . . that human society may make this planet uncomfortably warm with degraded energy long before it runs out of high-grade energy to consume.

Occasionally it is suggested erroneously that the process of biological evolution represents a violation of the second law of thermodynamics. After all, the development of complicated living organisms from primordial chemical precursors, and the growing structure and complexity of the biosphere over the eons, do appear to be the sort of spontaneous increases in order excluded by the second law. The catch is that Earth is not an isolated system; the process of evolution has been powered by the sun, and the decrease in entropy on Earth represented by the growing structure of the biosphere is more than counterbalanced by the increase in the entropy of the sun. . . .

It is often asked whether a revolutionary development in physics, such as Einstein's theory of relativity, might not open the way to circumvention of the laws of thermodynamics. Perhaps it would be imprudent to declare that in no distant corner of the universe or hitherto-unexplored compartment of subatomic matter will any exception ever turn up, even though our intrepid astrophysicists and particle physicists have not yet found a single one. But to wait for the laws of thermodynamics to be overturned as descriptions of everyday experiences on this planet is, literally, to wait for the day when beer refrigerates itself in hot weather and squashed cats on the freeway spontaneously reassemble themselves and trot away.

3 The Entropy Law and the Economic Problem

Nicholas Georgescu-Roegen

I

A curious event in the history of economic thought is that, years after the mechanistic dogma has lost its supremacy in physics and its grip on the philosophical world, the founders of the neoclassical school set out to erect an economic science after the pattern of mechanics—in the words of Jevons, as *"the mechanics of utility and self-interest."*[1] And while economics has made great strides since, nothing has happened to deviate economic thought from the mechanistic epistemology of the forefathers of standard economics. A glaring proof is the standard textbook representation of the economic process by a circular diagram, a pendulum movement between production and consumption within a completely closed system.[2] The situation is not different with the analytical pieces that adorn the standard economic literature; they, too, reduce the economic process to a self-sustained mechanical analogue. The patent fact that between the economic process and the material environment there exists a continuous mutual influence which is history making carries no weight with the standard economist. And the same is true of Marxist economists, who swear by Marx's dogma that everything nature offers man is a spontaneous gift.[3] In Marx's famous diagram of reproduction, too, the economic process is represented as a completely circular and self-sustaining affair.[4]

Earlier writers, however, pointed in another direction, as did Sir William Petty in arguing that labor is the father and nature is the mother of wealth.[5] The entire economic history of mankind proves

Appeared previously in The University of Alabama Distinguished Lecture Series, no. 1, 1971; reprinted by permission of the author and The University of Alabama.

beyond question that nature, too, plays an important role in the economic process as well as in the formation of economic value. It is high time, I believe, that we should accept this fact and consider its consequences for the economic problem of mankind. For, as I shall endeavor to show in this paper, some of these consequences have an exceptional importance for the understanding of the nature and the evolution of man's economy.

II

Some economists have alluded to the fact that man can neither create nor destroy matter or energy[6]—a truth which follows from the principle of conservation of matter-energy, alias the first law of thermodynamics. Yet no one seems to have been struck by the question—so puzzling in the light of this law—"what then does the economic process do?" All that we find in the cardinal literature is an occasional remark that man can produce only utilities, a remark which actually accentuates the puzzle. How is it possible for man to produce something material, given the fact that he cannot produce either matter or energy?

To answer this question, let us consider the economic process as a whole and view it only from the purely physical viewpoint. What we must note first of all is that this process is a partial process which, like all partial processes, is circumscribed by a boundary across which matter and energy are exchanged with the rest of the material universe.[7] The answer to the question of what this *material* process does is simple: it neither produces nor consumes matter-energy; it only absorbs matter-energy and throws it out continuously. This is what pure physics teaches us. However, economics—let us say it high and loud—is not pure physics, not even physics in some other form. We may trust that even the fiercest partisan of the position that natural resources have nothing to do with value will admit in the end that there is a difference between what goes into the economic process and what comes out of it. To be sure, this difference can be only qualitative.

An unorthodox economist—such as myself—would say that what goes into the economic process represents *valuable natural resources* and what is thrown out of it is *valueless waste*. But this qualitative difference is confirmed, albeit in different terms, by a particular (and peculiar) branch of physics known as thermodynamics. From the viewpoint of thermodynamics, matter-energy enters the economic process in a state of *low entropy* and comes out of it in a state of *high entropy*.[8]

To explain in detail what entropy means is not a simple task. The notion is so involved that, to trust an authority on thermodynamics, it is "not easily understood even by physicists."[9] To make matters worse not only for the layman but for everyone else as well, the term now circulates with several meanings, not all associated with a physical coordinate.[10] The 1965 edition of *Webster's Collegiate Dictionary* has three entries under "entropy." Moreover, the definition pertaining to the meaning relevant for the economic process is likely to confuse rather than enlighten the reader: "a measure of unavailable energy in a closed thermodynamic system so related to the state of the system that a change in the measure varies with change in the ratio of the increment of heat taken in the absolute temperature at which it is absorbed." But (as if intended to prove that not all progress is for the better) some older editions supply a more intelligible definition. "A measure of the unavailable energy in a thermodynamic system"—as we read in the 1948 edition—cannot satisfy the specialist but would do for general purposes. To explain (again in broad lines) what unavailable energy means is now a relatively simple task.

Energy exists in two qualitative states—*available* or *free* energy, over which man has almost complete command, and *unavailable* or *bound* energy, which man cannot possibly use. The chemical energy contained in a piece of coal is free energy because man can transform it into heat or, if he wants, into mechanical work. But the fantastic amount of heat-energy contained in the waters of the seas, for example, is bound energy. Ships sail on top of this energy, but to do so they need the free energy of some fuel or of the wind.

When a piece of coal is burned, its chemical energy is neither decreased nor increased. But the initial free energy has become so dissipated in the form of heat, smoke, and ashes that man can no longer use it. It has been degraded into bound energy. Free energy means energy that displays a differential level, as exemplified most simply by the difference of temperatures between the inside and the outside of a boiler. Bound energy is, on the contrary, chaotically dissipated energy. This difference may be expressed in yet another way. Free energy implies some ordered structure, comparable with that of a store in which all meat is on one counter, vegetables on another, and so on. Bound energy is energy dissipated in disorder, like the same store after being struck by a tornado. This is why entropy is also defined as a measure of disorder. It fits the fact that a copper sheet represents a lower entropy than the copper ore from which it was produced.

The distinction between free and bound energy is certainly an anthropomorphic one. But this fact need not trouble a student of man, nay, even a student of matter in its simple form. Every element by which man seeks to get in mental contact with actuality can be but anthropomorphic. Only, the case of thermodynamics happens to be more striking. The point is that it was the economic distinction between things having an economic value and waste which prompted the thermodynamic distinction, not conversely. Indeed, the discipline of thermodynamics grew out of a memoir in which the French engineer Sadi Carnot (1824) studied for the first time the *economy* of heat engines. Thermodynamics thus began as a physics of economic value and has remained so in spite of the numerous subsequent contributions of a more abstract nature.

III

Thanks to Carnot's memoir, the elementary fact that heat moves by itself only from the hotter to the colder body acquired a place among the truths recognized by physics. Still more important was the consequent recognition of the additional truth that once the heat of a closed system has diffused itself so that the temperature has become uniform throughout the system, the movement of the heat cannot be reversed without external intervention. The ice cubes in a glass of water, once melted, will not form again by themselves. In general, the free heat-energy of a closed system continuously and irrevocably degrades itself into bound energy. The extension of this property from heat-energy to all other kinds of energy led to the second law of thermodynamics, alias the entropy law. This law states that the entropy (i.e., the amount of bound energy) of a closed system continuously increases or that the order of such a system steadily turns into disorder.

The reference to a closed system is crucial. Let us visualize a closed system, a room with an electric stove and a pan of water that has just been boiled. What the entropy law tells us is, first, that the heat of the boiled water will continuously dissipate into the system. Ultimately, the system will attain thermodynamic equilibrium—a state in which the temperature is uniform throughout (and all energy is bound). This applies to every kind of energy in a closed system. The free chemical energy of a piece of coal, for instance, will ultimately become degraded into bound energy even if the coal is left in the ground. Free energy will do so in any case.

The law also tells us that once thermodynamic equilibrium is reached, the water will not start boiling by itself.[11] But, as everyone knows, we can make it boil again by turning on the stove. This does not mean, however, that we have defeated the entropy law. If the entropy of the room has been decreased as the result of the temperature differential created by boiling the water, it is only because some low entropy (free energy) was brought into the system from the outside. And if we include the electric plant in the system, the entropy of this new system must have decreased, as the entropy law states. This means that the decrease in the entropy of the room has been obtained only at the cost of a greater increase in entropy elsewhere.

Some writers, impressed by the fact that living organisms remain almost unchanged over short periods of time, have set forth the idea that life eludes the entropy law. Now, life may have properties that cannot be accounted for by the natural laws, but the mere thought that it may violate some law of matter (which is an entirely different thing) is sheer nonsense. The truth is that every living organism strives only to maintain its own entropy constant. To the extent to which it achieves this, it does so by sucking low entropy from the environment to compensate for the increase in entropy to which, like every material structure, the organism is continuously subject. But the entropy of the entire system—consisting of the organism and its environment—must increase. Actually, the entropy of a system must increase faster if life is present than if it is absent. The fact that any living organism fights the entropic degradation of its own material structure may be a characteristic property of life, not accountable by material laws, but it does not constitute a violation of these laws.

Practically all organisms live on low entropy in the form found immediately in the environment. Man is the most striking exception: he cooks most of his food and also transforms natural resources into mechanical work or into various objects of utility. Here again, we should not let ourselves be misled. The entropy of copper metal is lower than the entropy of the ore from which it was refined, but this does not mean that man's *economic* activity eludes the entropy law.

The refining of the ore causes a more than compensating increase in the entropy of the surroundings. Economists are fond of saying that we cannot get something for nothing. The entropy law teaches us that the rule of biological life and, in man's case, of its economic continuation is far harsher. In entropy terms, the cost of any biological or economic enterprise is always greater than the product. In entropy terms, any such activity necessarily results in a deficit.

IV

The statement made earlier—that, from a purely physical viewpoint, the economic process only transforms valuable natural resources (low entropy) into waste (high entropy)—is thus completely vindicated. But the puzzle of why such a process should go on is still with us. And it will remain a puzzle as long as we do not see that the true economic output of the economic process is not a material flow of waste, but an immaterial flux: the enjoyment of life. If we do not recognize the existence of this flux, we are not in the economic world. Nor do we have a complete picture of the economic process if we ignore the fact that this flux—which, as an entropic feeling, must characterize life at all levels—exists only as long as it can continuously feed itself on environmental low entropy. And if we go one step further, we discover that every object of economic value—be it a fruit just picked from a tree, or a piece of clothing, or furniture, etc.—has a highly ordered structure, hence, a low entropy.[12]

There are several lessons to be derived from this analysis. The first lesson is that man's economic struggle centers on environmental low entropy. Second, environmental low entropy is scarce in a different sense than Ricardian land. Both Ricardian land and the coal deposits are available in limited amounts. The difference is that a piece of coal can be used only once. And, in fact, the entropy law is the reason why an engine (even a biological organism) ultimately wears out and must be replaced by a *new* one, which means an additional tapping of environmental low entropy.

Man's continuous tapping of natural resources is not an activity that makes no history. On the contrary, it is the most important long-run element of mankind's fate. It is because of the irrevocability of the entropic degradation of matter-energy that, for instance, the peoples from the Asian steppes, whose economy was based on sheep raising, began their Great Migration over the entire European continent at the beginning of the first millennium. The same element—the pressure on natural resources—had, no doubt, a role in other migrations, including that from Europe to the New World. The fantastic efforts made for reaching the moon may also reflect some vaguely felt hope of obtaining access to additional sources of low entropy. It is also because of the particular scarcity of environmental low entropy that ever since the dawn of history man has continuously sought to invent means for sifting low entropy better. In most (though not in all) of man's inventions one can definitely see a progressively better economy of low entropy.

Nothing could, therefore, be further from the truth than the notion that the economic process is an isolated, circular affair—as Marxist and standard analysis represent it. The economic process is solidly anchored to a material base which is subject to definite constraints. It is because of these constraints that the economic process has a unidirectional irrevocable evolution. In the economic world only money circulates back and forth between one economic sector and another (although, in truth, even the bullion slowly wears out and its stock must be continuously replenished from the mineral deposits). In retrospect it appears that the economists of both persuasions have succumbed to the worst economic fetishism—money fetishism.

V

Economic thought has always been influenced by the economic issues of the day. It also has reflected—with some lag—the trend of ideas in the natural sciences. A salient illustration of this correlation is the very fact that, when economists began ignoring the natural environment in representing the economic process, the event reflected a turning point in the temper of the entire scholarly world. The unprecedented achievements of the Industrial Revolution so amazed everyone with what man might do with the aid of machines that the general attention became confined to the factory. The landslide of spectacular scientific discoveries triggered by the new technical facilities strengthened this general awe for the power of technology. It also induced the literati to overestimate and, ultimately, to oversell to their audiences the powers of science. Naturally, from such a pedestal one could not even conceive that there is any real obstacle inherent in the human condition.

The sober truth is different. Even the lifespan of the human species represents just a blink when compared with that of a galaxy. So, even with progress in space travel, mankind will remain confined to a speck of space. Man's biological nature sets other limitations as to what he can do. Too high or too low a temperature is incompatible with his existence. And so are many radiations. It is not only that he cannot reach up to the stars, but he cannot even reach down to an individual elementary particle, nay, to an individual atom.

Precisely because man has felt, however unsophisticatedly, that his life depends on scarce, irretrievable low entropy, man has all along nourished the hope that he may eventually discover a self-perpetuating force. The discovery of electricity enticed many to believe that the hope was actually fulfilled. Following the strange marriage of thermodynamics with mechanics, some began seriously thinking about schemes to unbind bound energy.[13] The discovery of atomic energy spread another wave of sanguine hopes that, this time, we have truly gotten hold of a self-perpetuating power. The shortage of electricity which plagues New York and is gradually extending to other cities should suffice to sober us up. Both the nuclear theorists and the operators of atomic plants vouch that it all boils down to a problem of cost, which in the perspective of this paper means a problem of a balance sheet in entropy terms.

With natural sciences preaching that science can do away with all limitations felt by man and with the economists following suit in not relating the analysis of the economic process to the limitations of man's material environment, no wonder that no one realized that we cannot produce "better and bigger" refrigerators, automobiles, or jet planes without producing also "better and bigger" waste. So, when everyone (in the countries with "better and bigger" industrial production) was, literally, hit in the face by pollution, scientists as well as economists were taken by surprise. But even now no one seems to see that the cause of all this is that we have failed to acknowledge the entropic nature of the economic process. A convincing proof is that the various authorities on pollution now try to sell us, on the one hand, the idea of machines and chemical reactions that produce no waste, and, on the other, salvation through a perpetual recycling of waste. There is no denial that, in principle at least, we can recycle even the gold dispersed in the sand of the seas just as we can recycle the boiling water in my earlier example. But in both cases we must use an additional amount of low entropy much greater than the decrease in the entropy of what is recycled. There is no free recycling just as there is no wasteless industry.

VI

The globe to which the human species is bound floats, as it were, within the cosmic store of free energy, which may be even infinite. But for the reasons mentioned in the preceding section, man cannot have access to all this fantastic amount, nor to all possible forms of free energy. Man cannot, for example, tap directly the immense thermonuclear energy of the sun. The most important impediment (valid also for the industrial use of the "hydrogen bomb") is that no material container can resist the temperature of massive thermonuclear reactions. Such reactions can occur only in free space.

The free energy to which man can have access comes from two distinct sources. The first source is a *stock,* the stock of free energy of the mineral deposits in the bowels of the earth. The second source is a *flow,* the flow of solar radiation intercepted by the earth. Several differences between these two sources should be well marked. Man has almost complete command over the terrestrial dowry; conceivably, we may use it all within a single year. But, for all practical purposes, man has no control over the flow of solar radiation. Neither can he use the flow of the future *now.* Another asymmetry between the two sources pertains to their specific roles. Only the terrestrial source provides us with the low-entropy materials from which we manufacture our most important implements. On the other hand, solar radiation is the primary source of all life on earth, which begins with chlorophyll photosynthesis. Finally, the terrestrial stock is a paltry source in comparison with that of the sun. In all probability, the active life of the sun—during which the earth will receive a flow of solar energy of significant intensity—will last another five billion years.[14] But hard to believe though it may be, the entire terrestrial stock could only yield a few days of sunlight.[15]

All this casts a new light on the population problem, which is so topical today. Some students are alarmed at the possibility that the world population will reach seven billion by 2000 A.D.—the level predicted by United Nations demographers. On the other side of the fence, there are those who, like Colin Clark, claim that with a proper administration of resources the earth may feed as many as forty-five billion people.[16] Yet no population expert seems to have raised the far more vital question for mankind's future: How long can a given world population—be it of one billion or of forty-five billion—be maintained? Only if we raise this question can we see how complicated the population problem is. Even the analytical concept of optimum population, on which many population studies have been erected, emerges as an inept fiction.

What has happened to man's entropic struggle over the last two hundred years is a telling story in this respect. On the one hand, thanks to the spectacular progress of science man has achieved an almost miraculous level of economic development. On the other hand, this development has forced man to push his tapping of terrestrial sources to a staggering degree (witness offshore oil drilling). It has also sustained a population growth which has accentuated the struggle for food and, in some areas, brought this pressure to critical levels. The solution, advocated unanimously, is an increased mechanization of agriculture. But let us see what this solution means in terms of entropy.

In the first place, by eliminating the traditional partner of the farmer—the draft animal—the mechanization of agriculture allows the entire land area to be allocated to the production of food (and to fodder only to the extent of the need for meat). But the ultimate and the most important result is a shift of the low-entropy input from the solar to the terrestrial source. The ox or the water buffalo—which derive their mechanical power from the solar radiation caught by chlorophyll photosynthesis—is replaced by the tractor—which is produced and operated with the aid of terrestrial low entropy. And the same goes for the shift from manure to artificial fertilizers. The upshot is that the mechanization of agriculture is a solution which, though inevitable in the present impasse, is antieconomical in the long run. Man's biological existence is made to depend in the future more and more upon the scarcer of the two sources of low entropy. There is also the risk that mechanized agriculture may trap the human species in a cul-de-sac because of the possibility that some of the biological species involved in the other method of farming will be forced into extinction.

Actually, the problem of the economic use of the terrestrial stock of low entropy is not limited to the mechanization of agriculture only: it is the main problem for the fate of the human species. To see this, let S denote the present stock of terrestrial low entropy and let r be some average annual amount of depletion. If we abstract (as we can safely do here) from the slow degradation of S, the *theoretical* maximum number of years until the complete exhaustion of that stock is S/r. This is also the number of years until the *industrial* phase in the evolution of mankind will forcibly come to its end. Give the fantastic disproportion between S and the flow of solar energy that reaches the globe annually, it is beyond question that, even with a very parsimonious use of S, the industrial phase of man's evolution will end long before the sun will cease to shine. What will happen then (if the extinction of the human species is not brought about earlier by some totally resistant bug or some insidious chemical) is hard to say. Man could continue to live by reverting to the stage of a berry-picking species—as he once was. But, in the light of what we know about evolution, such an evolutionary reversal does not seem probable. Be that as it may, the fact remains that the higher the degree of economic development, the greater must be the annual depletion r and, hence, the shorter becomes the expected life of the human species.

VII

The upshot is clear. Every time we produce a Cadillac, we irrevocably destroy an amount of low entropy that could otherwise be used for producing a plow or a spade. In other words, every time we produce a Cadillac, we do it at the cost of decreasing the number of human lives in the future. Economic development through industrial abundance may be a blessing for us now and for those who will be able to enjoy it in the near future, but it is definitely against the interest of the human species as a whole, if its interest is to have a lifespan as long as is compatible with its dowry of low entropy. In this paradox of economic development we can see the price man has to pay for the unique privilege of being able to go beyond the biological limits in his struggle for life.

Biologists are fond of repeating that natural selection is a series of fantastic blunders since future conditions are not taken into account. The remark, which implies that man is wiser than nature and should take over her job, proves that man's vanity and the scholar's self-confidence will never know their limits. For the race of economic development that is the hallmark of modern civilization leaves no doubt about man's lack of foresight. It is only because of his biological nature (his inherited instincts) that man cares for the fate of only some of his immediate descendants, generally not beyond his great-grandchildren. And there is neither cynicism nor pessimism in believing that, even if made aware of the entropic problem of the human species, mankind would not be willing to give up its present luxuries in order to ease the life of those humans who will live ten thousand or even one thousand years from now. Once man expanded his biological powers by means of industrial artifacts, he became *ipso facto* not only dependent on a very scarce source of life support but also addicted to industrial luxuries. It is as if the human species were determined to have a short but exciting life. Let the less ambitious species have a long but uneventful existence.

Issues such as those discussed in these pages pertain to long-run forces. Because these forces act extremely slowly we are apt to ignore their existence or, if we recognize them, to belittle their importance. Man's nature is such that he is always interested in what will happen until tomorrow, not in thousands of years from now. Yet it is the slow-acting forces that are the more fateful in general. Most people die not because of some quickly acting force—such as pneumonia or an automobile accident—but because of the slow-acting forces that cause aging. As a Jain philosopher remarked, man begins to die at birth. The point is that it would not be hazardous to venture some thoughts about the distant future of man's economy any more than it would be to predict in broad lines the life of a newly born child. One such thought is that the increased pressure on the stock of mineral resources created by the modern fever of industrial development, together with the mounting problem of making pollution less noxious (which places additional demands on the same stock), will necessarily concentrate man's attention on ways to make greater use of solar radiation, the more abundant source of free energy.

Some scientists now proudly claim that the food problem is on the verge of being completely solved by the imminent conversion on an industrial scale of mineral oil into food protein—an inept thought in view of what we know about the entropic problem. The logic of this problem justifies instead the prediction that, under the pressure of necessity, man will ultimately turn to the contrary conversion, of vegetable products into gasoline (if he will still have any use for it).[17] We may also be quasi-certain that, under the same pressure, man will discover means by which to transform solar radiation into motor power directly. Certainly, such a discovery will represent the greatest possible breakthrough for man's entropic problem, for it will bring under his command also the more abundant source of life support. Recycling and pollution purification would still consume low entropy, but not from the rapidly exhaustible stock of our globe.

Notes

1. W. Stanley Jevons, *The Theory of Political Economy* (4th ed., London, 1924), p. 21.

2. For examples, R. T. Bye, *Principles of Economics* (5th ed., New York, 1956), p. 253; G. L. Bach, *Economics* (2d ed., Englewood Cliffs, N.J., 1957), p. 60; J. H. Dodd, C. W. Hasek, T. J. Hailstones, *Economics* (Cincinnati, 1957), p. 125; R. M. Havens, J. S. Henderson, D. L. Cramer, *Economics* (New York, 1966), p. 49; Paul A. Samuelson, *Economics* (8th ed., New York, 1970), p. 42.

3. Karl Marx, *Capital* (3 vols., Chicago, 1906–1933), 1: 94, 199, 230, and passim.

4. Ibid., vol. 2, ch. 20.

5. *The Economic Writings of Sir William Petty*, ed. C. H. Hull (2 vols., Cambridge, Eng., 1899), 2: 377. Curiously, Marx went along with Petty's idea; but he claimed that nature only "helps to create use value without contributing to the formation of exchange value." Karl Marx, *Capital*, 1: 227. See also ibid., p. 94.

6. For example, Alfred Marshall, *Principles of Economics* (8th ed., New York, 1924), p. 63.

7. On the problem of the analytical representation of a process, see N. Georgescu-Roegen, *The Entropy Law and the Economic Process* (Cambridge, Mass., 1971), pp. 211–231.

8. This distinction together with the fact that no one would exchange some natural resources for waste disposes of Marx's assertion that "no chemist has ever discovered exchange value in a pearl or a diamond." Karl Marx, *Capital*, 1: 95.

9. D. ter Harr, "The Quantum Nature of Matter and Radiation," in *Turning Points in Physics*, ed. R. J. Blin-Stoyle et al. (Amsterdam, 1959), p. 37.

10. One meaning that has recently made the term extremely popular is "the amount of information." For an argument that this term is misleading and for a critique of the alleged connection between information and physical entropy, see N. Georgescu-Roegen, *The Entropy Law and the Economic Process*, appendix B.

11. This position calls for some technical elaboration. The opposition between the entropy law—with its unidirectional qualitative change—and mechanics—where everything can move either forward or backward while remaining self-identical—is accepted without reservation by every physicist and phi-

losopher of science. However, the mechanistic dogma retained (as it still does) its grip on scientific activity even after physics recanted it. The result was that mechanics was soon brought into thermodynamics in the company of randomness. This is the strangest possible company, for randomness is the very antithesis of the deterministic nature of the laws of mechanics. To be sure, the new edifice (known as statistical mechanics) could not include mechanics under its roof and, at the same time, exclude reversibility. So, statistical mechanics must teach that a pail of water may start boiling by itself, a thought which is slipped under the rug by the argument that the miracle has not been observed because of its extremely small probability. This position has fostered the belief in the possibility of converting bound into free energy or, as P. W. Bridgman wittily put it, of bootlegging entropy. For a critique of the logical fallacies of statistical mechanics and of the various attempts to patch them, see N. Georgescu-Roegen, *The Entropy Law and the Economic Process,* ch. 6.

12. This does not mean that everything of low entropy necessarily has economic value. Poisonous mushrooms, too, have a low entropy. The relation between low entropy and economic value is similar to that between economic value and price. An object can have a price only if it has economic value, and it can have economic value only if its entropy is low. But the converse is not true.

13. See note 11, above.

14. George Gamow, *Matter, Earth, and Sky* (Englewood Cliffs, N.J., 1958), pp. 493f.

15. Four days, according to Eugene Ayres, "Power from the Sun," *Scientific American,* August 1950, p. 16. The situation is not changed even if we admit that the calculations might be in error by as much as one thousand times.

16. Colin Clark, "Agricultural Productivity in Relation to Population," in *Man and His Future,* ed. G. Wolstenholme (Boston, 1963), p. 35.

17. That the idea is not farfetched is proved by the fact that in Sweden, during World War II, automobiles were driven by the poor gas obtained by heating wood with wood.

4 Selections from "Energy and Economic Myths"

Nicholas Georgescu-Roegen

Myths about Mankind's Entropic Problem

Hardly anyone would nowadays openly profess a belief in the immortality of mankind. Yet many of us prefer not to exclude this possibility; to this end, we endeavor to impugn any factor that could limit mankind's life. The most natural rallying idea is that mankind's entropic dowry is virtually inexhaustible, primarily because of man's inherent power to defeat the Entropy Law in some way or another.

To begin with, there is the simple argument that, just as has happened with many natural laws, the laws on which the finiteness of accessible resources rests will be refuted in turn. The difficulty of this historical argument is that history proves with even greater force, first, that in a finite space there can be only a finite amount of low entropy and, second, that low entropy continuously and irrevocably dwindles away. The impossibility of perpetual motion (of both kinds) is as firmly anchored in history as the law of gravitation.

More sophisticated weapons have been forged by the statistical interpretation of thermodynamic phenomena—an endeavor to re-establish the supremacy of mechanics propped up this time by a *sui generis* notion of probability. According to this interpretation, the reversibility of high into low entropy is only a highly improbable, not a totally impossible event. And since the event *is possible*, we should be able by an ingenious device to cause the event to happen as often as we please, just as an adroit sharper may throw a "six" almost at will. The argument only brings to the surface the irreducible

Reprinted from *Southern Economic Journal* 41, no. 3 (January 1975), by permission of the author and *Southern Economic Journal*. Notes are renumbered from the original version.

contradictions and fallacies packed into the foundations of the statistical interpretation by the worshipers of mechanics [32, ch. 6]. The hopes raised by this interpretation were so sanguine at one time that P. W. Bridgman, an authority on thermodynamics, felt it necessary to write an article just to expose the fallacy of the idea that one may fill one's pockets with money by "bootlegging entropy" [11].

Occasionally and *sotto voce* some express the hope, once fostered by a scientific authority such as John von Neumann, that man will eventually discover how to make energy a free good, "just like the unmetered air" [3, p. 32]. Some envision a "catalyst" by which to decompose, for example, the sea water into oxygen and hydrogen, the combustion of which will yield as much available energy as we would want. But the analogy with the small ember which sets a whole log on fire is unavailing. The entropy of the log and the oxygen used in the combustion is lower than that of the resulting ashes and smoke, whereas the entropy of water is higher than that of the oxygen and hydrogen after decomposition. Therefore, the miraculous catalyst also implies entropy bootlegging.[1]

With the notion, now propagated from one syndicated column to another, that the breeder reactor produces more energy than it consumes, the fallacy of entropy bootlegging seems to have reached its greatest currency even among the large circles of literati, including economists. Unfortunately, the illusion feeds on misconceived sales talk by some nuclear experts who extol the reactors which transform fertile but nonfissionable material into fissionable fuel as the breeders that "produce more fuel than they consume" [81, p. 82]. The stark truth is that the breeder is in no way different from a plant which produces hammers with the aid of some hammers. According to the deficit principle of the Entropy Law . . . , even in breeding chickens a greater amount of low entropy is consumed than is contained in the product.[2]

Apparently in defense of the standard vision of the economic process, economists have set forth themes of their own. We may mention first the argument that "the notion of an absolute limit to natural resource availability is untenable when the definition of resources changes drastically and unpredictably over time. . . . A limit may exist, but it can be neither defined nor specified in economic terms" [3, pp. 7, 11]. We also read that there is no upper limit even for arable land because "arable is infinitely indefinable" [55, p. 22]. The sophistry of these arguments is flagrant. No one would deny that we cannot say *exactly* how much coal, for example, is accessible. Estimates of natural resources have constantly been shown to be too low. Also, the point that metals contained in the top mile of the earth's crust may be a million times as much as the present known reserves [4, p. 338; 58, p. 331] does not prove the inexhaustibility of resources, but, characteristically, it ignores both the issues of accessibility and disposability.[3] Whatever resources or arable land we may need at one time or another, they will consist of accessible low entropy and accessible land. *And since all kinds together are in finite amount, no taxonomic switch can do away with that finiteness.*

The favorite thesis of standard and Marxist economists alike, however, is that the power of technology is without limits [3; 4; 10; 49; 51; 69; 74]. We will always be able not only to find a substitute for a resource which has become scarce, but also to increase the *productivity* of any kind of energy and material. Should we run out of some resources, we will always think up something, just as we have continuously done since the time of Pericles [4, pp. 332–334]. Nothing, therefore, could ever stand in the way of an increasingly happier existence of the human species. One can hardly think of a more blunt form of linear thinking. By the same logic, no healthy young human should ever become afflicted with rheumatism or any other old-age ailments; nor should he ever die. Dinosaurs, just before they disappeared from this very same planet, had behind them not less than one hundred and fifty million years of truly prosperous existence. (And they did not pollute environment with industrial waste!) But the logic to be truly savored is Solo's [73, p. 516]. If entropic degradation is to bring mankind to its knees sometime in the future, it should have done so sometime after A.D. 1000. The old truth of Seigneur de La Palice has never been turned around—and in such a delightful form.[4]

In support of the same thesis, there also are arguments directly pertaining to its substance. First, there is the assertion that only a few kinds of resources are "so resistant to technological advance as to be incapable of eventually yielding extractive products at constant or declining cost" [3, p. 10].[5] More recently, some have come out with a specific law which, in a way, is the contrary of Malthus's law concerning resources. The idea is that technology improves exponentially [4, p. 236; 51, p. 664; 74, p. 45]. The superficial justification is that one technological advance induces another. This is true, only it does not work cumulatively as in population growth. And it is terribly wrong to argue, as Maddox does [59, p. 21], that to insist on the existence of a limit to technology means to deny man's power to influence progress. Even if technology continues to progress, it will not necessary exceed any limit; an increasing sequence may have an upper limit. In the case of technology this limit is set by the theoretical coefficient of efficiency. . . . If progress were indeed exponential, then the input i per unit of output would follow in time the law $i = i_0(1 + r)^{-t}$ and would constantly approach zero. Production would ultimately become incorporeal and the earth a new Garden of Eden.

Finally, there is the thesis which may be called the fallacy of endless substitution: "Few components of the earth's crust, including farm land, are so specific as to defy economic replacement; . . . nature imposes particular scarcities, not an inescapable general scarcity" [3, pp. 10f].[6] Bray's protest notwithstanding [10, p. 8], this is "an economist's conjuring trick." True, there are only a few "vitamin" elements which play a totally specific role such as phosphorus plays in living organisms. Aluminum, on the other hand, has replaced iron and copper in many, although not in all uses.[7] However, *substitution within a finite stock of accessible low entropy* whose irrevocable degradation is speeded up through use cannot possibly go on forever.

In Solow's hands, substitution becomes the key factor that supports technological progress even as resources become increasingly scarce. There will be, first, a substitution within the spectrum of consumer goods. With prices reacting to increasing scarcity, consumers will buy "fewer resource-intensive goods and more of other things" [74, p. 47].[8] More recently, he extended the same idea to production, too. We may, he argues, substitute "other factors for natural resources" [75, p. 11]. One must have a very erroneous view of the economic process as a whole not to see that there are no material factors other than natural resources. To maintain further that "the world can, in effect, get along without natural resources" is to ignore the difference between the actual world and the Garden of Eden.

More impressive are the statistical data invoked in support of some of the foregoing theses. The data adduced by Solow [74, pp. 44f] show that in the United States between 1950 and 1970 the consumption of a series of mineral elements per unit of GNP decreased substantially. The exceptions were attributed to substitution but were expected to get in line sooner or later. In strict logic, the data do not prove that during the same period technology necessarily progressed to a greater economy of resources. The GNP may increase more than any input of minerals even if technology remains the same, or even if it deteriorates. But we also know that during practically the same period, 1947–1967, the consumption per capita of basic materials increased in the United States. And in the world, during only one decade, 1957–1967, the consumption of steel per capita grew by 44 percent [12, pp. 198–200]. What matters in the end is not only the impact of technological progress on the consumption of resources per unit of GNP, but especially the increase in the rate of resource depletion, which is a side effect of that progress.

Still more impressive—as they have actually proved to be—are the data used by Barnett and Morse to show that, from 1870 to 1957, the ratios of labor and capital costs to net output decreased appreciably in agriculture and mining, both critical sectors as concerns depletion of resources [3, 8f, 167–178]. In spite of some arithmetical incongruities,[9] the picture emerging from these data cannot be repudiated. Only its interpretation must be corrected.

For the environmental problem it is essential to understand the typical forms in which technological progress may occur. A first group includes the *economy innovations,* which achieve a *net* economy of low entropy—be it by a more complete combustion, by decreasing friction, by deriving a more intensive light from gas or electricity, by substituting materials costing less in energy for others costing more, and so on. Under this heading we should also include the discovery of how to use new kinds of accessible low entropy. A second group consists of *substitution innovations,* which simply substitute physico-chemical energy for human energy. A good illustration is the innovation of gunpowder, which did away with the catapult. Such innovations generally enable us not only to do things better but also (and especially) to do things which could not be done before—to fly in airplanes, for example. Finally, there are the *spectrum innovations,* which bring into existence new consumer goods, such as the hat, nylon stockings, etc. Most of the innovations of this group are at the same time substitution innovations. In fact, most innovations belong to more than one category. But the classification serves analytical purposes.

Now, economic history confirms a rather elementary fact—the fact that the great strides in technological progress have generally been touched off by a discovery of how to use a new kind of accessible energy. On the other hand, a great stride in technological progress cannot materialize unless the corresponding innovation is followed by a great mineralogical expansion. Even a substantial increase in the efficiency of the use of gasoline as fuel would pale in comparison with a manifold increase of the known, rich oil fields.

This sort of expansion is what has happened during the last one hundred years. We have struck oil and discovered new coal and gas deposits in a far greater proportion than we could use during the same period. Still more important, all mineralogical discoveries have included a substantial proportion of *easily* accessible resources. This exceptional bonanza by itself has sufficed to lower the real cost of bringing mineral resources *in situ* to the surface. Energy of mineral source thus becoming cheaper, substitution innovations have caused the ratio of labor to net output to decline. Capital also must have evolved toward forms which cost less but use more energy to achieve the same result. What has happened during this period is a modification of the cost structure, the flow factors being increased and the fund factors decreased.[10] By examining, therefore, only the relative variations of the fund factors during a period of exceptional mineral bonanza, we cannot prove either that the unitary total cost will always follow a declining trend or that the continuous progress of technology renders accessible resources almost inexhaustible—as Barnett and Morse claim [3, p. 239].

Little doubt is thus left about the fact that the theses examined in this section are anchored in a deep-lying belief in mankind's immortality. Some of their defenders have even urged us to have faith in the human species: such faith will triumph over all limitations.[11] But neither faith nor assurance from some famous academic chair [4] could alter the fact that, according to the basic law of thermodynamics, mankind's dowry is finite. Even if one were inclined to believe in the possible refutation of these principles in the future, one still must not act on that faith now. We must take into account that evolution does not consist of a linear repetition, even though over short intervals it may fool us into the contrary belief.

A great deal of confusion about the environmental problem prevails not only among economists generally (as evidenced by the numerous cases already cited), but also among the highest intellectual circles simply because the sheer entropic nature of all happenings is ignored or misunderstood. Sir Macfarlane Burnet, a Nobelite, in a special lecture considered it imperative "to prevent the progressive destruction of the earth's irreplaceable resources" [quoted, 15, p. 1].

And a prestigious institution such as the United Nations, in its Declaration on the Human Environment (Stockholm, 1972), repeatedly urged everyone "to improve the environment." Both urgings reflect the fallacy that man can reverse the march of entropy. The truth, however unpleasant, is that the most we can do is to prevent any unnecessary depletion of resources and any unnecessary deterioration of the environment, but without claiming that we know the precise meaning of "unnecessary" in this context.

The Steady State: A Topical Mirage

Malthus, as we know, was criticized primarily because he assumed that population and resources grow according to some simple mathematical laws. But this criticism did not touch the real error of Malthus (which has apparently remained unnoticed). This error is the implicit assumption that population may grow beyond any limit both in number and time *provided that it does not grow too rapidly.*[12] An essentially similar error has been committed by the authors of *The Limits,* by the authors of the nonmathematical yet more articulate "Blueprint for Survival," as well as by several earlier writers. Because, like Malthus, they were set exclusively on proving the impossibility of growth, they were easily deluded by a simple, now widespread, but false syllogism: since exponential growth in a finite world leads to disasters of all kinds, ecological salvation lies in the stationary state [42; 47; 62, pp. 156–184; 6, pp. 3f, 8, 20].[13] H. Daly even claims that "the stationary state economy is, therefore, a necessity" [21, p. 5].

This vision of a blissful world in which both population and capital stock remain constant, once expounded with his usual skill by John Stuart Mill [64, bk. 4, ch. 6], was until recently in oblivion.[14] Because of the spectacular revival of this myth of ecological salvation, it is well to point out its various logical and factual snags. The crucial error consists in not seeing that not only growth, but also a zero-growth state, nay, even a declining state which does not converge toward annihilation, cannot exist forever in a finite environment. The error perhaps stems from some confusion between finite stock and finite flow rate, as the incongruous dimensionalities of several graphs suggest [62, pp. 62, 64f, 124ff; 6, p. 6]. And contrary to what some advocates of the stationary state claim [21, p. 15], this state does not occupy a privileged position vis-à-vis physical laws.

To get to the core of the problem, let S denote the actual amount of accessible resources in the crust of the earth. Let P_i and s_i be the population and the amount of depleted resources per person in the year i. Let the "amount of total life," measured in years of life, be defined by $L = \Sigma P_i$, from $i = 0$ to $i = \infty$. S sets an upper limit for L through the obvious constraint $\Sigma P_i s_i \leq S$. For although s_i is a historical variable, it cannot be zero or even negligible (unless mankind reverts sometime to a berry-picking economy). Therefore, $P = 0$ for i greater than some finite n, and $P_i > 0$ otherwise. That value of n *is* the maximum duration of the human species [31, pp. 12f; 32, p. 304].

The earth also has a so-called carrying capacity, which depends on a complex of factors, including the size of s_i.[15] This capacity sets a limit on any single P_i. But this limit does not render the other limits, of L and n, superfluous. It is therefore inexact to argue—as the Meadows group seems to do [62, pp. 91f]—that the stationary state can go on forever as long as P_i does not exceed that capacity. The proponents of salvation through the stationary state must admit that such a state can have only a finite duration—unless they are willing to join the "No Limit" Club by maintaining that S is inexhaustible or almost so—as the Meadows group does in fact [62, p. 172]. Alternatively, they must explain the puzzle of how a whole economy, stationary for a long era, all of a sudden comes to an end.

Apparently, the advocates of the stationary state equate it with an open *thermodynamic* steady state. This state consists of an *open* macrosystem which maintains its entropic structure constant through material exchanges with its "environment." As one would immediately guess, the concept constitutes a highly useful tool for the study of biological organisms. We must, however, observe that the concept rests on some special conditions introduced by L. Onsager [50, pp. 89–97]. These conditions are so delicate (they are called the principle of *detailed* balance) that in actuality they can hold only "within a deviation of a few percent" [50, p. 140]. For this reason, a steady state may exist in fact only in an approximated manner and over a finite duration. This impossibility of a macrosystem not in a state of chaos to be perpetually durable may one day be explicitly recognized by a new thermodynamic law just as the impossibility of perpetual motion once was. Specialists recognize that the present thermodynamic laws do not suffice to explain all nonreversible phenomena, including especially life processes.

Independently of these snags there are simple reasons against believing that mankind can live in a perpetual stationary state. The structure of such a state remains the same throughout; it does not contain in itself the seed of the inexorable death of all open macro-systems. On the other hand, a world with a stationary population would, on the contrary, be continually forced to change its technology as well as its mode of life in response to the inevitable decrease of resource accessibility. Even if we beg the issue of how capital may change qualitatively and still remain constant, we could have to assume that the unpredictable decrease in accessibility will be miraculously compensated by the right innovations at the right time. A stationary world may for a while be interlocked with the changing environment through a system of balancing feedbacks analogous to those of a living organism during one phase of its life. But as Bormann reminded us [7, p. 707], the miracle cannot last forever; sooner or later the balancing system will collapse. At that time, the stationary state will enter a crisis, which will defeat its alleged purpose and nature.

One must be cautioned against another logical pitfall, that of invoking the Prigogine principle in support of the stationary state. This principle states that the minimum of the entropy produced by an Onsager type of open thermodynamic system is reached when the system becomes steady [50, ch. 16]. It says nothing about how this last entropy compares with that produced by other open systems.[16]

The usual arguments adduced in favor of the stationary state are, however, of a different, more direct nature. It is, for example, argued that in such a state there is more time for pollution to be reduced by natural processes and for technology to adapt itself to the decrease of resource accessibility [62, p. 166]. It is plainly true that we could use much more efficiently today the coal we have burned in the past. The rub is that we might not have mastered the present efficient techniques if we had not burned all that coal "inefficiently." The point that in a stationary state people will not have to work additionally to accumulate capital (which in view of what I have said in the last paragraphs is not quite accurate) is related to Mill's claim that people could devote more time to intellectual activities. "The trampling, crushing, elbowing, and treading on each other's heel" will cease [64, p. 754]. History, however, offers multiple examples—the Middle Ages, for one—of quasi stationary societies where arts and sciences were practically stagnant. In a stationary state, too, people may be busy in the fields and shops all day long. Whatever the state, free time for intellectual progress depends on the intensity of the pressure of population on resources. Therein lies the main weakness of Mill's vision. Witness the fact that—as Daly explicitly admits [21, pp. 6–8]—its writ offers no basis for determining even in principle the optimum levels of population and capital. This brings to light the important, yet unnoticed point, that *the necessary conclusion of the arguments in favor of that vision is that the most desirable state is not a stationary, but a declining one.*

Undoubtedly, the current growth must cease, nay, be reversed. But anyone who believes that he can draw a blueprint for the ecological salvation of the human species does not understand the nature of evolution, or even of history—which is that of permanent struggle in continuously novel forms, not that of a predictable, controllable physico-chemical process, such as boiling an egg or launching a rocket to the moon.

Some Basic Bioeconomics[17]

Apart from a few insignificant exceptions, all species other than man use only *endosomatic* instruments—as Alfred Lotka proposed to call those instruments (legs, claws, wings, etc.) which belong to the individual organism *by birth*. Man alone came, in time, to use a club, which does not belong to him by birth, but which extended his endosomatic arm and increased its power. At that point in time, man's evolution transcended the biological limits to include also (and primarily) the evolution of *exosomatic* instruments, i.e., of instruments produced by man but not belonging to his body.[18] That is why man can now fly in the sky or swim under water even though his body has no wings, no fins, and no gills.

The exosomatic evolution brought down upon the human species two fundamental and irrevocable changes. The first is the irreducible social conflict which characterizes the human species [29, pp. 98–101; 32, pp. 306–315, 348f]. Indeed, there are other species which also live in society, but which are free from such conflict. The reason is that their "social classes" correspond to some clear-cut biological divisions. The periodic killing of a great part of the drones by the bees is a natural, biological action, not a civil war.

The second change is man's addiction to exosomatic instruments—a phenomenon analogous to that of the flying fish which became addicted to the atmosphere and mutated into birds forever. It is because of this addiction that mankind's survival presents a problem entirely different from that of all other species [31; 32, pp. 302–305]. It is neither only biological nor only economic. It is bioeconomic. Its broad contours depend on the multiple asymmetries existing among the three sources of low entropy which together constitute mankind's dowry—the free energy received from the sun, on the one hand, and the free energy and the ordered material structures stored in the bowels of the earth, on the other.

The *first* asymmetry concerns the fact that the terrestrial component is a *stock,* whereas the solar one is a flow. The difference needs to be well understood [32, pp. 226f]. Coal *in situ* is a stock because we are free to use it all today (conceivably) or over centuries. But at no time can we use any part of a future flow of solar radiation. Moreover, the flow rate of this radiation is wholly beyond our control; it is completely determined by cosmological conditions, including the size of our globe.[19] One generation, whatever it may do, cannot alter the share of solar radiation of any future generation. Because of the priority of the present over the future and the irrevocability of entropic degradation, the opposite is true for the terrestrial shares. These shares are affected by how much of the terrestrial dowry the past generations have consumed.

Second, since no practical procedure is available at human scale for transforming energy into matter . . . , accessible material low entropy is by far the most critical element from the bioeconomic viewpoint. True, a piece of coal burned by our forefathers is gone forever, just as is part of the silver or iron, for instance, mined by them. Yet future generations will still have their inalienable share of solar energy (which, as we shall see next, is enormous). Hence, they will be able, at least, to use each year an amount of wood equivalent to the annual vegetable growth. For the silver and iron dissipated by the earlier generations there is no similar compensation. This is why in bioeconomics we must emphasize that every Cadillac or every Zim—let alone any instrument of war—means fewer plowshares for some future generations, and implicitly, fewer future human beings, too [31, p. 13; 32, p. 304].

Third, there is an astronomical difference between the amount of the flow of solar energy and the size of the stock of terrestrial free energy. At the cost of a decrease in mass of 131×10^{12} tons, the sun radiates annually 10^{13} Q—one single Q being equal to 10^{18} BTU! Of this fantastic flow, only some 5,300 Q are intercepted at the limits of the earth's atmosphere, with roughly one half of that amount being reflected back into outer space. At our own scale, however, even this amount is fantastic; for the total world consumption of energy currently amounts to no more than 0.2 Q annually. From the solar energy that reaches the ground level, photosynthesis absorbs only 1.2 Q. From waterfalls we could obtain at most 0.08 Q, but we are now using only one tenth of that potential. Think also of the additional fact that the sun will continue to shine with practically the same intensity for another five billion years (before becoming a red giant which will raise the earth's temperature to 1,000°F). Undoubtedly, the human species will not survive to benefit from all this abundance.

Passing to the terrestrial dowry, we find that, according to the best estimates, the initial dowry of fossil fuel amounted to only 215 Q. The outstanding recoverable reserves (known and probable) amount to about 200 Q. These reserves, therefore, could produce only two weeks of sunlight on the globe.[20] If their depletion continues to increase at the current pace, these reserves may support man's industrial activity for just a few more decades. Even the reserves of uranium 235 will not last for a longer period if used in the ordinary reactors. Hopes are now set on the breeder reactor, which, with the aid of uranium 235, may "extract" the energy of the fertile but not fissionable elements, uranium 238 and thorium 232. Some experts claim that this source of energy is "essentially inexhaustible" [83, p. 412]. In the United States alone, it is believed, there are large areas covered with black shale and granite which contain 60 grams of natural uranium or thorium per metric ton [46, pp. 226f]. On this basis, Weinberg and Hammond [83, pp. 415f] have come out with a grand plan. By stripmining and crushing all these rocks, we could obtain enough nuclear fuel for some 32,000 breeder reactors distributed in 4,000 offshore parks and capable of supplying a population of twenty billion for millions of years with twice as much energy per capita as the current consumption rate in the USA. The grand plan is a typical example of linear thinking, according to which all that is needed for the existence of a population, even "considerably larger than twenty billion," is to increase all supplies proportionally.[21] Not that the authors deny that there also are nontechnical issues; only, they play them down with noticeable zeal [83, pp. 417f]. The most important issue, of whether a social organization compatible with the density of population and the nuclear manipulation at the grand level can be achieved, is brushed aside by Weinberg as "transscientific" [82].[22] Technicians are prone to forget that due to their own successes, nowadays it may be easier to move the mountain to Mohammed than to induce Mohammed to go to the mountain. For the time being, the snag is far more palpable. As responsible forums openly admit, even one breeder still presents substantial risks of nuclear catastrophes, and the problem of safe transportation of nuclear fuels and especially that of safe storage of the radioactive garbage still await a solution even for a moderate scale of operations [35; 36; especially 39 and 67].

There remains the physicist's greatest dream, controlled thermonuclear reaction. To constitute a real breakthrough, it must be the deuterium-deuterium reaction, the only one that could open up a formidable source of terrestrial energy for a long era.[23] However, because of the difficulties alluded to earlier . . . , even the experts working at it do not find reasons for being too hopeful.

For completion, we should also mention the tidal and geothermal energies, which, although not negligible (in all, 0.1 Q per year), can be harnessed only in very limited situations.

The general picture is now clear. The terrestrial energies on which we can rely effectively exist in very small amounts, whereas the use of those which exist in ampler amounts is surrounded by great risks and formidable technical obstacles. On the other hand, there is the immense energy from the sun which reaches us without fail. Its direct use is not yet practiced on a significant scale, the main reason being that the alternative industries are now much more efficient economically. But promising results are coming from various directions [37; 41]. What counts from the bioeconomic viewpoint is that the feasibility of using the sun's energy directly is not surrounded by risks or big question marks; it is a proven fact.

The conclusion is that mankind's entropic dowry presents another important differential scarcity. From the viewpoint of the extreme long run, the terrestrial free energy is far scarcer than that received from the sun. The point exposes the foolishness of the victory cry that we can finally obtain protein from fossil fuels! Sane reason tells us to move in the opposite direction, to convert vegetable stuff into hydrocarbon fuel—an obviously natural line already pursued by several researchers [22, pp. 311–313].[24]

Fourth, from the viewpoint of industrial utilization, solar energy has an immense drawback in comparison with energy of terrestrial origin. The latter is available in a concentrated form; in some cases, in a too concentrated form. As a result, it enables us to obtain almost instantaneously enormous amounts of work, most of which could not even be obtained otherwise. By great contrast, the flow of solar energy comes to us with an extremely low intensity, like a very fine rain, almost a microscopic mist. The important difference from true rain is that this radiation rain is not collected naturally into streamlets, then into creeks and rivers, and finally into lakes from where we could use it in a concentrated form, as is the case with waterfalls. Imagine the difficulty one would face if one tried to use *directly* the kinetic energy of some microscopic rain drops as they fall. The same difficulty presents itself in using solar energy directly (i.e., not through the chemical energy of green plants, or the kinetic energy of the wind and waterfalls). But as was emphasized a while ago, the difficulty does not amount to impossibility.[25]

Fifth, solar energy, on the other hand, has a unique and incommensurable advantage. The use of any terrestrial energy produces some noxious pollution, which, moreover, is irreducible and hence cumulative, be it in the form of thermal pollution alone. By contrast, any use of solar energy is *pollution-free*. For, whether this energy is used or not, its ultimate fate is the same, namely, to become the dissipated heat that maintains the thermodynamic equilibrium between the globe and outer space at a propitious temperature.[26]

The *sixth* asymmetry involves the elementary fact that the survival of every species on earth depends, directly or indirectly, on solar radiation (in addition to some elements of a superficial environmental layer). Man alone, because of his exosomatic addiction, depends on mineral resources as well. For the use of these resources man competes with no other species; yet his use of them usually endangers many forms of life, including his own. Some species have in fact been brought to the brink of extinction merely because of man's exosomatic needs or his craving for the extravagant. But nothing in nature compares in fierceness with man's competition for solar energy (in its primary or its by-product forms). Man has not deviated one bit from the law of the jungle; if anything, he has made it even more merciless by his sophisticated exosomatic instruments. Man has openly sought to exterminate any species that robs him of his food or feeds on him—wolves, rabbits, weeds, insects, microbes, etc.

But this struggle of man with other species for food (in ultimate analysis, for solar energy) has some unobtrusive aspects as well. And, curiously, it is one of these aspects that has some far-reaching consequences in addition to supplying a most instructive refutation of the common belief that every technological innovation constitutes a move in the right direction as concerns the economy of resources. The case pertains to the economy of modern agricultural techniques. . . .

Justus von Liebig observed that "civilization is the economy of power" [32, p. 304]. At the present hour, the economy of power in all its aspects calls for a turning point. Instead of continuing to be opportunistic in the highest degree and concentrating our research toward finding more economically efficient ways of tapping mineral energies—all in finite supply and all heavy pollutants—we should direct all our efforts toward improving the direct uses of solar energy—the only clean and essentially unlimited source. Already-known techniques should without delay be diffused among all people so that we all may learn from practice and develop the corresponding trade.

An economy based primarily on the flow of solar energy will also do away, though not completely, with the monopoly of the present over future generations, for even such an economy will still need to tap the terrestrial dowry, especially for materials. Technological innovations will certainly have a role in this direction. But it is high time for us to stop emphasizing exclusively—as all platforms have apparently done so far—the increase of supply. Demand can also play a role, an even greater and more efficient one in the ultimate analysis.

It would be foolish to propose a complete renunciation of the industrial comfort of the exosomatic evolution. Mankind will not return to the cave or, rather, to the tree. But there are a few points that may be included in a minimal bioeconomic program.

First, the production of all instruments of war, *not only of war itself,* should be prohibited completely. It is utterly absurd (and also hypocritical) to continue growing tobacco if, avowedly, no one intends to smoke. The nations which are so developed as to be the main producers of armaments should be able to reach a consensus over this prohibition without any difficulty if, as they claim, they also possess the wisdom to lead mankind. Discontinuing the production of all instruments of war will not only do away at least with the mass killings by ingenious weapons but will also release some tremendous productive forces for international aid without lowering the standard of living in the corresponding countries.

Second, through the use of these productive forces as well as by additional well-planned and sincerely intended measures, the underdeveloped nations must be aided to arrive as quickly as possible at a good (not luxurious) life. Both ends of the spectrum must effectively participate in the efforts required by this transformation and accept the necessity of a radical change in their polarized outlooks on life.[27]

Third, mankind should gradually lower its population to a level that could be adequately fed only by organic agriculture.[28] Naturally, the nations now experiencing a very high demographic growth will have to strive hard for the most rapid possible results in that direction.

Fourth, until either the direct use of solar energy becomes a general convenience or controlled fusion is achieved, all waste of energy—by overheating, overcooling, overspeeding, overlighting, etc.—should be carefully avoided, and if necessary, strictly regulated.

Fifth, we must cure ourselves of the morbid craving for extravagant gadgetry, splendidly illustrated by such a contradictory item as the golf cart, and for such mammoth splendors as *two-garage* cars. Once we do so, manufacturers will have to stop manufacturing such "commodities."

Sixth, we must also get rid of fashion, of "that disease of the human mind," as Abbot Fernando Galliani characterized it in his celebrated *Della moneta* (1750). It is indeed a disease of the mind to throw away a coat or a piece of furniture while it can still perform its specific service. To get a "new" car every year and to refashion the house every other is a bioeconomic crime. Other writers have already proposed that goods be manufactured in such a way as to be more durable [e.g., 43, p. 146]. But it is even more important that consumers should reeducate themselves to despise fashion. Manufacturers will then have to focus on durability.

Seventh, and closely related to the preceding point, is the necessity that durable goods be made still more durable by being designed so as to be repairable. (To put it in a plastic analogy, in many cases nowadays, we have to throw away a pair of shoes merely because one lace has broken.)

Eighth, in a compelling harmony with all the above thoughts we should cure ourselves of what I have been calling "the circumdrome of the shaving machine," which is to shave oneself faster so as to have more time to work on a machine that shaves faster so as to have more time to work on a machine that shaves still faster, and so on *ad infinitum*. This change will call for a great deal of recanting on the part of all those professions which have lured man into this empty infinite regress. We must come to realize that an important prerequisite for a good life is a substantial amount of leisure spent in an intelligent manner.

Considered on paper, in the abstract, the foregoing recommendations would on the whole seem reasonable to anyone willing to examine the logic on which they rest. But one thought has persisted in my mind ever since I became interested in the entropic nature of the economic process. Will mankind listen to any program that implies a constriction of its addiction to exosomatic comfort? Perhaps the destiny of man is to have a short but fiery, exciting, and extravagant life rather than a long, uneventful, and vegetative existence. Let other species—the amoebas, for example—which have no spiritual ambitions inherit an earth still bathed in plenty of sunshine.

Notes

1. A specific suggestion implying entropy bootlegging is Harry Johnson's: it envisages the possibility of reconstituting the stores of coal and oil "with enough ingenuity" [49, p. 8]. And if he means with enough energy as well, why should one wish to lose a great part of that energy through the transformation?

2. How incredibly resilient is the myth of energy breeding is evidenced by the very recent statement of Roger Revelle [70, p. 169] that "farming can be thought of as a kind of breeder reactor in which much more energy is produced than consumed." Ignorance of the main laws governing energy is widespread indeed.

3. Marxist economists also are part of this chorus. A Romanian review of [32], for example, objected that we have barely scratched the surface of the earth.

4. To recall the famous old French quatrain: "Seigneur de La Palice / fell in the battle for Pavia. / A quarter of an hour before his death / he was still alive." (My translation.) See *Grand Dictionnaire Universel du XIXe Siècle*, vol. 10, p. 179.

5. Even some natural scientists, e.g., [1], have taken this position. Curiously, the historical fact that some civilizations were unable "to think up something" is brushed aside with the remark that they were "relatively isolated" [13,

p. 6]. But is not mankind, too, a community completely isolated from any external cultural diffusion and one, also, which is unable to migrate?

6. Similar arguments can be found in [4, pp. 338f; 59, p. 102; 74, p. 45]. Interestingly, Kaysen [51, p. 661] and Solow [74, p. 43], while recognizing the finitude of mankind's entropic dowry, pooh-pooh the fact because it does not "lead to any very interesting conclusions." Economists, of all students, should know that the finite, not the infinite, poses extremely interesting questions. The present paper hopes to offer proof of this.

7. Even in this most cited case, substitution has not been as successful in every direction as we have generally believed. Recently, it has been discovered that aluminum electrical cables constitute fire hazards.

8. The pearl on this issue, however, is supplied by Maddox [59, p. 104]: "Just as prosperity in countries now advanced has been accompanied by an actual decrease in the consumption of bread, so it is to be expected that affluence will make societies less dependent on metals such as steel."

9. The point refers to the addition of capital (measured in *money terms)* and labor (measured in *workers employed)* as well as the computation of net output (by subtraction) from physical gross output [3, pp. 167f].

10. For these distinctions, see [27, pp. 512–519; 30, p. 4; 32, pp. 223–225].

11. See the dialogue between Preston Cloud and Roger Revelle quoted in [66, p. 416]. The same refrain runs through Maddox's complaint against those who point out mankind's limitations [59, pp. vi, 138, 280]. In relation to Maddox's chapter, "Manmade Men," see [32, pp. 348–359].

12. Joseph J. Spengler, a recognized authority in this broad domain, tells me that indeed he knows of no one who may have made the observation. For some very penetrating discussions of Malthus and of the present population pressure, see [76; 77]

13. The substance of the argument of *The Limits* beyond that of Mill's is borrowed from Boulding and Daly [8; 9; 20; 21].

14. In *International Encyclopedia of the Social Sciences*, for example, the point is mentioned only in passing.

15. Obviously, any increase in s_i will generally result in a decrease of L and of n. Also, the carrying capacity in any year may be increased by a greater use of terrestrial resources. These elementary points should be retained for further use. . . .

16. The point recalls Boulding's idea that the inflow from nature into the economic process, which he calls "throughput," is "something to be minimized rather than maximized" and that we should pass from an economy of flow to one of stock [8, pp. 9f; 9, pp. 359f]. The idea is more striking than enlightening. True, economists suffer from a flow complex [29; 55; 88]; also, they have little realized that the proper analytical description of a process must include *both flows and funds* [30; 32, pp. 219f, 228–234]. Entrepreneurs,

as far as Boulding's idea is concerned, have at all times aimed at minimizing the flow necessary to maintain their capital funds. If the present inflow from nature is incommensurate with the safety of our species, it is only because the population is too large and part of it enjoys excessive comfort. Economic decisions will always forcibly involve both flows and stocks. Is it not true that mankind's problem is to economize S (a stock) for as large an amount of life as possible, which implies to minimize s_j (a flow) for some "good life"?

17. I saw this term used for the first time in a letter from Jiří Zeman.

18. The practice of slavery, in the past, and the possible procurement, in the future, of organs for transplant are phenomena akin to the exosomatic evolution.

19. A fact greatly misunderstood: Ricardian land has economic value for the same reason as a fisherman's net. Ricardian land catches the most valuable energy, roughly in proportion to its total size [27, p. 508; 32, p. 232].

20. The figures used in this section have been calculated from the data of Daniels [22] and Hubbert [46]. Such data, especially those about reserves, vary from author to author but not to the extent that really matters. However, the assertion that "the vast oil shales which are to be found all over the world [would last] for no less than 40,000 years" [59, p. 99] is sheer fantasy.

21. In an answer to critics *(American Scientist* 58, no. 6, p. 610), the same authors prove, again linearly, that the agro-industrial complexes of the grand plan could easily feed such a population.

22. For a recent discussion of the social impact of industrial growth, in general, and of the social problems growing out of a large-scale use of nuclear energy, in particular, see [78], a monograph by Harold and Margaret Sprout, pioneers in this field.

23. One percent only of the deuterium in the oceans would provide 10^8 Q through that reaction, an amount amply sufficient for some hundred millions of years of very high industrial comfort. The reaction deuterium-tritium stands a better chance of success because it requires a lower temperature. But since it involves lithium 6, which exists in small supply, it would yield only about 200 Q in all.

24. It should be of interest to know that during World War II in Sweden, for one, automobiles were driven with the poor gas obtained by heating charcoal with kindlings in a container serving as a tank!

25. [Editors' note: Georgescu-Roegen's more recent writings are less sanguine about the prospects for direct use of solar energy. See his "Energy Analysis and Economic *Valuation," Southern Economic Journal,* April 1979.]

26. One necessary qualification: even the use of solar energy may disturb the climate if the energy is released in another place than where collected. The same is true for a difference in time, but this case is unlikely to have any practical importance.

27. At the Dai Dong Conference (Stockholm, 1972), I suggested the adoption of a measure which seems to me to be applicable with much less difficulty than dealing with installations of all sorts. My suggestion, instead, was to allow people to move freely from any country to any other country whatsover. Its reception was less than lukewarm. See [2, p. 72].

28. To avoid any misinterpretation, I should add that the present fad for organic foods has nothing to do with this proposal. . . .

References

[1] Abelson, Philip H. "Limits to Growth." *Science,* 17 March 1972, p. 1197.

[2] Artin, Tom. *Earth Talk: Independent Voices on the Environment.* New York: Grossman, 1973.

[3] Barnett, Harold J., and Chandler Morse. *Scarcity and Growth.* Baltimore: Johns Hopkins University Press, 1963.

[4] Beckerman, Wilfred. "Economists, Scientists, and Environmental Catastrophe." *Oxford Economic Papers* (November 1972), 327–344.

[5] Blin-Stoyle, R. J. "The End of Mechanistic Philosophy and the Rise of Field Physics." In *Turning Points in Physics,* edited by R. J. Blin-Stoyle et al. Amsterdam: North-Holland, 1959, pp. 5–29.

[6] "A Blueprint for Survival." *The Ecologist* (January 1972), 1–43.

[7] Bormann, F. H. "Unlimited Growth: Growing, Growing, Gone?" *BioScience* (December 1972), 706–709.

[8] Boulding, Kenneth. "The Economics of the Coming Spaceship Earth." In *Environmental Quality in a Growing Economy,* edited by Henry Jarrett. Baltimore: Johns Hopkins University Press, 1966, pp. 3–14.

[9] Boulding, Kenneth. "Environment and Economics." In [66], pp. 359–367.

[10] Bray, Jeremy. *The Politics of the Environment,* Fabian Tract 412. London: Fabian Society, 1972.

[11] Bridgman, P. W. "Statistical Mechanics and the Second Law of Thermodynamics." In *Reflections of a Physicist,* 2d ed. New York: Philosophical Library, 1955, pp. 236–268.

[12] Brown, Harrison. "Human Materials Production as a Process in the Biosphere." *Scientific American* (September 1970), 195–208.

[13] Brown, Lester R., and Gail Finsterbusch. "Man, Food and Environment." In [66], pp. 53–69.

[14] Cannon, James. "Steel: The Recyclable Material." *Environment* (November 1973), 11–20.

[15] Cloud, Preston, ed. *Resources and Man.* San Francisco: W. H. Freeman, 1969.

[16] Cloud, Preston. "Resources, Population, and Quality of Life." In *Is There an Optimum Level of Population?*, edited by S. F. Singer. New York: McGraw-Hill, 1971, pp. 8–31.

[17] Cloud, Preston. "Mineral Resources in Fact and Fancy." In [66], pp. 71–88.

[18] Commoner, Barry. *The Closing Circle*. New York: Knopf, 1971.

[19] Culbertson, John M. *Economic Development: An Ecological Approach*. New York: Knopf, 1971.

[20] Daly, Herman E. "Toward a Stationary-State Economy." In *Patient Earth*, edited by J. Harte and R. Socolow. New York: Holt, Rinehart and Winston, 1971, pp. 226–244.

[21] Daly, Herman E. *The Stationary-State Economy*. Distinguished Lecture Series no. 2, Department of Economics, University of Alabama, 1971.

[22] Daniels, Farrington. *Direct Use of the Sun's Energy*. New Haven: Yale University Press, 1964.

[23] Einstein, Albert, and Leopold Infeld. *The Evolution of Physics*. New York: Simon and Schuster, 1938.

[24] "The Fragile Climate of Spaceship Earth." *Intellectual Digest* (March 1972), 78–80.

[25] Georgescu-Roegen, Nicholas. "The Theory of Choice and the Constancy of Economic Laws." *Quarterly Journal of Economics* (February 1950), 125–138. Reprinted in [29], pp. 171–183.

[26] Georgescu-Roegen, Nicholas. "Toward a Partial Redirection of Econometrics," Part III. *Review of Economics and Statistics* 34 (August 1952), 206–211.

[27] Georgescu-Roegen, Nicholas. "Process in Farming versus Process in Manufacturing: A Problem of Balanced Development." In *Economic Problems of Agriculture in Industrial Societies*, edited by Ugo Papi and Charles Nunn. London: Macmillan; New York: St. Martin's Press, 1969, pp. 497–528.

[28] Georgescu-Roegen, Nicholas. "Further Thoughts on Corrado Gini's *Dellusioni dell' econometria*." *Metron* 25, no. 104 (1966), 265–279.

[29] Georgescu-Roegen, Nicholas. *Analytical Economics: Issues and Problems*. Cambridge: Harvard University Press, 1966.

[30] Georgescu-Roegen, Nicholas. "The Economics of Production." *American Economic Review* 40 (May 1970), 1–9.

[31] Georgescu-Roegen, Nicholas. "The Entropy Law and the Economic Problem." Distinguished Lecture Series no. 1, Department of Economics, University of Alabama, 1971. Reprinted in this volume.

[32] Georgescu-Roegen, Nicholas. *The Entropy Law and the Economic Process*. Cambridge: Harvard University Press, 1971.

[33] Georgescu-Roegen, Nicholas. "Process Analysis and the Neoclassical Theory of Production." *American Journal of Agricultural Economics* 54 (May 1972), 279–294.

[34] Gillette, Robert. "The Limits to Growth: Hard Sell for a Computer View of Doomsday." *Science,* 10 March 1972, pp. 1088–1092.

[35] Gillette, Robert. "Nuclear Safety: Damaged Fuel Ignites a New Debate in AEC." *Science,* 28 July 1972, pp. 330–331.

[36] Gillette, Robert. "Reactor Safety: AEC Concedes Some Points to Its Critics." *Science,* 3 November 1972, pp. 482–484.

[37] Glaser, Peter E. "Power from the Sun: Its Future." *Science,* 22 November 1968, pp. 857–861.

[38] Goeller, H. E. "The Ultimate Mineral Resource Situation." *Proceedings of the National Academy of Science, USA* (October 1972), 2991–2992.

[39] Gofman, John W. "Time for a Moratorium." *Environmental Action* (November 1972), 11–15.

[40] Haar, D. ter. "The Quantum Nature of Matter and Radiation." In *Turning Points in Physics,* edited by R. J. Blin-Stoyle et al. (Amsterdam: North-Holland, 1959), pp. 30–44.

[41] Hammond, Allen L. "Solar Energy: A Feasible Source of Power?" *Science,* 14 May 1971, p. 660.

[42] Hardin, Garrett. "The Tragedy of the Commons." *Science,* 13 December 1968, pp. 1234–1248.

[43] Hibbard, Walter R., Jr. "Mineral Resources: Challenge or Threat?" *Science,* 12 April 1968, pp. 143–145.

[44] Holdren, John, and Philip Herera. *Energy.* San Francisco: Sierra Club, 1971.

[45] Hotelling, Harold. "The Economics of Exhaustible Resources." *Journal of Political Economy* (March-April 1931), 137–175.

[46] Hubbert, M. King. "Energy Resources." In [15], pp. 157–242.

[47] Istock, Conrad A. "Modern Environmental Deterioration as a Natural Process." *International Journal of Environmental Studies* (1971), 151–155.

[48] Jevons, W. Stanley. *The Theory of Political Economy,* 2d ed. London: Macmillan, 1879.

[49] Johnson, Harry G. *Man and His Environment.* London: The British-North American Committee, 1973.

[50] Katchalsky, A., and Peter F. Curran. *Nonequilibrium Thermodynamics in Biophysics.* Cambridge, Mass.: Harvard University Press, 1965.

[51] Kaysen, Carl. "The Computer That Printed Out W*O*L*F*." *Foreign Affairs* (July 1972), 660–668.

[52] Kneese, Allen, and Ronald Ridker. "Predicament of Mankind." *Washington Post*, 2 March 1972.

[53] Laplace, Pierre Simon de. *A Philosophical Essay on Probability*. New York: Wiley, 1902.

[54] Leontief, Wassily. "Theoretical Assumptions and Nonobservable Facts." *American Economic Review* (March 1971), 1–7.

[55] "Limits to Misconception." *The Economist*, 11 March 1972, pp. 20–22.

[56] Lovering, Thomas S. "Mineral Resources from the Land." In [15], pp. 109–134.

[57] MacDonald, Gordon J. F. "Pollution, Weather and Climate." In [66], pp. 326–336.

[58] Maddox, John. "Raw Materials and the Price Mechanism." *Nature*, 14 April 1972, pp. 331–334.

[59] Maddox, John. *The Doomsday Syndrome*. New York: McGraw-Hill, 1972.

[60] Marshall, Alfred. *Principles of Economics*, 8th ed. London: Macmillan, 1920.

[61] Marx, Karl. *Capital*. 3 vols. Chicago: Charles H. Kerr, 1906–1933.

[62] Meadows, Donella H., et al. *The Limits to Growth*. New York: Universe Books, 1972.

[63] Metz, William D. "Fusion: Princeton Tokamak Proves a Principle." *Science*, 22 December 1972, p. 1274B.

[64] Mill, John Stuart. *Principles of Political Economy*. In *Collected Works*, edited by J. M. Robson, vols. 2–3. Toronto: University of Toronto Press, 1965.

[65] Mishan, E. J. *Technology and Growth: The Price We Pay*. New York: Praeger, 1970.

[66] Murdoch, William W., ed. *Environment: Resources, Pollution and Society*. Stamford, Conn.: Sinauer, 1971.

[67] Novick, Sheldon. "Nuclear Breeders." *Environment* (July-August 1974), 6–15.

[68] Pigou, A. C. *The Economics of Stationary States*. London: Macmillan, 1935.

[69] *Report on Limits to Growth*. Mimeographed. A Study of the Staff of the International Bank for Reconstruction and Development, Washington, D.C., 1972.

[70] Revelle, Roger. "Food and Population." *Scientific American* (September 1974), 161–170.

[71] Schrödinger, Erwin. *What Is Life?* Cambridge, England: The University Press, 1944.

[72] Silk, Leonard. "On the Imminence of Disaster " *New York Times,* 14 March 1972.

[73] Solo, Robert A. "Arithmomorphism and Entropy." *Economic Development and Cultural Change* (April 1974), 510–517.

[74] Solow, Robert M. "Is the End of the World at Hand?" *Challenge* (March-April 1973), 39–50.

[75] Solow, Robert M. "The Economics of Resources or the Resources of Economics." Richard T. Ely Lecture, *American Economic Review* (May 1974), 1–14.

[76] Spengler, Joseph J. "Was Malthus Right?" *Southern Economic Journal* (July 1966), 17–34.

[77] Spengler, Joseph J. "Homosphere, Seen and Unseen: Retreat from Atomism." *Proceedings of the Nineteenth Southern Water Resources and Pollution Control Conference,* 1970, pp. 7–16.

[78] Sprout, Harold, and Margaret Sprout. *Multiple Vulnerabilities.* Mimeographed. Research Monograph No. 40, Center of International Studies, Princeton University, 1974.

[79] Summers, Claude M. "The Conversion of Energy." *Scientific American* (September 1971), 149–160.

[80] Wallich, Henry C. "How to Live with Economic Growth." *Fortune* (October 1972), 115–122.

[81] Weinberg, Alvin M. "Breeder Reactors." *Scientific American* (January 1960), 82–94.

[82] Weinberg, Alvin M. "Social Institutions and Nuclear Energy." *Science,* 7 July 1972, pp. 27–34.

[83] Weinberg, Alvin M., and R. Philip Hammond. "Limits to the Use of Energy." *American Scientist* (July-August 1970), 412–418.

5

Exponential Growth as a Transient Phenomenon in Human History

M. King Hubbert

In this bicentennial year of American history, it is useful for us to reflect that the two-hundred-year period from 1776 to 1976 marks the emergence of an entirely new phase in human history. This is the period during which our industrial civilization has arisen and developed. It is also the period during which there has occurred a transition from a social state whose material and energy requirements were satisfied mainly from renewable resources to our present state that is overwhelmingly dependent upon nonrenewable resources. In 1776 our material requirements for food, housing, clothing, and industrial equipment were principally satisfied by renewable vegetable and animal products. Nonrenewable mineral products, clay products, lime, sand, and metals, were used in such small amounts that the available supplies, at that rate of consumption, seemed almost inexhaustible.

The energy requirements two centuries ago were likewise met principally by renewable resources. Vegetable and animal products were used for food and warmth; human and animal labor and wind and water power for mechanical work. The only nonrenewable energy source then used was coal, which was consumed in such small amounts per year that the total supply at this rate would likewise have seemed almost inexhaustible.

During the ensuing two centuries, the development of the world's present highly industrialized society has occurred. The magnitude and significance of this transition can most readily be appreciated if we consider the graphs showing the growth in the world's annual

Reprinted from *Societal Issues, Scientific Viewpoints* (New York: American Institute of Physics), edited by Margaret A. Strom, pp. 75–84, by permission of the author.

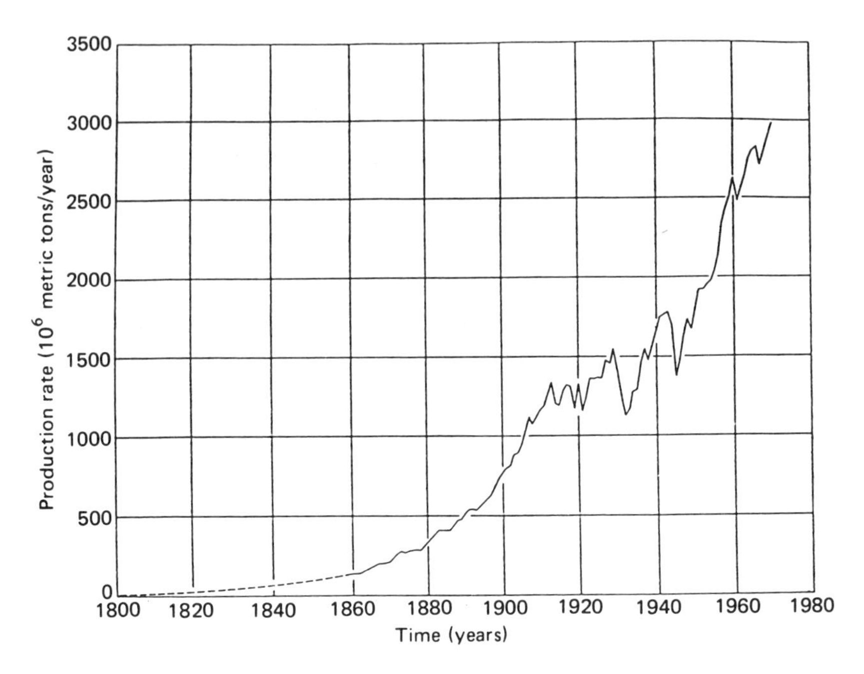

Figure 5.1
World production of coal and lignite (Hubbert, 1974a, fig. 3).

production of the principal sources of industrial energy, coal and petroleum.

Figure 5.1 shows the annual world production of coal and lignite from 1860 to 1970; figure 5.2 shows the corresponding growth in the annual world production of crude oil from 1880 and 1970.

The mining of coal as a continuous industrial enterprise began nine centuries ago near the town of Newcastle upon Tyne in northeast England. Annual statistics of world coal production are difficult to assemble before 1860, but by that time the annual production rate had reached 138 million metric tons per year. From the earlier history of coal mining and from scattered statistics it can be estimated that during the eight centuries from 1060 to 1860, the average growth rate in annual coal production was about 2.3 percent per year with an average doubling period of about 30 years. From this it can be estimated that the production of coal in 1776 was about 20 million metric tons. As of this year, the annual coal production rate has reached about 3.3 billion metric tons—a 165-fold increase during the two centuries since 1776. This has been at an average growth rate of

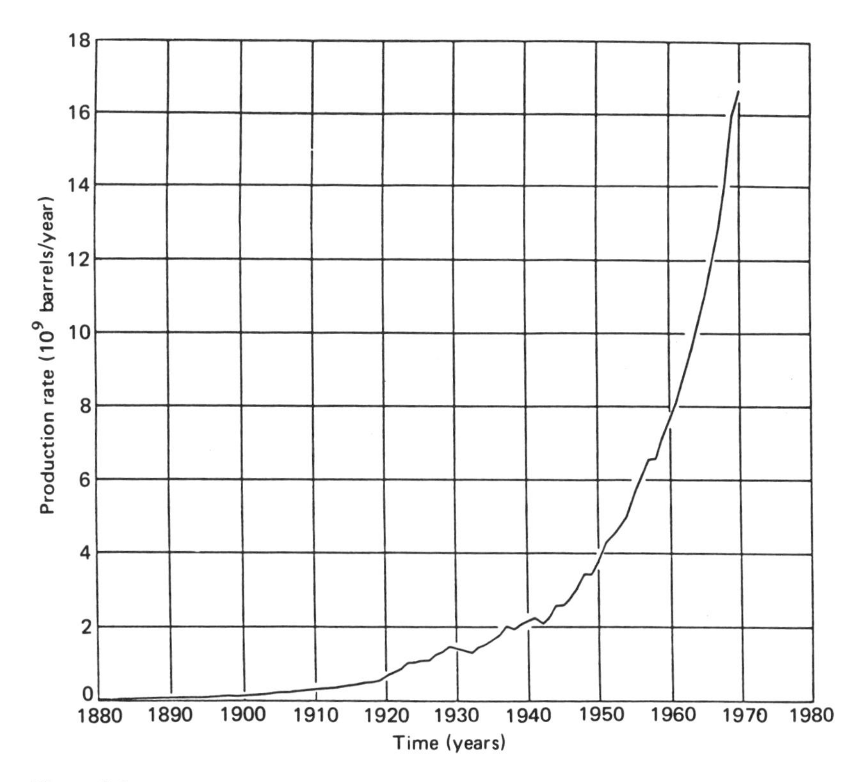

Figure 5.2
World production of crude oil (Hubbert, 1974a, fig. 5).

about 2.55 percent per year, or an average period of doubling of 27 years.

Although very small amounts of oil were produced in China and Burma at earlier times, the world's production of crude oil as a continuous industrial enterprise was begun in Romania in 1857 and in the United States two years later. As figure 5.2 shows, from 1880 until 1970 the growth in annual crude oil production increased smoothly and spectacularly. During this period the growth rate averaged 7.04 percent per year and the production rate doubled, on the average, every 9.8 years. The cumulative production also doubled about every 10 years. For example, the amount of oil produced during the decade 1960–1970 was almost exactly equal to all the oil produced from 1857 to 1960.

The increase in the consumption of and dependence upon the industrial metals during the last two centuries is comparable to the increased consumption of energy from the fossil fuels. Consider iron. In 1776 the amount of pig iron produced in the world was about 360,000 metric tons. By 1976 this had increased by a factor of 1556 times to a present annual production of about 560 million metric tons. This corresponds to an average growth rate in the annual production of 3.67 percent per year, with the production rate doubling, on the average, every 18.9 years.

In 1776 the world's human population was approximately 790 million. It has increased to 4.24 billion. The per capita consumption of iron in 1776 amounted to only 0.46 kilogram. This has increased to 132 kilograms, a 287-fold increase during these two centuries.

These figures are illustrative of the profound changes in human affairs that have occurred during the last two centuries. Our present concern, however, is with the future. Is it possible that such rates of growth can be maintained during the next two centuries, or do the industrial and demographic growth rates experienced during the last two centuries represent a transient and ephemeral epoch in the longer span of human history?

Answers to such questions can be obtained by considering the physical nature of various growth phenomena. Consider first a renewable phenomenon such as a food supply or the development of water power. Human food supply is derived almost entirely from plant or animal products. Therefore, the problem of the increase of food supply, or of the human population itself, reduces to the basic problem of how much the population of any biologic species can be increased on the earth. The basics of this problem are well understood. Biologists discovered a couple of centuries ago that the population of any biologic species, plant or animal, if given a favorable environment, will increase exponentially with time. That is, the population will double and redouble in the successive ratios of 1, 2, 4, 8, etc. during successive equal intervals of time. The period required for the population to double is different for different species. For elephants and humans the doubling period is a few decades, but for some bacteria it is as short as 20 minutes. Such a manner of growth obviously cannot continue indefinitely, but the significant question is this: About how many successive doublings on a finite earth are possible?

An approximate answer can be obtained when we consider the magnitudes obtained by successive doublings. Consider, for example, the classical problem of placing one grain of wheat on the first square of a chessboard, two on the second square, four on the third, and continuing the doubling for each successive square of the board. On the *n*th square—the last, or 64th—the number of grains will be 2^{n-1} or 2^{63}. How much wheat will this amount to? On the last square alone, the amount of wheat would be equal to approximately 1,000 times the world's present annual wheat crop; for the whole board, it would be twice this amount.

From one point of view this is merely a trivial problem in arithmetic; from another it is of profound significance, for it tells us that the earth itself will not tolerate the doubling of 1 grain of wheat 64 times.

Similar results are obtained when we consider the successive doublings of other biologic populations, or of industrial activities. The present world human population, were it to have descended from a single pair, say Adam and Eve, would have been generated by only 31 doublings, and a total of 46 doublings would yield a population density of one person per square meter over all the land areas of the earth. For an industrial example, the world's population of automobiles is also doubling repeatedly. If we apply the chessboard arithmetic to the automobile population, beginning with one car and doubling this 64 times, and then let the resulting quantity of automobiles be stacked uniformly on all the land surfaces of the earth, how deep would be the layer formed? About 1,200 miles or 2,000 kilometers.

From such considerations it becomes evident that the maximum number of doublings that any biological population or industrial component can experience is but a few tens. Therefore, any rapid rate of growth of such a component must be a transient phenomenon of temporary duration. The normal state of a biologic population, when averaged over a few years, must be one of an extremely slow rate of change—a near steady state. For example, the maximum possible number of times the human population could have doubled during the last million years is 31. Consequently, the minimum value of the average period of doubling during that time would have been 32,000 years, as contrasted with the present period of but 32 years.

The growth curve that characterizes a rapid increase of any biologic population or the exploitation of any renewable nonbiologic resource such as water power must accordingly be similar to that shown in figure 5.3. Beginning at a near steady state of a constant biologic population, or at zero as in the case of water power, the quantity

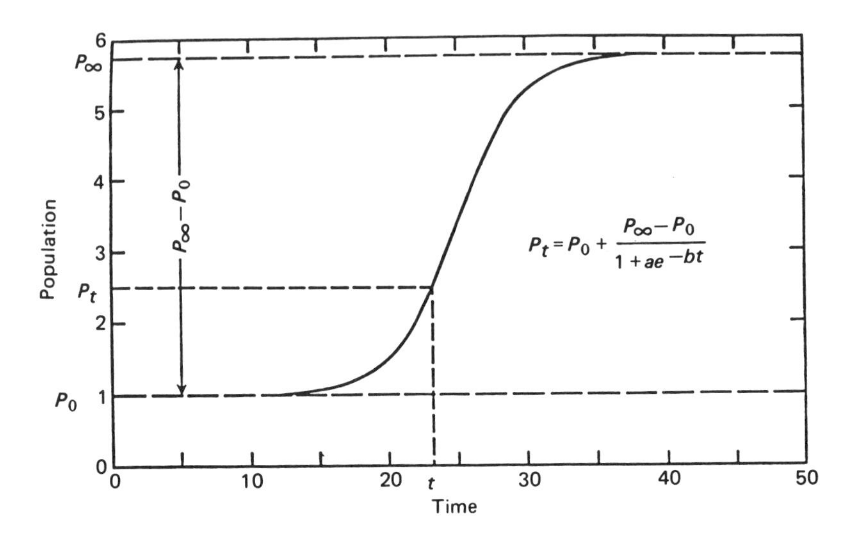

Figure 5.3
Logistic curve of population growth.

considered increases exponentially—that is, it doubles in equal intervals of time—for a while, and then, in response to retarding influences, gradually levels off to a constant figure, characterized by a state of nongrowth.

In the biologic case the inverse can also happen. The disturbance may be unfavorable and the population may undergo a decline or a negative growth. It then must stabilize at a lower level, or else become extinct.

The long-term behavior of the exploitation of the nonrenewable resources such as the fossil fuels or the ores of metals differs from that of the renewable resources in a very fundamental respect. Nonrenewable resources are absolutely exhaustible. Figure 5.4 is a flow diagram of the mining and utilization of a fossil fuel. The fuel is taken from the earth, and, in burning, is combined chemically with atmospheric oxygen. The materials, in the form of gaseous compounds CO_2, H_2O, and SO_2 and of asheous solids, remain on the earth, but the energy content released initially as heat, after undergoing successive degradations, eventually leaves the earth as spent low-temperature radiation.

The exploitation of metallic ores is illustrated in figure 5.5. Here, an ore body consisting of a naturally occurring rock containing an

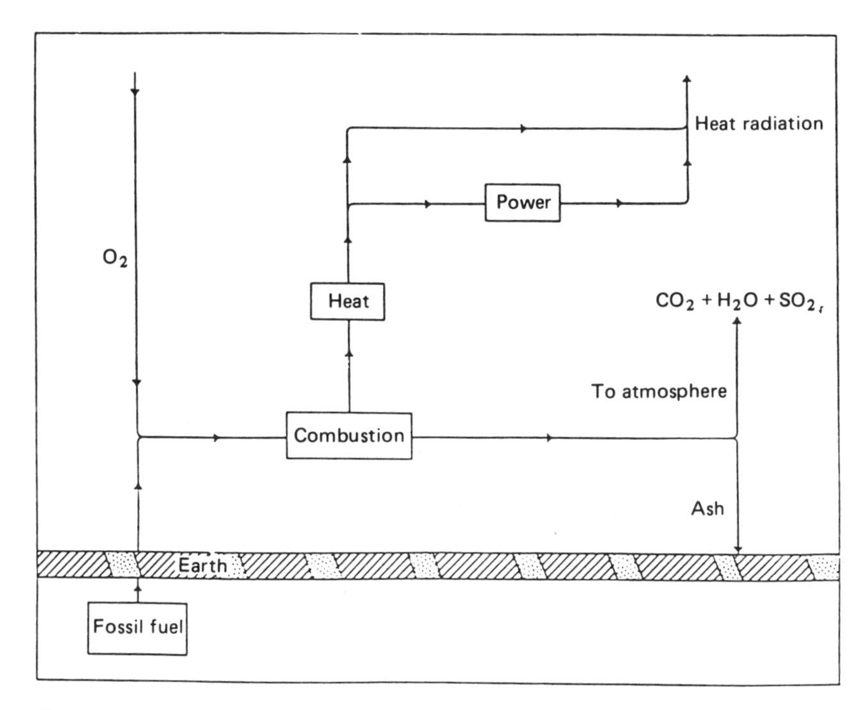

Figure 5.4
Flow diagram of matter and energy in combustion of fossil fuel for power production (Hubbert, 1974a, fig. 15).

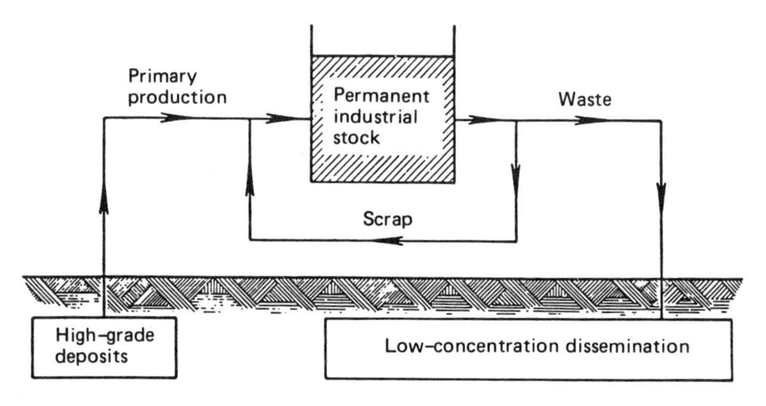

Figure 5.5
Flow diagram for the production and use of an industrial metal (Hubbert, 1972, fig. 26).

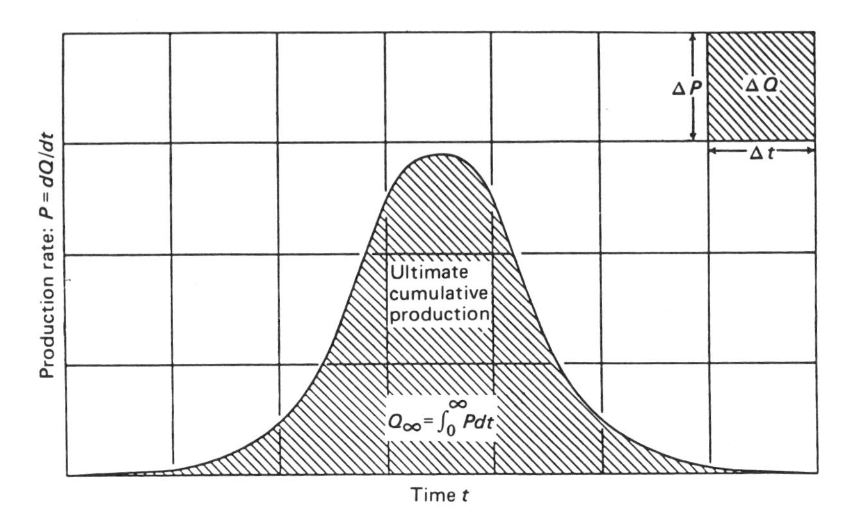

Figure 5.6
Mathematical properties of complete-cycle curve of production of an exhaustible resource (Hubbert, 1974a, fig. 18).

abnormally high concentration of a given metal is mined and the metal extracted by smelting. This then goes into the industrial stock from which it eventually is removed—lead in gasoline, for example, is irretrievably scattered.

Thus the production history of the fossil fuels or of the metallic ores, in any given region and eventually for the entire earth, is characterized by a complete-cycle curve such as that shown in figure 5.6, where annual production is plotted graphically as a function of time, as has been done for coal and oil in figures 5.1 and 5.2. The characteristics of such a curve are that the production, beginning initially at zero, increases, usually exponentially, for a period. The growth rate then slows down and the rate of production passes one or more maxima and finally, as the resource becomes exhausted, goes into a long negative-exponential decline.

The area beneath such a curve is also a graphical measure of cumulative production. If from geological mapping, drilling, and other means, the total recoverable quantity of the resource considered can be estimated in the earlier stages of exploitation, the future of the production history can be estimated, because the complete-cycle curve can only encompass a total area corresponding to the estimate made.

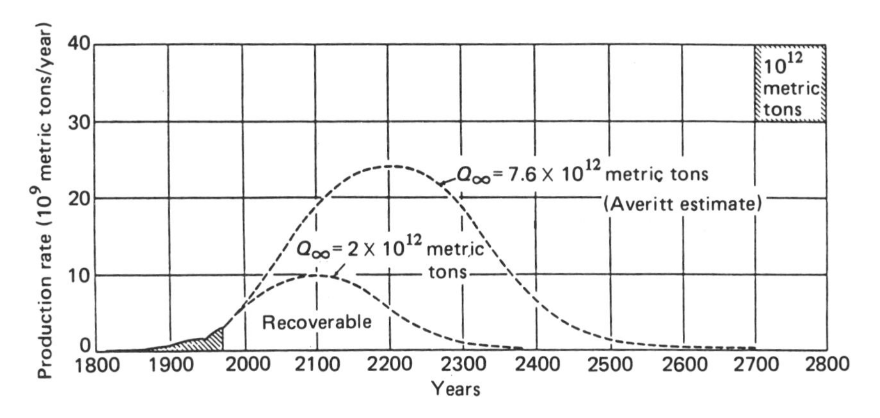

Figure 5.7
Complete-cycle curves of world coal production for higher figure of 7.6×12^{12} and lower figure of 2×10^{12} metric tons for ultimate production (Hubbert, 1974a, fig. 21).

By this means we are able to gain a reasonably reliable estimate of the time during which the fossil fuels can serve as major sources of the world's industrial energy. Figure 5.7 shows two interpretations of the complete cycle of world coal production, based upon a high estimate for the ultimate amount of coal to be produced of 7.6 trillion metric tons, and a low figure of 2 trillion. The high figure is based upon estimates, assuming 50 percent recovery, of beds as thin as 1 foot (0.3 meters) and to depths as great as 1,200 meters. The lower figure is for coal beds not thinner than 0.7 meters and not deeper than 300 meters.

Two aspects of such curves are particularly significant, the approximate date of the peak rate of production, and the period of time during which the cumulative production increases from 10 to 90 percent of the ultimate cumulative production—the time span required to produce the middle 80 percent. For the high estimate of 7.6 trillion tons the estimated date of peak production would be about 150 to 200 years hence, and the time period required to produce the middle 80 percent would be that from about the year 2000 to 2300 or 2400. For the lower figure of 2 trillion tons of ultimate production the peak in the production rate would occur earlier, about the year 2100, and the middle 80 percent of the ultimate coal would be consumed during the approximately 200 years from about 2000 to 2200.

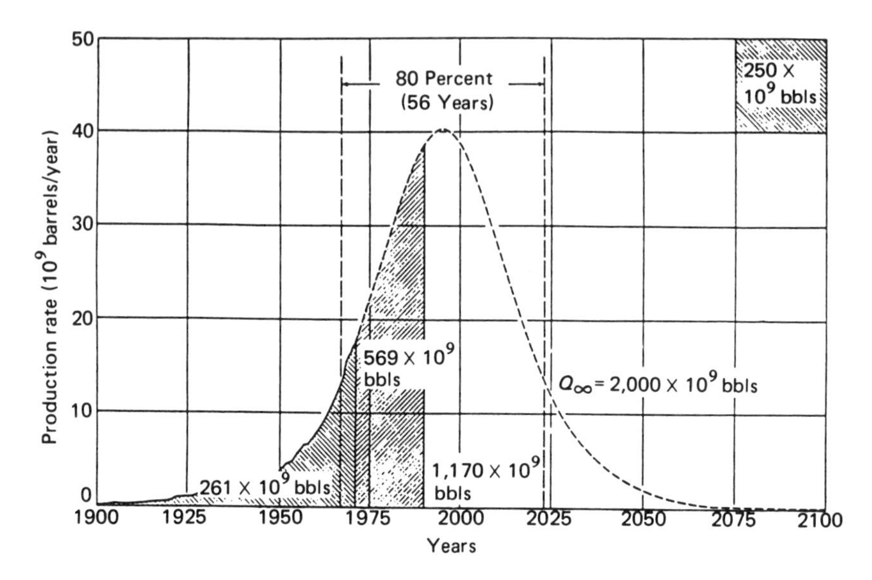

Figure 5.8
Complete-cycle curve for world crude oil production (Hubbert, 1974a, fig. 68).

The estimate of the complete cycle for the world production of crude oil is shown in figure 5.8. This is based upon an estimate of 2,000 billion barrels for the ultimate production, which is somewhat higher than the average of 15 estimates published since 1958 by international oil companies and leading international petroleum geologists. Using this figure, and assuming an orderly evolution of petroleum consumption, the peak in the production rate will probably occur during the decade 1990 to 2000, and the middle 80 percent will be consumed within the 60-year period from 1965 to 2025. Hence, children born within the last 10 years will see the world consume most of its oil during their lifetimes.

It is possible that oil production could be curtailed by the exporting nations to somewhat near the present rate. Were that to occur, the upper area under the curve in figure 5.8 would be displaced to the back slope and thus prolong the middle 80 percent period. Even so, this would be only about 10 to 15 years.

In parallel with the energy resources, modern industry, as we have seen, is also dependent upon the industrial metals. One of the most abundant of these is iron, the known ores of which are estimated to amount to about 250 trillion metric tons and the ultimate amount to

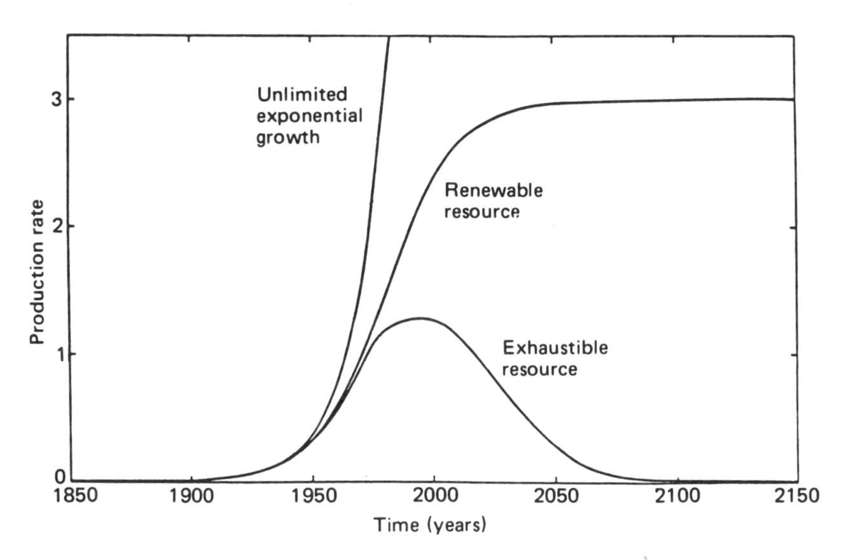

Figure 5.9
Three types of growth (Hubbert, 1974b, fig. 1).

possibly as much as 750 trillion. Based upon these figures in conjunction with the record of past production, the time span for iron production would be about the same as that for coal, the peak in the production rate occurring one or two centuries hence, and the middle 80 percent span extending about two to four centuries.

The known and estimated ores of most other industrial metals, copper, tin, lead, zinc, and others, are in very much shorter supply, with the time until the peak production rate occurs and the time span for the middle 80 percent being measurable in decades rather than in centuries.

Three types of growth phenomena with which an industrial society must deal are shown graphically in figure 5.9. Here, the lower curve represents the rise, culmination, and decline in the production rate of any nonrenewable resource such as the fossil fuels or the ores of metals. The middle curve represents the rise and leveling off of the production of a renewable resource such as water power or a biological product. The third curve is simply the mathematical curve of exponential growth. No physical quantity can follow this curve for more than a brief period of time. However, a sum of money, being of a nonphysical nature and growing according to the rules of com-

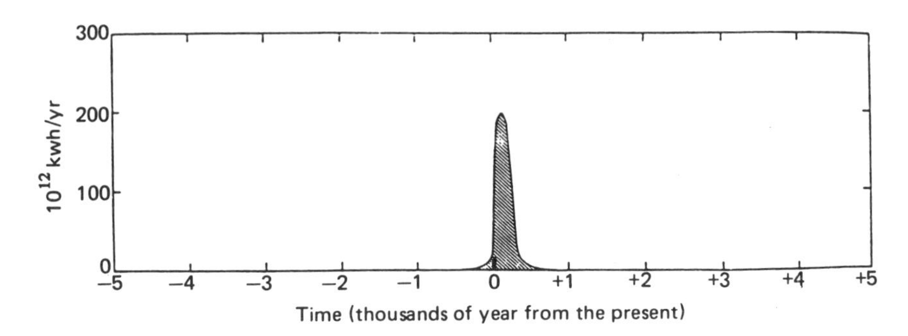

Figure 5.10
Epoch of fossil fuel exploitation in human history during the period from 5,000 years ago to 5,000 years in the future (Hubbert, 1974a, fig. 69).

pound interest at a fixed interest rate, can follow that curve indefinitely.

In their initial phases, the curves for each of these types of growth are indistinguishable from one another, but as industrial growth approaches maturity, the separate curves begin to diverge from one another. In its present state the world industrial system has already entered the divergence phase of these curves but is still somewhat short of the culmination of the curve of nonrenewable resources.

A better appreciation of the brevity and exceptional character of the epoch of the fossil fuels can be gained if we view it in the perspective of a longer time span of human history that we have considered heretofore. In figure 5.10 the complete cycle of exploitation of the world's total supply of fossil fuels, coal and petroleum, is shown on a time scale extending from 5,000 years in the past to 5,000 years in the future. On such a scale, the Washington Monument–like spike in the middle of this range, with a middle 80 percent spread of about three centuries, represents the period of exploitation of the fossil fuels in the much longer span of human history. Brief as this period is, having arisen, as we have seen, principally within the last century, it has already exercised one of the most disturbing influences ever experienced by the human species in its entire biological existence.

The position in which human society now finds itself in this longer span of history is depicted in figure 5.11. What we see there is that we are now in a period of transition between a past characterized by a much smaller population, a low level of technology, of energy

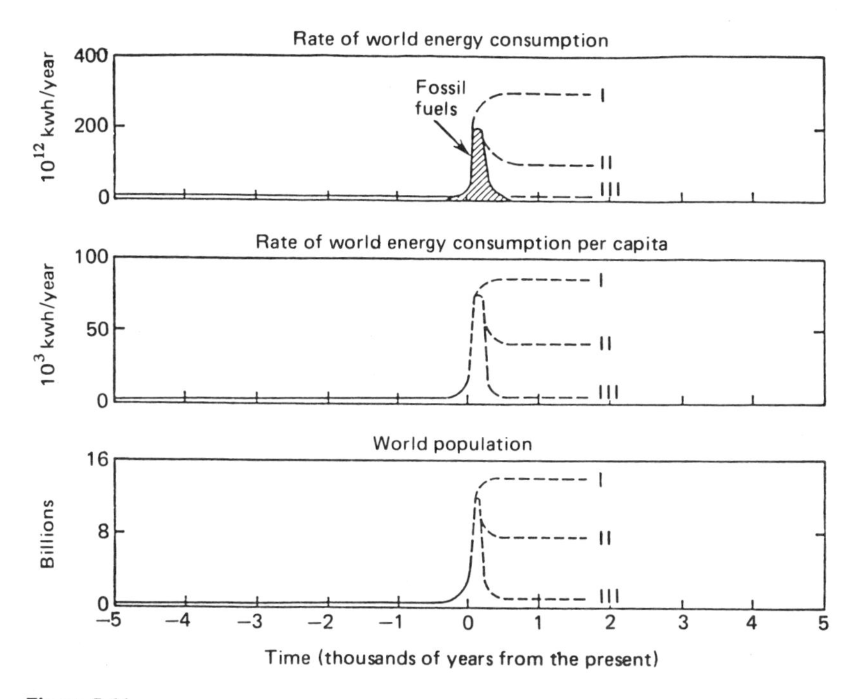

Figure 5.11
Human affairs in historical perspective from 5,000 years in the past to 5,000 years in the future (after Hubbert, 1962, fig. 61).

consumption, and of dependence upon nonrenewable resources, and by very slow rates of change, and a future also characterized by slow rates of change, but by means of utilization of the world's largest source of energy, that of inexhaustible sunshine, capable of sustaining a population of optimum size at a very comfortable standard of living for a prolonged period of time.

It appears therefore that one of the foremost problems confronting humanity today is how to make the transition from the precarious state that we are now in to this optimum future state by a least catastrophic progression. Our principal impediments at present are neither lack of energy or material resources nor of essential physical and biological knowledge. Our principal constraints are cultural. During the last two centuries we have known nothing but exponential growth and in parallel we have evolved what amounts to an exponential-growth culture, a culture so heavily dependent upon the continuance of exponential growth for its stability that it is incapable of reckoning with problems of nongrowth.

Since the problems confronting us are not intrinsically insoluble, it behooves us, while there is yet time, to begin a serious examination of the nature of our cultural constraints and of the cultural adjustments necessary to permit us to deal effectively with the problems rapidly arising. Provided this can be done before unmanageable crises arise, there is promise that we could be on the threshold of achieving one of the greatest intellectual and cultural advances in human history.

References

Hubbert, M. King. 1962. "Energy Resources: A Report to the Committee on Natural Resources." National Academy of Sciences–National Research Council, Publication 1000-D, 141 pp.

Hubbert, M. King. 1972. "Man's Conquest of Energy: Its Ecological and Human Consequences." In *The Environmental and Ecological Forum 1970–1971*, edited by Burt Kline, pp. 1–50. U.S. Atomic Energy Commission Office of Information Services. Available as TID 25857, National Technical Information Service, U.S. Department of Commerce, Springfield, Virginia 22151.

Hubbert, M. King. 1974a. "U.S. Energy Resources: A Review as of 1972," Part I. Committee on Interior and Insular Affairs, U.S. Senate, Serial no. 93–40 (92–75). U.S. Government Printing Office, 267 pp.

Hubbert, M. King. 1974b. "Statement on the Relations between Industrial Growth, the Monetary Interest Rate, and Price Inflation." Hearings before the Subcommittee on the Environment, Committee on Interior and Insular Affairs, House of Representatives, June 4, 6, and 10, July 19 and 26, 1974. Serial no. 93–55, pp. 58–77. U.S. Government Printing Office.

6 The Tragedy of the Commons

Garrett Hardin

At the end of a thoughtful article on the future of nuclear war, Wiesner and York concluded that:

> Both sides in the arms race are . . . confronted by the dilemma of steadily increasing military power and steadily decreasing national security. It *is our considered professional judgment that this dilemma has no technical solution*. If the great powers continue to look for solutions in the area of science and technology only, the result will be to worsen the situation.[1]

I would like to focus your attention not on the subject of the article (national security in a nuclear world) but on the kind of conclusion they reached, namely that there is no technical solution to the problem. An implicit and almost universal assumption of discussions published in professional and semipopular scientific journals is that the problem under discussion has a technical solution. A technical solution may be defined as one that requires a change only in the techniques of the natural sciences, demanding little or nothing in the way of change in human values or ideas of morality.

In our day (though not in earlier times) technical solutions are always welcome. Because of previous failures in prophecy, it takes courage to assert that a desired technical solution is not possible. Wiesner and York exhibited this courage; publishing in a science journal, they insisted that the solution to the problem was not to be found in the natural sciences. They cautiously qualified their statement with the phrase "It is our considered professional judgment." Whether they were right or not is not the concern of the present article. Rather, the concern here is with the important concept of a

From *Science* 162, 13 December 1968, pp. 1243–1248.

class of human problems which can be called "no technical solution problems," and, more specifically, with the identification and discussion of one of these.

It is easy to show that the class is not a null class. Recall the game of tick-tack-toe. Consider the problem "How can I win the game of tick-tack-toe?" It is well known that I cannot, if I assume (in keeping with the conventions of game theory) that my opponent understands the game perfectly. Put another way, there is no "technical solution" to the problem. I can win only by giving a radical meaning to the word "win." I can hit my opponent over the head; or I can drug him; or I can falsify the records. Every way in which I "win" involves, in some sense, an abandonment of the game, as we intuitively understand it. (I can also, of course, openly abandon the game—refuse to play it. This is what most adults do.)

The class of "no technical solution problems" has members. My thesis is that the "population problem," as conventionally conceived, is a member of this class. How it is conventionally conceived needs some comment. It is fair to say that most people who anguish over the population problem are trying to find a way to avoid the evils of overpopulation without relinquishing any of the privileges they now enjoy. They think that farming the seas or developing new strains of wheat will solve the problem—technologically. I try to show here that the solution they seek cannot be found. The population problem cannot be solved in a technical way, any more than can the problem of winning the game of tick-tack-toe.

What Shall We Maximize?

Population, as Malthus said, naturally tends to grow "geometrically," or, as we would now say, exponentially. In a finite world this means that the per capita share of the world's goods must steadily decrease. Is ours a finite world?

A fair defense can be put forward for the view that the world is infinite; or that we do not know that it is not. But, in terms of the practical problems that we must face in the next few generations with the foreseeable technology, it is clear that we will greatly increase human misery if we do not, during the immediate future, assume that the world available to the terrestrial human population is finite. "Space" is no escape.[2]

A finite world can support only a finite population; therefore, population growth must eventually equal zero. (The case of perpetual wide fluctuations above and below zero is a trivial variant that need not be discussed.) When this condition is met, what will be the situation of mankind? Specifically, can Bentham's goal of "the greatest good for the greatest number" be realized?

No—for two reasons, each sufficient by itself. The first is a theoretical one. It is not mathematically possible to maximize for two (or more) variables at the same time. This was clearly stated by von Neumann and Morgenstern,[3] but the principle is implicit in the theory of partial differential equations, dating back at least to D'Alembert (1717–1783).

The second reason springs directly from biological facts. To live, any organism must have a source of energy (for example, food). This energy is utilized for two purposes: mere maintenance and work. For man, maintenance of life requires about 1,600 kilocalories a day ("maintenance calories"). Anything that he does over and above merely staying alive will be defined as work, and is supported by "work calories" which he takes in. Work calories are used not only for what we call work in common speech; they are also required for all forms of enjoyment, from swimming and automobile racing to playing music and writing poetry. If our goal is to maximize population it is obvious what we must do: We must make the work calories per person approach as close to zero as possible. No gourmet meals, no vacations, no sports, no music, no literature, no art. . . . I think that everyone will grant, without argument or proof, that maximizing population does not maximize goods. Bentham's goal is impossible.

In reaching this conclusion I have made the usual assumption that it is the acquisition of energy that is the problem. The appearance of atomic energy has led some to question this assumption. However, given an infinite source of energy, population growth still produces an inescapable problem. The problem of the acquisition of energy is replaced by the problem of its dissipation, as J. H. Fremlin has so wittily shown.[4] The arithmetic signs in the analysis are, as it were, reversed; but Bentham's goal is still unobtainable.

The optimum population is, then, less than the maximum. The difficulty of defining the optimum is enormous; so far as I know, no one has seriously tackled this problem. Reaching an acceptable and stable solution will surely require more than one generation of hard analytical work—and much persuasion.

We want the maximum good per person; but what is good? To one person it is wilderness, to another it is ski lodges for thousands. To one it is estuaries to nourish ducks for hunters to shoot; to another it is factory land. Comparing one good with another is, we usually say, impossible because goods are incommensurable. Incommensurables cannot be compared.

Theoretically this may be true; but in real life incommensurables *are* commensurable. Only a criterion of judgment and a system of weighting are needed. In nature the criterion is survival. Is it better for a species to be small and hidable, or large and powerful? Natural selection commensurates the incommensurables. The compromise achieved depends on a natural weighting of the values of the variables.

Man must imitate this process. There is no doubt that in fact he already does, but unconsciously. It is when the hidden decisions are made explicit that the arguments begin. The problem for the years ahead is to work out an acceptable theory of weighting. Synergistic effects, nonlinear variation, and difficulties in discounting the future make the intellectual problem difficult, but not (in principle) insoluble.

Has any cultural group solved this practical problem at the present time, even on an intuitive level? One simple fact proves that none has: there is no prosperous population in the world today that has, and has had for some time, a growth rate of zero. Any people that has intuitively identified its optimum point will soon reach it, after which its growth rate becomes and remains zero.[5]

Of course, a positive growth rate might be taken as evidence that a population is below its optimum. However, by any reasonable standards, the most rapidly growing populations on earth today are (in general) the most miserable. This association (which need not be invariable) casts doubt on the optimistic assumption that the positive growth rate of a population is evidence that it has yet to reach its optimum.

We can make little progress in working toward optimum population size until we explicitly exorcize the spirit of Adam Smith in the field of practical demography. In economic affairs, *The Wealth of Nations* (1776) popularized the "invisible hand," the idea that an individual who "intends only his own gain" is, as it were, "led by an invisible hand to promote . . . the public interest."[6] Adam Smith did not assert that this was invariably true, and perhaps neither did any of his followers. But he contributed to a dominant tendency of thought that has ever since interfered with positive action based on rational analysis, namely, the tendency to assume that decisions reached individually will, in fact, be the best decisions for an entire society. If this assumption is correct it justifies the continuance of our present policy of laissez-faire in reproduction. If it is correct we can assume that men will control their individual fecundity so as to produce the optimum population. If the assumption is not correct, we need to reexamine our individual freedoms to see which ones are defensible.

The Tragedy of Freedom in a Commons

The rebuttal to the invisible hand in population control is to be found in a scenario first sketched in a little-known pamphlet in 1833 by a mathematical amateur named William Forster Lloyd (1794–1852).[7] We may well call it "the tragedy of the commons," using the word "tragedy" as the philosopher Whitehead used it: "The essence of dramatic tragedy is not unhappiness. It resides in the solemnity of the remorseless working of things." He then goes on to say, "This inevitableness of destiny can only be illustrated in terms of human life by incidents which in fact involve unhappiness. For it is only by them that the futility of escape can be made evident in the drama."[8]

The tragedy of the commons develops in this way. Picture a pasture open to all. It is to be expected that each herdsman will try to keep as many cattle as possible on the commons. Such an arrangement may work reasonably satisfactorily for centuries because tribal wars, poaching, and disease keep the numbers of both man and beast well below the carrying capacity of the land. Finally, however, comes the day of reckoning, that is, the day when the long-desired goal of social stability becomes a reality. At this point, the inherent logic of the commons remorselessly generates tragedy.

As a rational being, each herdsman seeks to maximize his gain. Explicitly or implicitly, more or less consciously, he asks, "What is the utility to me of adding one more animal to my herd?" This utility has one negative and one positive component.

1. The positive component is a function of the increment of one animal. Since the herdsman receives all the proceeds from the sale of the additional animal, the positive utility is nearly +1.

2. The negative component is a function of the additional overgrazing created by one more animal. Since, however, the effects of overgrazing are shared by all the herdsmen, the negative utility for any particular decision-making herdsman is only a fraction of −1.

Adding together the component partial utilities, the rational herdsman concludes that the only sensible course for him to pursue is to add another animal to his herd. And another; and another. . . . But this is the conclusion reached by each and every rational herdsman sharing a commons. Therein is the tragedy. Each man is locked into a system that compels him to increase his herd without limit—in a world that is limited. Ruin is the destination toward which all men rush, each pursuing his own best interest in a society that believes in the freedom of the commons. Freedom in a commons brings ruin to all.

Some would say that this is a platitude. Would that it were! In a sense, it was learned thousands of years ago, but natural selection favors the forces of psychological denial.[9] The individual benefits as an individual from his ability to deny the truth even though society as a whole, of which he is a part, suffers. Education can counteract the natural tendency to do the wrong thing, but the inexorable succession of generations requires that the basis for this knowledge be constantly refreshed.

A simple incident that occurred a few years ago in Leominster, Massachusetts, shows how perishable the knowledge is. During the Christmas shopping season the parking meters downtown were covered with plastic bags that bore tags reading: "Do not open until after Christmas. Free parking courtesy of the mayor and city council." In other words, facing the prospect of an increased demand for already scarce space, the city fathers reinstituted the system of the commons. (Cynically, we suspect that they gained more votes than they lost by this retrogressive act.)

In an approximate way, the logic of the commons has been understood for a long time, perhaps since the discovery of agriculture or the invention of private property in real estate. But it is understood mostly only in special cases which are not sufficiently generalized. Even at this late date, cattlemen leasing national land on the western ranges demonstrate no more than an ambivalent understanding, in constantly pressuring federal authorities to increase the head count to the point where overgrazing produces erosion and weed dominance. Likewise, the oceans of the world continue to suffer from the survival of the philosophy of the commons. Maritime nations still respond automatically to the shibboleth of the "freedom of the seas." Professing to believe in the "inexhaustible resources of the oceans," they bring species after species of fish and whales closer to extinction.[10]

The national parks present another instance of the working out of the tragedy of the commons. At present, they are open to all, without limit. The parks themselves are limited in extent—there is only one Yosemite Valley—whereas population seems to grow without limit. The values that visitors seek in the parks are steadily eroded. Plainly, we must soon cease to treat the parks as commons or they will be of no value to anyone.

What shall we do? We have several options. We might sell them off as private property. We might keep them as public property, but allocate the right to enter them. The allocation might be on the basis of wealth by the use of an auction system. It might be on the basis of merit, as defined by some agreed-upon standards. It might be by lottery. Or it might be on a first-come, first-served basis, administered to long queues. These, I think, are all the reasonable possibilities. They are all objectionable. But we must choose—or acquiesce in the destruction of the commons that we call our national parks.

Pollution

In a reverse way, the tragedy of the commons reappears in problems of pollution. Here it is not a question of taking something out of the commons, but of putting something in—sewage, or chemical, radioactive, and heat wastes into water; noxious and dangerous fumes into the air; and distracting and unpleasant advertising signs into the light of sight. The calculations of utility are much the same as before. The rational man finds that his share of the cost of the wastes he discharges into the commons is less than the cost of purifying his wastes before releasing them. Since this is true for everyone, we are locked into a system of "fouling our own nest," so long as we behave only as independent, rational, free enterprisers.

The tragedy of the commons as a food basket is averted by private property, or something formally like it. But the air and waters surrounding us cannot readily be fenced, and so the tragedy of the commons as a cesspool must be prevented by different means, by coercive laws or taxing devices that make it cheaper for the polluter to treat his pollutants than to discharge them untreated. We have not progressed as far with the solution of this problem as we have with the first. Indeed, our particular concept of private property, which deters us from exhausting the positive resources of the earth, favors pollution. The owner of a factory on the bank of a stream—whose property extends to the middle of the stream—often has difficulty seeing why it is not his natural right to muddy the waters flowing past his door. The law, always behind the times, requires elaborate stitching and fitting to adapt it to this newly perceived aspect of the commons.

The pollution problem is a consequence of population. It did not much matter how a lonely American frontiersman disposed of his waste. "Flowing water purifies itself every ten miles," my grandfather used to say, and the myth was near enough to the truth when he was a boy, for there were not too many people. But as population became denser, the natural chemical and biological recycling processes became overloaded, calling for a redefinition of property rights.

How to Legislate Temperance?

Analysis of the pollution problem as a function of population density uncovers a not generally recognized principle of morality, namely: *the morality of an act is a function of the state of the system at the time it is performed.*[11] Using the commons as a cesspool does not harm the general public under frontier conditions, because there is no public; the same behavior in a metropolis is unbearable. A hundred and fifty years ago a plainsman could kill an American bison, cut out only the tongue for his dinner, and discard the rest of the animal. He was not in any important sense being wasteful. Today, with only a few thousand bison left, we would be appalled at such behavior.

In passing, it is worth noting that the morality of an act cannot be determined from a photograph. One does not know whether a man killing an elephant or setting fire to the grassland is harming others until one knows the total system in which his act appears. "One picture is worth a thousand words," said an ancient Chinese; but it may take 10,000 words to validate it. It is as tempting to ecologists as it is to reformers in general to try to persuade others by way of the photographic shortcut. But the essence of an argument cannot be photographed: it must be presented rationally—in words.

That morality is system-sensitive escaped the attention of most codifiers of ethics in the past. "Thou shalt not . . ." is the form of traditional ethical directives which make no allowance for particular circumstances. The laws of our society follow the pattern of ancient ethics, and therefore are poorly suited to governing a complex, crowded, changeable world. Our epicyclic solution is to augment statutory law with administrative law. Since it is practically impossible to spell out all the conditions under which it is safe to burn trash in the back yard or to run an automobile without smog control, by law we delegate the details to bureaus. The result is administrative law, which is rightly feared for an ancient reason—*Quis custodiet ipsos custodes?*—"Who shall watch the watchers themselves?" John Adams said that we must have "a government of laws and not men." Bureau administrators, trying to evaluate the morality of acts in the total system, are singularly liable to corruption, producing a government by men, not laws.

Prohibition is easy to legislate (though not necessarily to enforce); but how do we legislate temperance? Experience indicates that it can be accomplished best through the mediation of administrative law. We limit possibilities unnecessarily if we suppose that the sentiment of *Quis custodiet* denies us the use of administrative law. We should rather retain the phrase as a perpetual reminder of fearful dangers we cannot avoid. The great challenge facing us now is to invent the corrective feedbacks that are needed to keep custodians honest. We must find ways to legitimate the needed authority of both the custodians and the corrective feedbacks.

Freedom to Breed Is Intolerable

The tragedy of the commons is involved in population problems in another way. In a world governed solely by the principle of "dog eat dog"—if indeed there ever was such a world—how many children a family had would not be a matter of public concern. Parents who bred too exuberantly would leave fewer descendants, not more, because they would be unable to care adequately for their children. David Lack and others have found that such a negative feedback demonstrably controls the fecundity of birds.[12] But men are not birds, and have not acted like them for millenniums, at least.

If each human family were dependent only on its own resources; *if* the children of improvident parents starved to death; *if*, thus, overbreeding brought its own "punishment" to the germ line—*then* there would be no public interest in controlling the breeding of families. But our society is deeply committed to the welfare state,[13] and hence is confronted with another aspect of the tragedy of the commons.

In a welfare state, how shall we deal with the family, the religion, the race, or the class (or indeed any distinguishable and cohesive group) that adopts overbreeding as a policy to secure its own aggrandizement?[14] To couple the concept of freedom to breed with the belief that everyone born has an equal right to the commons is to lock the world into a tragic course of action.

Unfortunately this is just the course of action that is being pursued by the United Nations. In late 1967, some thirty nations agreed to the following:

> The Universal Declaration of Human Rights describes the family as the natural and fundamental unit of society. It follows that any choice and decision with regard to the size of the family must irrevocably rest with the family itself, and cannot be made by someone else.[15]

It is painful to have to deny categorically the validity of this right; denying it, one feels as uncomfortable as a resident of Salem, Massachusetts, who denied the reality of witches in the seventeenth century. At the present time, in liberal quarters, something like a taboo acts to inhibit criticism of the United Nations. There is a feeling that the United Nations is "our last and best hope," that we shouldn't find fault with it; we shouldn't play into the hands of the archconservatives. However, let us not forget what Robert Louis Stevenson said: "The truth that is suppressed by friends is the readiest weapon of the enemy." If we love the truth we must openly deny the validity of the Universal Declaration of Human Rights, even though it is promoted by the United Nations. We should also join with Kingsley Davis[16] in attempting to get Planned Parenthood–World Population to see the error of its ways in embracing the same tragic ideal.

Conscience Is Self-Eliminating

It is a mistake to think that we can control the breeding of mankind in the long run by an appeal to conscience. Charles Galton Darwin made this point when he spoke on the centennial of the publication of his grandfather's great book. The argument is straightforward and Darwinian.

People vary. Confronted with appeals to limit breeding, some people will undoubtedly respond to the plea more than others. Those who have more children will produce a larger fraction of the next generation than those with more susceptible consciences. The difference will be accentuated, generation by generation.

In C. G. Darwin's words: "It may well be that it would take hundreds of generations for the progenitive instinct to develop in this way, but if it should do so, nature would have taken her revenge, and the variety *Homo contracipiens* would become extinct and would be replaced by the variety *Homo progenitivus*."[17]

The argument assumes that conscience or the desire for children (no matter which) is hereditary—but hereditary only in the most general formal sense. The result will be the same whether the attitude is transmitted through germ cells, or exosomatically, to use A. J. Lotka's term. (If one denies the latter possibility as well as the former, then what's the point of education?) The argument has here been stated in the context of the population problem, but it applies equally well to any instance in which society appeals to an individual exploiting a commons to restrain himself for the general good—by means of his conscience. To make such an appeal is to set up a selective system that works toward the elimination of conscience from the race.

Pathogenic Effects of Conscience

The long-term disadvantage of an appeal to conscience should be enough to condemn it; but it has serious short-term disadvantages as well. If we ask a man who is exploiting a commons to desist "in the name of conscience," what are we saying to him? What does he hear?—not only at the moment but also in the wee small hours of the night when, half asleep, he remembers not merely the words we used but also the nonverbal communication cues we gave him unawares? Sooner or later, consciously or subconsciously, he senses that he has received two communications, and that they are contradictory: (1, the intended communication) "If you don't do as we ask, we will openly condemn you for not acting like a responsible citizen"; (2, the unintended communication) "If you *do* behave as we ask, we will secretly condemn you for a simpleton who can be shamed into standing aside while the rest of us exploit the commons."

Everyman then is caught in what Bateson has called a "double bind." Bateson and his co-workers have made a plausible case for viewing the double bind as an important causative factor in the genesis of schizophrenia.[18] The double bind may not always be so damaging, but it always endangers the mental health of anyone to whom it is applied. "A bad conscience," said Nietzsche, "is a kind of illness."

To conjure up a conscience in others is tempting to anyone who wishes to extend his control beyond the legal limits. Leaders at the highest level succumb to this temptation. Has any President during the past generation failed to call on labor unions to moderate voluntarily their demands for higher wages, or to steel companies to honor voluntary guidelines on prices? I can recall none. The rhetoric used on such occasions is designed to produce feelings of guilt in noncooperators.

For centuries it was assumed without proof that guilt was a valuable, perhaps even an indispensable, ingredient of the civilized life. Now, in this post-Freudian world, we doubt it. Paul Goodman speaks from the modern point of view when he says: "No good has ever come from feeling guilty, neither intelligence, policy, nor compassion. The guilty do not pay attention to the object but only to themselves, and not even to their own interests, which might make sense, but to their anxieties."[19]

One does not have to be a professional psychiatrist to see the consequences of anxiety. We in the Western world are just emerging from a dreadful two-centuries-long Dark Ages of Eros that was sustained partly by prohibition laws, but perhaps more effectively by the anxiety-generating mechanisms of education. Alex Comfort has told the story well in *The Anxiety Makers;*[20] it is not a pretty one.

Since proof is difficult, we may even concede that the results of anxiety may sometimes, from certain points of view, be desirable. The larger question we should ask is whether, as a matter of policy, we should ever encourage the use of a technique the tendency (if not the intention) of which is psychologically pathogenic. We hear much talk these days of responsible parenthood; the coupled words are incorporated into the titles of some organizations devoted to birth control. Some people have proposed massive propaganda campaigns to instill responsibility into the nation's (or the world's) breeders. But what is the meaning of the word responsibility in this context? Is it not merely a synonym for the word conscience? When we use the word responsibility in the absence of substantial sanctions are we not trying to browbeat a free man in a commons into acting against his own interest? Responsibility is a verbal counterfeit for a substantial *quid pro quo*. It is an attempt to get something for nothing.

If the word responsibility is to be used at all, I suggest that it be in the sense Charles Frankel uses it.[21] "Responsibility," says this philosopher, "is the product of definite social arrangements." Notice that Frankel calls for social arrangements—not propaganda.

Mutual Coercion Mutually Agreed Upon

The social arrangements that produce responsibility are arrangements that create coercion, of some sort. Consider bank robbing. The man who takes money from a bank acts as if the bank were a commons. How do we prevent such action? Certainly not by trying to control his behavior solely by a verbal appeal to his sense of responsibility. Rather than rely on propaganda we follow Frankel's lead and insist that a bank is not a commons; we seek the definite social arrangements that will keep it from becoming a commons. That we thereby infringe on the freedom of would-be robbers we neither deny nor regret.

The morality of bank robbing is particularly easy to understand because we accept complete prohibition of this activity. We are willing to say "Thou shalt not rob banks," without providing for exceptions. But temperance also can be created by coercion. Taxing is a good coercive device. To keep downtown shoppers temperate in their use of parking space we introduce parking meters for short periods, and traffic fines for longer ones. We need not actually forbid a citizen to park as long as he wants to; we need merely make it increasingly expensive for him to do so. Not prohibition, but carefully biased options are what we offer him. A Madison Avenue man might call this persuasion; I prefer the greater candor of the word coercion.

Coercion is a dirty word to most liberals now, but it need not forever be so. As with the four-letter words, its dirtiness can be cleansed away by exposure to the light, by saying it over and over without apology or embarrassment. To many, the word coercion implies arbitrary decisions of distant and irresponsible bureaucrats; but this is not a necessary part of its meaning. The only kind of coercion I recommend is mutual coercion, mutually agreed upon by the majority of the people affected.

To say that we mutually agree to coercion is not to say that we are required to enjoy it, or even to pretend we enjoy it. Who enjoys taxes? We all grumble about them. But we accept compulsory taxes because we recognize that voluntary taxes would favor the conscienceless. We institute and (grumblingly) support taxes and other coercive devices to escape the horror of the commons.

An alternative to the commons need not be perfectly just to be preferable. With real estate and other material goods, the alternative we have chosen is the institution of private property coupled with legal inheritance. Is this system perfectly just? As a genetically trained biologist I deny that it is. It seems to me that, if there are to be differences in individual inheritance, legal possession should be perfectly correlated with biological inheritance—that those who are biologically more fit to be the custodians of property and power should legally inherit more. But genetic recombination continually makes a mockery of the doctrine of "like father, like son" implicit in our laws of legal inheritance. An idiot can inherit millions, and a trust fund can keep his estate intact. We must admit that our legal system of private property plus inheritance is unjust—but we put up with it because we are not convinced, at the moment, that anyone has invented a better system. The alternative of the commons is too horrifying to contemplate. Injustice is preferable to total ruin.

It is one of the peculiarities of the warfare between reform and the status quo that it is thoughtlessly governed by a double standard. Whenever a reform measure is proposed it is often defeated when its opponents triumphantly discover a flaw in it. As Kingsley Davis has pointed out, worshippers of the status quo sometimes imply that no reform is possible without unanimous agreement, an implication contrary to historical fact.[22] As nearly as I can make out, automatic rejection of proposed reforms is based on one of two unconscious assumptions: (1) that the status quo is perfect; or (2) that the choice we face is between reform and no action; if the proposed reform is imperfect, we presumably should take no action at all, while we wait for a perfect proposal.

But we can never do nothing. That which we have done for thousands of years is also action. It also produces evils. Once we are aware that the status quo is action, we can then compare its discoverable advantages and disadvantages with the predicted advantages and disadvantages of the proposed reform, discounting as best we can for our lack of experience. On the basis of such a comparison, we can make a rational decision which will not involve the unworkable assumption that only perfect systems are tolerable.

Recognition of Necessity

Perhaps the simplest summary of this analysis of man's population problems is this: the commons, if justifiable at all, is justifiable only under conditions of low population density. As the human population has increased, the commons has had to be abandoned in one aspect after another.

First we abandoned the commons in food gathering, enclosing farm land and restricting pastures and hunting and fishing areas. These restructions are still not complete throughout the world.

Somewhat later we saw that the commons as a place for waste disposal would also have to be abandoned. Restrictions on the disposal of domestic sewage are widely accepted in the Western world; we are still struggling to close the commons to pollution by automobiles, factories, insecticide sprayers, fertilizing operations, and atomic energy installations.

In a still more embryonic state is our recognition of the evils of the commons in matters of pleasure. There is almost no restriction on the propagation of sound waves in the public medium. The shopping public is assaulted with mindless music, without its consent. Our government is paying out billions of dollars to create supersonic transport which will disturb 50,000 people for every one person who is whisked from coast to coast three hours faster. Advertisers muddy the airwaves of radio and television and pollute the view of travelers. We are a long way from outlawing the commons in matters of pleasure. Is this because our Puritan inheritance makes us view pleasure as something of a sin, and pain (that is, the pollution of advertising) as the sign of virtue?

Every new enclosure of the commons involves the infringement of somebody's personal liberty. Infringements made in the distant past are accepted because no contemporary complains of a loss. It is the newly proposed infringements that we vigorously oppose; cries of "rights" and "freedom" fill the air. But what does "freedom" mean? When men mutually agreed to pass laws against robbing, mankind became more free, not less so. Individuals locked into the logic of the commons are free only to bring on universal ruin; once they see the necessity of mutual coercion, they become free to pursue other goals. I believe it was Hegel who said, "Freedom is the recognition of necessity."

The most important aspect of necessity that we must now recognize is the necessity of abandoning the commons in breeding. No technical solution can rescue us from the misery of overpopulation. Freedom to breed will bring ruin to all. At the moment, to avoid hard decisions many of us are tempted to propagandize for conscience and responsible parenthood. The temptation must be resisted, because an appeal to independently acting consciences selects for the disappearance of all conscience in the long run, and an increase in anxiety in the short.

The only way we can preserve and nurture other and more precious freedoms is by relinquishing the freedom to breed, and that very soon. "Freedom is the recognition of necessity"—and it is the role of education to reveal to all the necessity of abandoning the freedom to breed. Only so can we put an end to this aspect of the tragedy of the commons.

Notes

1. J. B. Wiesner and H. F. York, *Scientific American* 211, no. 4 (1964), 27. Offprint 319.

2. G. Hardin, *Journal of Heredity* 50 (1959), 68; S. von Hoernor, *Science* 137 (1962), 18.

3. J. von Neumann and O. Morgenstern, *Theory of Games and Economic Behavior* (Princeton, N.J.: Princeton University Press, 1947), p. 11.

4. J. H. Fremlin, *New Science*, no. 415 (1964), 285.

5. [Editors' note: Several European countries now have stable or declining populations.]

6. A. Smith, *The Wealth of Nations* (New York: Modern Library, 1937), p. 423.

7. W. F. Lloyd, *Two Lectures on the Checks to Population* (Oxford: Oxford University Press, 1833), reprinted (in part) in G. Hardin, ed., *Population, Evolution, and Birth Control*, 2d ed. (San Francisco: W. H. Freeman and Company, 1969), p. 28.

8. A. N. Whitehead, *Science and the Modern World* (New York: Mentor, 1948), p. 17.

9. Hardin, *Population, Evolution, and Birth Control*, p. 46.

10. S. McVay, *Scientific American* 216, no. 8 (1966), 13. Offprint 1046.

11. J. Fletcher, *Situation Ethics* (Philadelphia: Westminster, 1966).

12. D. Lack, *The Natural Regulation of Animal Numbers* (Oxford: Clarendon Press, 1954).

13. H. Girvetz, *From Wealth to Welfare* (Stanford, Calif.: Stanford University Press, 1950).

14. G. Hardin, *Perspectives in Biology and Medicine* 6 (1963), 366.

15. U Thant, *International Planned Parenthood News*, no. 168 (February 1968), 3.

16. K. Davis, *Science* 158 (1967), 730.

17. S. Tax, ed., *Evolution after Darwin* (Chicago: University of Chicago Press, 1960), 2:469.

18. G. Bateson, D. D. Jackson, J. Haley, J. Weakland, *Behavioral Science* 1 (1956), 251.

19. P. Goodman, *New York Review of Books* 10, no. 8 (23 May 1968), 22.

20. A. Comfort, *The Anxiety Makers* (London: Nelson, 1967).

21. C. Frankel, *The Case for Modern Man* (New York: Harper, 1955), p. 203.

22. J. D. Roslansky, *Genetics and the Future of Man* (New York: Appleton-Century-Crofts, 1966), p. 177.

7 Second Thoughts on "The Tragedy of the Commons"

Garrett Hardin

A scholar who ventures outside his own field can confidently expect soon to be informed that his pronouncements are not as original as he has supposed. So much have I learned in publishing "The Tragedy of the Commons." Though officially trained only in microbiology (and in only a micropart of that vast field), in writing "Tragedy" I did not hesitate to make assertions that impinged on the fields of economics, political science, human behavior, and ethics. I was soon informed that there was a considerable literature on "common pool resources" in economics and that Aristotle long ago had said, "What is common to the greatest number gets the least amount of care." So what is new in my essay?

Just this, I think: the *emphasis* on the tragedy of the situation. Aristotle's statement is as bland as a bureaucrat's: It hardly impels one to take action. Yet action is what we must have now that the world is overcrowded—action in the form of rejecting the commons as a distribution system. Failing that, a tragic end is our fate.

No fate is intrinsically inevitable; rather, each fate is contingent upon the mechanism that produces it. How overpopulation leads to a disastrous end was clearly set forth by an English geologist and minister, Joseph Townsend (1739–1816), in *A Dissertation on the Poor Laws,* published in 1786, a decade after Adam Smith's *Wealth of Nations*. The essential passage follows:

> In the South Seas there is an island, which from the first discoverer is called Juan Fernandez. In this sequestered spot, John Fernando placed a colony of goats, consisting of one male, attended by his female. This happy couple finding pasture in abundance, could readily obey the first commandment, to increase and multiply, till in process of time they had replenished their

---

This essay was written especially for the preceding edition of this book.

little island. In advancing to this period they were strangers to misery and want, and seemed to glory in their numbers: but from this unhappy moment they began to suffer hunger; yet continuing for a time to increase their numbers, had they been endued with reason, they must have apprehended the extremity of famine. In this situation the weakest first gave way, and plenty was again restored. Thus they fluctuated between happiness and misery, and either suffered want or rejoiced in abundance, according as their numbers were diminished or increased; never at a stay, yet nearly balancing at all times their quantity of food. This degree of aequipoise was from time to time destroyed, either by epidemical diseases or by the arrival of some vessel in distress. On such occasions their numbers were considerably reduced; but to compensate for this alarm, and to comfort them for the loss of their companions, the survivors never failed immediately to meet returning plenty. They were no longer in fear of famine: they ceased to regard each other with an evil eye; all had abundance, all were contented, all were happy. Thus, what might have been considered as misfortunes, proved a source of comfort; and, to them at least, partial evil was universal good.

When the Spaniards found that the English privateers resorted to this island for provisions, they resolved on the total extirpation of the goats, and for this purpuse they put on shore a greyhound dog and bitch. These in their turn increased and multiplied, in proportion to the quantity of the food they met with; but in consequence, as the Spaniards had foreseen, the breed of goats diminished. Had they been totally destroyed, the dogs likewise must have perished. But as many of the goats retired to the craggy rocks, where the dogs could never follow them, descending only for short intervals to feed with fear and circumspection in the vallies, few of these, besides the careless and the rash, became a prey; and none but the most watchful, strong, and active of the dogs could get a sufficiency of food. Thus a new kind of balance was established. The weakest of both species were among the first to pay the debt of nature; the most active and vigorous preserved their lives. It is the quantity of food which regulates the numbers of the human species. In the woods, and in the *savage state,* there can be few inhabitants; but of these there will be only a proportionable few to suffer want. As long as food is plenty they will continue to increase and multiply; and every man will have ability to support his family, or to relieve his friends, in proportion to his activity and strength. The weak must depend upon the precarious bounty of the strong; and, sooner or later, the lazy will be left to suffer the natural consequence of their indolence. Should they introduce a community of goods, and at the same time leave every man at liberty to marry, they would at first increase their numbers, but not the sum total of their happiness, till by degrees, all by being equally reduced to want and misery, the weakly would be the first to perish.

Ashley Montagu, commenting on this passage, says: "Townsend recounts the story of what is purported to have happened on Robinson Crusoe's island." The word *purported* belittles the moral of the story. Whether or not Townsend's tale is historically correct, the moral is nonetheless supported by empirical facts. In our time, David Klein's study of the reindeer on St. Matthew Island tells essentially the same story and points to the same moral. A population of 29 animals, minus predators, released on this remote island off the coast of Alaska, multiplied to 6,000 animals in 19 years and then, through starvation, "crashed" to 42 in three years. The 42 were in miserable condition and probably all sterile. This may well be the end of the story, though it is possible that the population might later "fluctuate between happiness and misery," as supposed by Townsend. In any case, a predator-free population, multiplying freely and enjoying "a community of goods," can know only transient happiness as it races toward overpopulation.

Stability and happiness are possible only if the positive feedback of natural biological reproduction—interest earning interest, *compound interest* in economic terms—finds a countervailing force in negative feedback. To use other terms, *runaway feedback* must be opposed by *corrective feedback* to produce a stable system. Wolves could have been the corrective feedback on St. Matthew Island, but the human beings who stocked the island did not think to include predators. Bacteria and other disease organisms can act as micropredators and hence as negative feedbacks. So also can starvation, when worst comes to worst. How much fluctuation there is in the size of the population of the *propositus* population (reindeer, in this case) depends on the particularities of the life cycles in the organisms involved—prey, macropredators, micropredators, and food organisms (lichens). If the time constants of the interacting species are wrong, even a system with negative feedback may collapse. Complexities and discontinuities in the environment—the "craggy rocks" of Townsend's model—may also be required for persistence of the demographic system. Real demography is not simple, not even among nonhuman species.

The human species has freed itself of all macropredators and bids fair soon to do the same for all micropredators. What effective controls remain? There are only three substantial possibilities: the "misery and vice" (the "positive checks") of the first edition of Malthus's essay—starvation and human conflict, in which man is his own predator—and the "preventive checks" of the second edition. The only acceptable preventive checks, in Malthus's mind, fell under the heading of "moral restraint"—delay in the age of marriage and considerable sexual continence in marriage (hopefully coupled with complete continence outside). We latter-day Malthusians (though not Malthus himself) also regard contraception as a genuinely moral restraint. Whatever its definition, moral restraint raises political problems of great difficulty.

It is in the nature of biological reproduction that it anticipates no limit: In every species the exponential curve of unopposed population growth soars off toward infinity. Population is prevented from reaching this goal by countervailing forces that are extrinsic to the propositus species—predators and the like. The coexistence of all the forces produces a more or less precise population equilibrium. By removing the countervailing forces of predators and disease from the population of our own species, we have created a disequilibrium that can persist for only a moment, as time is reckoned geologically. If utter misery is to be avoided, *Homo sapiens* must create and maintain new countervailing forces that are intrinsic to his own species—contraception and continence, to mention the most acceptable. (Less acceptably, war and genocide can perform the same function.) The adoption and enforcement of intrinsic controls necessarily involve policy. Who is to be controlled? By what means? What do we do with noncooperators? How do we superintend education so as to minimize noncooperation and the problems it creates? These are all political questions.

It is astonishing how the inescapably political nature of human population control is ignored by the political establishment. The United States government annually finances several hundred million dollars of research that is labeled *population research* in official summaries. From the titles of the projects I would estimate that at least 98 percent of them are either pedestrian surveys of the most unproductive sort or attempts to improve the technology of contraception. But we already have a nearly perfect system of birth control, and a perfect system of birth control merely permits people to have the number of children they want. It is an empirical fact that in every country in the world the number of children wanted by the average family is greater than the number needed to produce population equilibrium in that nation. Theoretically, this is perfectly understandable. Human psychology has evolved in an environment that included extrinsic countervailing forces which human ingenuity has only lately removed. Some mating couples internalize effective control forces, but not all do so. Population problems are created by those who cannot, or will not, internalize the necessary controls.

In the absence of extrinsic population controls, a totally voluntary control system selects for its own failure. Even a partial abandonment of voluntarism in the United States will require difficult political adjustments. With government money, it is perilous even to study the possibilities of change. How can a going political concern finance its own (partial) dissolution? Our exceptional Constitution makes change more practicable than it is for most nations, but even for us it is not easy.

Before significant political change can be instituted, there must be a fundamental improvement in the theory and practice of economics, which needs to be firmly tied down to the sort of conservation laws that have proven essential to the progress of the natural sciences. The idea of a conservation law is much older than science. Basically it is this—that in any equation that stands for before-and-after, what is on the left side of the equation must equal what is on the right side. This is the spirit behind double-entry bookkeeping, known to exist (in Genoa) as far back as the year 1340. In the investigation of nature, a lack of the discipline of balancing income and outgo kept alchemy from achieving any significant insights into chemical processes. It was Lavoisier, on the eve of the French Revolution, who changed alchemy to chemistry by the discipline of the mandatory balanced equation. A similar advance was made in physics in the middle of the nineteenth century with the explicit statement of the law of the conservation of energy. This signal advance was momentarily imperiled when in 1896 Becquerel discovered the spontaneous fluorescence of uranium, which seemed to mean that we could get something (radiant energy) for nothing (no known energy input). The anomaly was resolved nine years later when Einstein wrote an equation that conserved matter and energy jointly. Since then, conservation principles have reigned supreme in physics and chemistry. It is hard to see how these sciences could have progressed otherwise.

What about economics? There is some ambivalence here. On the one hand, economists constantly remind their students that there is no such thing as a free lunch. On the other, in many public areas, particularly when politically pressed, some economists act on the nonconservative assumption that demand creates supply. In a limited economic context there is a quasi-truth in this doctrine, but not in any rigorous sense. Unfortunately, the rapid progress of technology during the past two centuries has repeatedly expanded the area of known supplies, thus making it difficult to denote the explicit limits of conservation in everyday affairs. Faced with threats of limited supplies, politicians and economists, like Dickens's Mr. Micawber, optimistically intone "Something will turn up." Unfortunately, under the impetus of science, something quite often does, and this undeserved good fortune encourages economist-advisers to sound more like alchemists than scientists, or even good bookkeepers.

The basic conservative concept of ecologists is the idea of *carrying capacity,* which defines that population of the propositus species that can be supported by a given territory year after year without degrading the environment—that is, without lowering its carrying capacity subsequently. Transgressing the carrying capacity even for a short time can set in train degradative processes (such as soil erosion) that operate by the rule of positive feedback (runaway feedback). For this reason, transgressing the carrying capacity even momentarily is an error of the most serious sort. The degraded environment of the southern and eastern shores of the Mediterranean is a permanent monument to transgressions committed two thousand years ago. Because carrying capacity varies seasonally in a predictable way, and unpredictably over longer periods of time, conservative management dictates that the operating figure for carrying capacity always be kept well below the momentary actual carrying capacity by an amount that we can call the *margin of safety.*

Even with these refinements, determining the carrying capacity for the animal populations that people exploit is fairly straightforward; for people themselves this is not so. We must first face the reality that this carrying capacity has, for a long time, been slowly moved upward by technology. Should we base the political and economic decisions of today on the increase in carrying capacity anticipated for tomorrow? The true conservative would say *no;* but few economists in the twentieth century have been conservatives. Conservatives are satisfied to postpone the spending of tomorrow's (possible) income until tomorrow; that way they do not risk transgressing the carrying capacity.

For human beings, carrying capacity is ineluctably tied to the desired standard of living, which is a political decision. Some people want to include meat in their diet, recreational vehicles in their life, and the delights of wilderness and uncrowded areas in which "unimproved" land creates a "waste" of natural resources. (The quotation marks surround the implicit value judgments of those of the opposite persuasion.) By political decisions, such high-standard people define the carrying capacity of their land at a lower figure than that agreed upon by those who want food only to live and are willing to sacrifice all recreational values to the support of more human lives. The contradiction between these two carrying capacities is a political contradiction, not a fact of natural science. If the two political views are allowed to compete with each other under laissez-faire rules, those people who accept lower standards will displace those who desire (but will not fight for) higher standards, in a sort of Gresham's Law of Living Standards. But whatever the standards—whatever the quantitative definition of carrying capacity—the transgression of carrying capacity must not be allowed because it leads to ruin.

In retrospect, I see the revival of Lloyd's image of the commons and the use of this metaphor in the formulation of conservation laws as necessary measures in putting the policy sciences on the path toward a rigorous grounding in conservation principles. To the extent that we achieve this goal, we will discover that much of the vagueness and prolixity that now afflict the policy sciences will disappear. The goal is worth pursuing.

References

Hardin, Garrett. "Carrying Capacity as an Ethical Concept." *Soundings* 59, no. 1 (1976), 120–137. Reprinted in Hardin, *Stalking the Wild Taboo*, 2d ed. Los Altos, Calif.: Kaufmann, 1978.

Klein, David R. "The Introduction, Increase, and Crash of Reindeer on St. Matthew Island." *Journal of Wildlife Management* 32 (1968), 350–367.

Townsend, Joseph. *A Dissertation on the Poor Laws, by a Well-Wisher to Mankind*. Berkeley: University of California Press, 1971, pp. 8, 36–38 (originally published 1786).

II Ethics: The Ultimate End and Value Constraints

The western world is now suffering from the limited moral outlook of the three previous generations. . . . The two evils are: one, the ignoration of the true relation of each organism to its environment; and the other, the habit of ignoring the intrinsic worth of the environment which must be allowed its weight in any consideration of final ends.

Alfred North Whitehead, *Science and the Modern World*, 1925

Introduction

Herman E. Daly and
Kenneth N. Townsend

Humanity, craving for the infinite, has been corrupted by the temptation to satisfy an insatiable hunger in the material realm. Turn these stones into bread, urges Satan, and modern man sets to it, even to the extent of devising energy-intensive schemes for grinding up ordinary rock for materials—to eat the spaceship itself! But Jesus' answer to the same temptation was more balanced: man does not live by bread alone. The proper object of economic activity is to have *enough* bread, not infinite bread, not a world turned into bread, not even vast storehouses full of bread. The infinite hunger of man, his moral and spiritual hunger, is not to be satisfied, is indeed exacerbated, by the current demonic madness of producing more and more things for more and more people. Afflicted with an infinite itch, modern man is scratching in the wrong place, and his frenetic clawing is drawing blood from the life-sustaining circulatory systems of his spaceship, the biosphere.

It is important to be very clear on the paramount importance of the moral issue. We could opt to scratch ourselves to death, destroying the spaceship in an orgy of procreation and consumption. The *only* arguments against doing this are religious and ethical: the obligation of stewardship for God's creation, the extension of brotherhood to future generations, and of some lesser degree of brotherhood to the nonhuman world.

The essays in this section give particular, though not exclusive, attention to the upper or ends segment of the ends-means spectrum (figure I.2). In "The Age of Plenty," E. F. Schumacher discusses the Ultimate End from an explicitly Christian perspective, taking St. Ignatius Loyola's "Foundation" as a basic principle by which sufficiency may be distinguished from excess of worldly goods. At what point does our use of the things of the earth cease to aid and begin to hinder us in the attainment of the end for which we were created? Those who do not accept the Christian view of the end for which humanity was created must nevertheless face a similar question. In his article "Buddhist Economics," Schumacher arrived at much the same conclusions starting from a Buddhist view of the purpose of human life. Schumacher himself was a Christian, and it is somewhat ironic that his article on Buddhist economics has attracted so much more attention than his economic writings from an explicitly Christian perspective.

Schumacher is not the only economist of the "modern period," extending roughly from the late eighteenth century with the works of Adam Smith up to the present, who has wrestled with the basic question of the end for which humanity was created, with its concomitant concern for what should govern the extent to which people exploit the physical environment in the attainment of that end. A historical perspective is provided by Gerald Alonzo Smith, who, in "The Purpose of Wealth: A Historical Perspective," reviews the ideas of the great critics of the industrial growth-oriented economy: Sismondi, Ruskin, Hobson, and Tawney. These thinkers refused to follow the trend of the emerging "positive economics" in divorcing themselves from the larger questions of purpose and insisted upon an explicitly teleological view of wealth. Well-defined ends never require infinite means for their fulfillment, whereas ill-defined ends seem, in their vagueness, always to require more resources. In terms of the ends-means spectrum, the thinkers surveyed by Smith are concerned mainly with the upper half of the spectrum, arguing from the top down.

In "Ecology, Ethics, and Theology," theologian John Cobb considers the extension of brotherhood to the nonhuman world and develops the thesis that our ethics should be based on feeling and on a perception of the hierarchy of feeling in the biotic pyramid.[1] Nonhuman feeling, Cobb contends, must be considered valuable, though its value is not on an equal footing with human feeling. Human beings may be the apex of the biotic pyramid from which they evolved, but an apex with no pyramid underneath is a dimensionless point.

The final essay, "The Abolition of Man," by C. S. Lewis, the late British professor of literature, lay theologian, and writer of science fiction, was first published in 1947, long before the wave of environmental concern had swelled. When, as in this case, the relevance of the article increases with time, it is a good indication that the author has based his arguments upon something very solid. Controlling nature, as Lewis shows, becomes after some point a very dangerous undertaking, and if carried to the limit, the whole enterprise blows up in our face—"Man's conquest of Nature turns out, in the moment of its consummation, to be Nature's conquest of Man."

Note

1. The reader further interested in dimensions of this brotherhood in terms of its extension to presently existing and future people, as well as the changes in political and economic thinking required to establish a sustainable future for the community of mankind, may wish to consult *For The Common Good: Redirecting the Economy toward Community, the Environment, and a Sustainable Future,* by Herman E. Daly and John B. Cobb, Jr. (Boston: Beacon Press, 1989).

8 The Age of Plenty: A Christian View

E. F. Schumacher

On all sides, the future of industrial society is being called into question. Not many years ago we were told that we had never had it so good; as time moved on we would have it better still. And the same would hold for all the world's people, particularly those who, for one reason or another, had been left behind in mankind's onward march into the Age of Plenty. The backward countries were politely and optimistically named the "developing countries," and the nineteen-sixties were to be known as the "Development Decade." If there was one thing on which everybody was expected to agree it was this: that at long last the problem of production had been solved. Modern science and technology had done it; western civilisation had done it; and the unique and dazzling achievements of this civilisation were now destined to spread across the globe in a very short time.

The Glory of the Modern Achievement

We may recall the claims that were made and indeed are still being made in some quarters—for the power and glory of the modern achievement. Theologians told us that God was dead: there was now no more need for this hypothesis; and high academics announced that man, the naked ape, emancipated and come-of-age, had now, at long last, taken possession of His seat of omnipotence. This was said to be the meaning of the Second Industrial Revolution, of the Nuclear Age, the Space Age, the Age of Automation. Energy would be forever cheap and plentiful; man was no longer confined to this miserable planet called the Earth; nor would he continue to have to

Originally published in pamphlet form by The Saint Andrew Press, Edinburgh, 1974. Reprinted by permission of Verena Schumacher.

labour for his living: automation and cybernetics would finally remove the curse allegedly put on Adam after the Fall. If there was a remaining problem, it was the problem of how to educate ourselves for endless leisure.

How is it that the mood has changed so suddenly and so profoundly, even though hardly a week passes without announcements of new, astonishing scientific or technological breakthroughs? The late President Kennedy set a target: that by the end of the sixties man would visit the moon—*and he did*. The week of this fantastic feat was referred to by President Nixon as "the greatest week since Genesis." It cannot be said that science and technology have suddenly lost their power. But somehow the glory is gone. At the height of success there is a smell of bankruptcy.

You may accuse me of overdramatisation—and you are quite right. The materialistic optimism of ten or twenty years ago did not possess all the people all the time. There was the awesome threat of the Atom Bomb; but we were advised to keep our cool; if we learned to "live with the Bomb" all would be well. Also, most of our artists did seem rather unhappy. The painters insisted on painting pictures which suggested the dissolution, if not the abolition, of man. The most gifted writers and composers did much the same in words and sounds. But this was interpreted as "experimentation in new art forms," indicating liberation from outdated restraints and an upsurge of creative vitality. Warning voices were raised also by people involved in agriculture, industry, and general economic affairs; but these were minority opinions, easily decried as "unbalanced," "irresponsible," or "unrealistic."

Today, however, those minority opinions of a few years ago are having a wide impact even on official utterances. The limits-to-growth debate is in full swing. The possibility of severe fuel shortages in many parts of the world, which only a few years ago was laughed out of court, is becoming a reality. Concern over environmental degradation and the dangers of ecological breakdown is no longer confined to a few minority groups. Most important of all, many people are beginning to take an overall view of the condition and prospects of life in industrial society and to feel that we may be moving into a real crisis of survival.

It is not easy to take an overall view, because such a view can be obtained only from a considerable height. But what is higher and what is lower? How could we measure this kind of height, and, if it cannot be measured has it any meaning? We cannot get an overall view merely by assembling more and more facts. By themselves, facts mean nothing, prove nothing, and lead to no conclusions. Facts need to be evaluated, that is to say fitted into a value system, to be of use. What, then, is our value system? Only from the height of a value system can we obtain a meaningful overall view.

A Christian View

That is why I have written into my subject title the words: "a Christian view." This is meant to indicate the value system which, as I think, rules what I have to say. Unfortunately, there is no unanimity today as to what constitutes a Christian point of view when it comes to such mundane matters as our economic life. I resort, therefore, to what a great Christian saint called "The Foundation." This is what he said:

> Man was created to praise, reverence, and serve God our Lord, and by this means to save his soul;
> And the other things on the face of the earth were created for man's sake, and in order to aid him in the prosecution of the end for which he was created.
> *Whence it follows*
> That man ought to make use of them just so far as they help him to attain his end,
> And that he ought to withdraw himself from them just so far as they hinder him.[1]

The logic of this statement is unshakeable; it is in fact the kind of logic we invariably try to apply in our everyday affairs, whether it be business, or science, or engineering, or politics. We first try to clarify what we want to achieve; we then study the means at our disposal; and we then use those means *just so far* as—in our judgment—they help us to attain our objective, and when it appears that we are overdoing things we withdraw from these means *just so far* as they hinder us.

When applied to mankind's present-day economic situation, the statement also seems eminently realistic. It implies that where people do not have enough means to attain their ends they should have more, and where they have more than enough they should "withdraw" from that which is excessive.

We can say therefore that the statement—"The Foundation"—is distinguished by both implacable logic and genial common sense. Anybody who is prepared to accept the two premises cannot possibly refuse to accept the conclusions. The question is: Does a Christian accept the two premises, namely, first, that man was created to "save his soul" and, second, that all the other things were created in order to aid him?

Obviously, some of us might wish to formulate these propositions slightly differently; but it hardly seems possible that a Christian should reject their essential meaning.

All the same, the more we look at this statement, and the more deeply we ponder the meaning of its artless syllogism, offered as "The Foundation" on which all our thinking, policy making, and acting should be built, the more incredible, remote, impractical, irritating it must seem to us; something completely out of keeping with what we actually think, plan, and do. We might feel this is something only for hermits or monks, but not for ordinary Christian householders like ourselves.

Are we then saying that salvation is only for hermits and monks? Surely not. How can we extricate ourselves from this disturbing logic? We might say: This statement, called "The Foundation," is indeed irrefutable as an ideal; but as ordinary mortals we cannot attain the ideal. Fair enough. The validity of an ideal, or goal, or objective depends on its inherent Truth, not on the number of people who actually attain it or live up to it. And those who accept it, while they cannot promise attainment, undertake to walk in one particular direction and not in any other. It would appear, therefore, that the Christian, as far as the goods of this world are concerned, is called upon to *strive* to use them *just so far* as they help him to attain salvation, and that he should *strive* to withdraw himself from them *just so far* as they hinder him.

If this is "The Foundation" from which the Christian obtains his overall view, we may now proceed to a consideration of the future of industrial society.

A Christian Looks at the Future

From this point of view it can be seen right away that we cannot be the least bit interested in purely quantitative concepts, such as economic growth or nongrowth applied to that mysterious aggregate,

GNP (Gross National Product) or, even more mysterious, GWP (Gross World Product). That which is good and helpful ought to be growing and that which is bad and hindering ought to be diminishing; whether aggregation of these two processes yields a higher grand total or a lower one is of no interest whatever. In fact, the aggregation itself is quite meaningless. We therefore need, above all else, qualitative concepts: concepts that enable us to choose the right direction of our movement and not merely to measure its speed.

The future of industrial society will depend on the development and adoption of such qualitative concepts. This, of course, is a somewhat arid way of expressing oneself. Let us look at some of the living details of which the overall picture is made up. We all know that, economically speaking, mankind is not a reasonably homogeneous group but is most unreasonably split between, on the one hand, about one quarter who (as conventionally measured) are immensely rich, and, on the other hand, about three-quarters who (similarly measured) are immensely poor. The gap between the rich and the poor, worldwide, is quite enormously large, and there are not many people situated, as it were, in the middle. A *normal* distribution curve would show most people in the middle and relatively few at the two extremes. But this distribution curve, classifying people in accordance with their income, is highly "abnormal," almost an inverted image of a normal distribution curve: with the extremes—of great poverty and of great affluence—heavily populated and the middle almost empty. This striking abnormality suggests that there is something seriously wrong. What ought to be—and used to be—one world has broken up into two worlds, neither of which is normal and healthy. This split grievously afflicts many countries internally. They have lost all internal cohesion, reflecting the world situation in miniature: "dual societies" with the extremes of poverty and of wealth heavily represented and the middle relatively empty, again almost an inverted image of normal distribution.

From the point of view of "The Foundation" this indicates a seriously pathological condition, because the rich evidently use too much and the poor use too little, and the golden mean is being achieved only by a minority.

The Rich Society Norm

What makes the world situation particularly abnormal and even alarming is the almost universally held idea that the rich societies set the norm and demonstrate what can and ought to be achieved by everybody. It does not require deep insight, nor extensive factual knowledge, to realise that this idea is mischievous and unsupportable. Let us take the allegedly most "advanced" society of the modern age, the United States of America, which President Nixon calls "the hope of the world." It contains 5.6 percent of the world's population and absorbs something approaching 40 percent of the world's output of raw materials, many of which are non-renewable. To obtain its supplies, the United States, in spite of the vastness of its superbly endowed territory, has had to extend its commercial tentacles into every corner of the globe.

How, then, could a life style that makes such exorbitant demands upon world resources serve as a norm or a model for development? The very idea that "the problem of production" has been solved by the achievements of modern science and technology is based on a most astonishing oversight, namely, that the whole edifice of modern industry is built on non-renewable energy resources.

In short, the poor countries, which assuredly need development to regain some kind of economic health, have to evolve a life-style for which America or for that matter Japan or any other "advanced" country cannot serve as a model, and the so-called advanced countries have the even more difficult task of achieving some basis of existence which is compatible with peace and permanence.

As I have mentioned already, it is not only the problem of resources that has, suddenly and somewhat belatedly, moved into the centre of discussion; there is also the problem of pollution and of ecological breakdown. In addition, we do not need to look far afield to realize that modern industrial society is involved in some kind of human crisis which manifests itself in inflation, various types of unrest, rising crime rates, drug addiction, and so on. All this suggests that there is something wrong at the root of things—as indeed a Christian would be inclined to think. What is being called into question, so it seems, is not our *technical* competence but our value system and the very aims and objects we are pursuing.

However, we cannot leave the matter there. Many people, particularly among the young, are only too ready to agree with any criticism that can be made of modern society's aims and objects; they are only too anxious to adopt less materialistic and (shall we say) more idealistic aims for their own lives; but they are at a loss what to do. "The system," they find, is geared to a certain set of values, it can produce "growth" but it cannot produce justice; it can improve the quality of goods but not the quality of people; it can find money for the development of Concorde, moon rockets, and heart transplants, but not for adequate housing, public transport, school or hospital building. As Professor Galbraith complained about the richest country in the world, it produces private affluence and public squalor. Are they then to "drop out of the system"? But they have nothing sensible to drop into. It appears that any substantial re-setting of aims and objects implies or even presupposes a re-setting of the "system." How is that to be done and who can do it?

Efficient Production

Etienne Gilson once said that only the professors of philosophy speak of ideas; a true philosopher speaks always of things.[2] So let us try to be true philosophers. The "things" we might speak about are the methods of production, both technical and social, which the modern world mainly employs. We may note that these methods are chosen and developed primarily from the point of view of efficiency, and this is of course as it should be. No one in his senses favours inefficiency. The concept of efficiency, however, has become quite uncannily narrow and exclusive: it relates only to the material side of things and only to profit. It certainly does not relate to people, the actual persons involved in the process of production. If I said: "This process is efficient, because it makes the worker a happy man," I should be accused of talking sentimental nonsense, unless I could demonstrate that the worker's happiness actually led to increased output, better quality output, and above all to more profitable output. What the work does to the worker is not recognised as a decisive criterion of efficiency. Among all the machines engaged in the productive process there are also workers to fill certain gaps of mechanisation or automation. Pope Pius XI described the situation thus:

> With the leaders of business abandoning the true path, it was easy for the working class also to fall at times into the same abyss; all the more so because very many employers treated their workmen as mere tools, without the slightest thought of spiritual things. . . . And so bodily labour, which even after original sin was decreed by Providence for the good of man's body and soul, is in many instances changed into an instrument of perversion; for from the factory dead matter goes out improved, whereas men there are corrupted and degraded.[3]

This is the outcome of a concept of efficiency which relates to goods and not to people. And the damage cannot be undone by paying the worker higher wages or treating him with respect or in some other way trying to compensate him. Compensation never compensates: it can never undo the damage that has been done, but can merely try to mitigate some of the consequences.

From the point of view of this kind of material efficiency, there always appear to be so-called "economies of scale," and every increase in scale offers opportunities for the introduction of more specialised equipment, while at the same time forcing the worker into a greater division of labour. So the production units become ever bigger, more complex, more expensive, and also in a special sense more violent. Although it is, of course, society that produces the production system, once a particular system has come into existence it begins to mould society: it, as it were, insists that the members of society respect the immanent logic of the system and adapt to it by accepting its implicit aims as their own. Man then becomes the captive of the system whether he approves of its aims or not, *and he cannot effectively adopt different aims or values unless he takes steps to alter the system of production.*

New Concepts Required

In other words, ideas can change the world only by some process of "incarnation." The prevailing concept of efficiency rules the modern world not by itself, but by the type of technology and organisation it has produced. A mere change of the concept remains wishful thinking until new technologies and types of organisation have been evolved.

This is of decisive importance. It shows that appeals for good behaviour and the teaching of ethical or spiritual principles, necessary as they always are, invariably stay, as it were, *inside* the system and are powerless to alter it: unless and until the preaching leads to significant *new types of work* in the physical world.

It is true, indeed, that in the beginning is the word. But the word remains ineffectual unless it comes into this world; in the words of the Gospel: unless it is made flesh and dwells among us. It may then, of course, not be recognised and accepted by the established order; it may be looked upon as impractical and subversive. Whether it will eventually succeed in changing the world will depend not just on its truth, but on the work it manages to get done "in the flesh."

What kind of work? Since the prevailing system has been shaped by a technology that drives it into giantism, infinite complexity, vast expensiveness, and violence, it would seem to follow that we should engage the best of our intelligence to devise a technology that moves in the opposite direction—towards smallness, simplicity, cheapness, and non-violence. With the help of such a technology it would be possible, I am sure, to create an economic system to serve man, in the place of the present system which enslaves him. But this is said from a Christian point of view, where "serving man" means something different from what it may mean when the point of view is modern materialism.

Back to the Human Scale

To strive for smallness means to try to bring organisations and units of production back to a human scale. This is not a field in which precise definitions are possible, nor are they needed. When we feel that things have become too large for human comfort, let's see if we cannot make them smaller. There are many reasons for favoring smallness. Small units of production can use small resources—a very important point when concentrated, large resources are becoming scarce or inaccessible. Small units are ecologically sounder than big ones: the pollution or damage they may cause has a better chance of fitting into nature's tolerance margins. Small units can be used for decentralised production, leading to a more even distribution of the population, a better use of space, the avoidance of congestion and of monster transport. Most important of all: small units, of which there can be a great number, enable more people "to do their own thing" than large units of which there can only be a few. Smallness is also conductive to simplicity.

Simplicity, from a Christian point of view, is a value in itself. Making a living should not absorb all or most of a man's attention, energy, or time, as if it were the primary purpose of his existence on earth. Complexity forces people to become so highly specialised that it is virtually impossible for them to attain to wisdom or higher understanding. As Thomas Aquinas said: "The smallest knowledge of the highest things is more to be desired than the most certain knowledge of the lower things." A life-style full of complexity and specialisation, while conducive to the acquisition of knowledge of the lower things, normally involves such agitation and constant strain that it tends to act as a complete barrier against the acquisition of any higher knowledge.

Giantism and complexity bring it about that the capital required for production grows to fantastic proportions. Only people already rich and powerful can gain access to it. All other people are excluded; they cannot create a job for themselves, but must try to find one created by the rich. The effect is that, as has been said, capital employs labour, instead of labour employing capital. If the rich have not created enough job opportunities to employ all job seekers, the latter have no practicable opportunities for self-help and self-reliance. Not surprisingly, the poor countries, caught in the net of gigantic, complex, immensely expensive technology, used for even the simplest tasks, are saddled with mass unemployment, which affects mainly school leavers and other young people and makes all exhortations to self-reliance a mockery.

Non-Violence

From a Christian point of view, non-violence is also a value in itself. It flows from man forming a true view of himself: seeing himself not as creator but as a creature which has been sent into life for a purpose. The smallest mosquito, as St. Thomas Aquinas said, is more wonderful than anything man has produced and will ever produce. So man must never lose his sense of the marvellousness of the world around and inside him—a world which he has not made and which, assuredly, has not made itself. Such an attitude engenders a spirit of non-violence, which is a form or aspect of wisdom. With all the great powers man has recently acquired through his science and technology, it seems certain that he is now far too clever to be able to survive without wisdom.

As with so many other things, perfect non-violence may not be attainable in this imperfect world. But it *does* make a difference in which direction we *strive*. A system of production and a style of living, or a concept of efficiency, which advance steadily in the direction of violence, which refuse to recognize non-violence as a valid criterion of success, move on a disaster course. And the warning signals are appearing all around us. We call them pollution, environmental degradation, ugliness, intolerable noise, rapid exhaustion of resources, social disintegration and so forth. In other words, I do not think of violence only in the context of man's relation to other men, but in the context of all his relations including those with animate and inanimate nature.

It is sometimes said that modern man's ruthlessness vis-à-vis the rest of creation stems from the teaching in the first book of the Bible according to which God gave man "dominion" over all creatures of the earth. This is an excessively superficial view. Man, the noblest of all creatures, has indeed been given "dominion" over the rest, but he is not entitled to forget that *noblesse oblige*.

In more mundane terms: what is non-violence? We can say, for instance, that biologically and ecologically sound farming systems, with "good husbandry" and the careful observance of the Law of Return (recycling of all organic materials) represent a non-violent approach, whereas the ever intensifying warfare against nature of highly chemicalised, industrialised, computerised farming systems represents violence. Some people say: "The choice is between these violent systems and hunger. Look how productive, how efficient, these systems are. We need them to feed the growing populations of the world." The question is: Is this true? An immensity of R & D (research and development) expenditure has gone into the development of these violent systems, which completely depend on a vast chemical and pharmaceutical industry, which in turn completely depends on non-renewable oil. How much R & D has gone into the development of non-violent systems? Apart from a few private efforts, such as those of the Soil Association, hardly any. Even so, there are thousands of farmers around the world who are obtaining excellent yields and making a good living without resort to chemical fertilizers, insecticides, herbicides, fungicides, etc. Would it not be right to take these alternatives seriously and support them consciously, instead of putting all our eggs into the basket of violence?

There are many other directions in which the idea of non-violence can and should be developed. In medicine, we can say that prevention is essentially non-violent, compared with cure. Somebody once asked the question: "If an ancestor of long ago visited us today, what would he be more astonished at, the skill of our dentists or the rottenness of our teeth?" We should not need the violent interferences we get from our skillful dentists if we had maintained the health of our teeth the way other peoples have.

Possibilities of Change

It will be apparent that the four criteria of smallness, simplicity, capital-cheapness, and non-violence are closely interconnected, and I do not need to spell out all the interconnections. Can they become criteria for *action?* Indeed they can. It would, of course, be a violent approach if one suggested that all of a sudden the direction of progress should be changed one hundred percent. The "withdrawal symptoms" would be too severe, but there can be experimentation and gradual change. Why this prevailing immobilism? If only a few percent of our scientific, technical, intellectual, and financial effort and resources were diverted into a systematic search for

smallness

simplicity

capital saving and

non-violence

as concerns our industrial and farming systems—and many other fields of activity as well—it would emerge that a viable future can be attained. One does not even have to be a Christian to countenance and promote such a marginal diversion of effort. Do we not all, as good householders, use a small proportion of our income for insurance against calamities, some of which might never happen?

Smallness, simplicity, capital saving, and non-violence, *as a direction of conscious striving,* by means of very practical, down-to-earth work, commend themselves also from a social and political point of view. Good human relations, in my experience, are extremely difficult, if not impossible, to attain in large units, whether these are schools, universities, offices, or factories. Participation, so rightly demanded in industry and elsewhere, cannot become a reality when units are so large and complex that people cannot know each other as people and the minds of ordinary men and women cannot encompass the meaning and the ramifications of the whole.

People Do Matter

The future of industrial, technological society must be a future in which every man and woman, even "the least among my brethren," can be *persons,* can see themselves and be seen by their children as real people, not as cogs in vast machines and gap-fillers in automated processes, employed solely because, occasionally, the human machine is calculated to be a cheaper "means of production" than a mindless device.

It seems to me that, when looking at the future of industrial society from a Christian point of view, one is looking at it from a realistic point of view—as against a certain "crackpot realism" (as, I believe, Veblen called it) which is based on the implicit assumptions that people really do not matter; that we are masters of nature which can be ravaged and mutilated with impunity; that some Divine Improvidence has endowed a finite world with infinite material resources; and that consumption is the be-all and end-all of human life on earth. And it also seems to me that we *can* get off the hook of crackpot realism if we, as it were, *remember ourselves,* if we remember that we have a purpose in life that goes beyond the material; in other words, if we remember "The Foundation" which I have quoted.

As more and more people realise the predicament of modern technological society and the dangers it is facing, I can see the formation of a new battle line. On the one side, there will be what we might call the people of the forward stampede, with the slogan: "A breakthrough a day keeps the crisis at bay." On the other side there will be—what shall we call them?—the homecomers: people striving to lead things back to their proper place and function, realising that when it is said that man has dominion over the rest of creatures the reference is to man as a child of God, not to man as a higher animal. They believe that the spiritual has dominion over the material, which it is called upon to use *just so far* as it is needed for the attainment of spiritual ends, and no further.

It must be admitted that the people of the forward stampede, like the devil, have all the best—or at least all the most catchy—tunes. But the homecomers have the most exalted texts on which to base their patient and painstaking *practical* work; texts such as this one: "Seek ye first the Kingdom of God and all these other things—which you also need—will be added unto you."

The extraordinary thing about our period is the *great convergence.* The language of spiritual wisdom can now be understood also as the language of practical sanity, showing the road to survival in this world as well as to salvation in the next.

Notes

1. W. H. Longridge, *The Spiritual Exercises of St. Ignatius of Loyola* (London and Oxford: A. R. Mowbray, 1922).

2. Etienne Gilson, *God and Philosophy* (New Haven: Yale University Press, 1941).

3. Pius IX, *Quadragesimo Anno*, 134.

9 Buddhist Economics

E. F. Schumacher

"Right Livelihood" is one of the requirements of the Buddha's Eightfold Path. It is clear, therefore, that there must be such a thing as Buddhist Economics.

Buddhist countries, at the same time, have often stated that they wish to remain faithful to their heritage. So Burma: "The New Burma sees no conflict between religious values and economic progress. Spiritual health and material well-being are not enemies: they are natural allies."[1] Or: "We can blend successfully the religious and spiritual values of our heritage with the benefits of modern technology."[2] Or: "We Burmans have a sacred duty to conform both our dreams and our acts to our faith. This we shall ever do."[3]

All the same, such countries invariably assume that they can model their economic development plans in accordance with modern economics, and they call upon modern economists from so-called advanced countries to advise them, to formulate the policies to be pursued, and to construct the grand design for development, the Five-Year Plan or whatever it may be called. No one seems to think that a Buddhist way of life would call for Buddhist economics, just as the modern materialist way of life has brought forth modern economics.

Economists themselves, like most specialists, normally suffer from a kind of metaphysical blindness, assuming that theirs is a science of absolute and invariable truths, without any presuppositions. Some go as far as to claim that economic laws are as free from "metaphysics" or "values" as the law of gravitation. We need not, however, get involved in arguments of methodology. Instead, let us take some

Reprinted from *Resurgence* 1, no. 11 (January-February 1968), by permission of the author and the publisher.

fundamentals and see what they look like when viewed by a modern economist and a Buddhist economist.

There is universal agreement that the fundamental source of wealth is human labor. Now, the modern economist has been brought up to consider labor or work as little more than a necessary evil. From the point of view of the employer, it is in any case simply an item of cost, to be reduced to a minimum if it cannot be eliminated altogether, say, by automation. From the point of view of the workman, it is a "disutility": to work is to make a sacrifice of one's leisure and comfort, and wages are a kind of compensation for the sacrifice. Hence the ideal from the point of view of the employer is to have output without employees, and the ideal from the point of view of the employee is to have income without employment.

The consequences of these attitudes both in theory and in practice are, of course, extremely far-reaching. If the ideal with regard to work is to get rid of it, every method that "reduces the work load" is a good thing. The most potent method, short of automation, is the so-called division of labor and the classical example is the pin factory eulogized in Adam Smith's *Wealth of Nations*. Here it is not a matter of ordinary specialization, which mankind has practised from time immemorial, but of dividing up every complete process of production into minute parts, so that the final product can be produced at great speed without anyone having had to contribute more than a totally insignificant and, in most cases, unskilled movement of his limbs.

Work

The Buddhist point of view takes the function of work to be at least threefold: to give a man a chance to utilize and develop his faculties; to enable him to overcome his ego-centredness by joining with other people in a common task; and to bring forth the goods and services needed for a becoming existence. Again, the consequences that flow from this view are endless. To organize work in such a manner that it becomes meaningless, boring, stultifying, or nerve-racking for the worker would be little short of criminal; it would indicate a greater concern with goods than with people, an evil lack of compassion and a soul-destroying degree of attachment to the most primitive side of this worldly existence. Equally, to strive for leisure as an alternative to work would be considered a complete misunderstanding of one of the basic truths of human existence, namely, that work and leisure are complementary parts of the same living process and cannot be separated without destroying the joy of work and the bliss of leisure. From the Buddhist point of view, there are therefore two types of mechanization which must be clearly distinguished: one that enhances a man's skill and power and one that turns the work of man over to a mechanical slave, leaving man in a position of having to serve the slave. How to tell the one from the other? "The craftsman himself," says Ananda Coomaraswamy, a man equally competent to talk about the Modern West as the Ancient East, "the craftsman himself can always, if allowed to, draw the delicate distinction between the machine and the tool. The carpet loom is a tool, a contrivance for holding warp threads at a stretch for the pile to be woven round them by the craftsman's fingers; but the power loom is a machine, and its significance as a destroyer of culture lies in the fact that it does the essentially human part of the work."[4] It is clear, therefore, that Buddhist economics must be very different from the economics of modern materialism, since the Buddhist sees the essence of civilization not in a multiplication of wants but in the purification of human character. Character, at the same time, is formed primarily by a man's work. And work, properly conducted in conditions of human dignity and freedom, blesses those who do it and equally their products. The Indian philosopher and economist J. C. Kumarappa sums the matter up as follows:

> If the nature of the work is properly appreciated and applied, it will stand in the same relation to the higher faculties as food is to the physical body. It nourishes and enlivens the higher man and urges him to produce the best he is capable of. It directs his freewill along the proper course and disciplines the animal in him into progressive channels. It furnishes an excellent background for man to display his scale of values and develop his personality.[5]

If a man has no chance of obtaining work he is in a desperate position, not simply because he lacks an income but because he lacks this nourishment and enlivening factor of disciplined work which nothing can replace. A modern economist may engage in highly sophisticated calculations on whether full employment "pays" or whether it might be more "economic" to run an economy at less than full employment so as to ensure a greater mobility of labor, a better stability of wages, and so forth. His fundamental criterion of success is simply the total quantity of goods produced during a given period of time. "If the marginal urgency of goods is low," says Professor

Galbraith in *The Affluent Society,* "then so is the urgency of employing the last man or the last million men in the labor force." And again: "If . . . we can afford some unemployment in the interest of stability—a proposition, incidentally, of impeccably conservative antecedents—then we can afford to give those who are unemployed the goods that enable them to sustain their accustomed standard of living."[6]

From a Buddhist point of view, this is standing the truth on its head by considering goods as more important than people and consumption as more important than creative activity. It means shifting the emphasis from the worker to the product of work, that is, from the human to the subhuman, a surrender to the forces of evil. The very start of Buddhist economic planning would be a planning for full employment, and the primary purpose of this would in fact be employment for everyone who needs an "outside" job: it would not be the maximization of employment nor the maximization of production. Women, on the whole, do not need an outside job, and the large-scale employment of women in offices or factories would be considered a sign of serious economic failure. In particular, to let mothers of young children work in factories while the children run wild would be as uneconomic in the eyes of a Buddhist economist as the employment of a skilled worker as a soldier in the eyes of a modern economist.

While the materialist is mainly interested in goods, the Buddhist is mainly interested in liberation. But Buddhism is 'the Middle Way' and therefore in no way antagonistic to physical well-being. It is not wealth that stands in the way of liberation but the attachment to wealth; not the enjoyment of pleasurable things but the craving for them. The keynote of Buddhist economics, therefore, is simplicity and nonviolence. From an economist's point of view, the marvel of the Buddhist way of life is the utter rationality of its pattern—amazingly small means leading to extraordinarily satisfactory results.

The Standard of Living

For the modern economist this is very difficult to understand. He is used to measuring the standard of living by the amount of annual consumption, assuming all the time that a man who consumes more is "better off" than a man who consumes less. A Buddhist economist would consider this approach excessively irrational: since consumption is merely a means to human well-being, the aim should be to obtain the maximum of well-being with the minimum of consumption. Thus, if the purpose of clothing is a certain amount of temperature comfort and an attractive appearance, the task is to attain this purpose with the smallest possible effort, that is, with the smallest annual destruction of cloth and with the help of designs that involve the smallest possible input of toil. The less toil there is, the more time and strength is left for artistic creativity. It would be highly uneconomic, for instance, to go in for complicated tailoring, like the modern West, when a much more beautiful effect can be achieved by the skillful draping of uncut material. It would be the height of folly to make material so that it should wear out quickly and the height of barbarity to make anything ugly, shabby, or mean. What has just been said about clothing applies equally to all other human requirements. The ownership and the consumption of goods is a means to an end, and Buddhist economics is the systematic study of how to attain given ends with the minimum means.

Modern economics, on the other hand, considers consumption to be the sole end and purpose of all economic activity, taking the factors of production—land, labor, and capital—as the means. The former, in short, tries to maximize human satisfactions by the optimal pattern of consumption, while the latter tries to maximize consumption by the optimal pattern of productive effort. It is easy to see that the effort needed to sustain a way of life which seeks to attain the optimal pattern of consumption is likely to be much smaller than the effort needed to sustain a drive for maximum consumption. We need not be surprised, therefore, that the pressure and strain of living is very much less in, say, Burma than it is in the United States, in spite of the fact that the amount of labor-saving machinery used in the former country is only a minute fraction of the amount used in the latter.

The Pattern of Consumption

Simplicity and nonviolence are obviously closely related. The optimal pattern of consumption, producing a high degree of human satisfaction by means of a relatively low rate of consumption, allows people to live without great pressure and strain and to fulfill the primary injunction of Buddhist teaching: "Cease to do evil; try to do good." As physical resources are everywhere limited, people satisfying their needs by means of a modest use of resources are obviously less likely to be at each other's throats than people depending upon a high rate of use. Equally, people who live in highly self-sufficient local communities are less likely to get involved in large-scale violence than people whose existence depends on worldwide systems of trade.

From the point of view of Buddhist economics, therefore, production from local resources for local needs is the most rational way of economic life, while dependence on imports from afar and the consequent need to produce for export to unknown and distant peoples is highly uneconomic and justifiable only in exceptional cases and on a small scale. Just as the modern economist would admit that a high rate of consumption of transport services between a man's home and his place of work signifies a misfortune and not a high standard of life, so the Buddhist economist would hold that to satisfy human wants from far-away sources rather than from sources nearby signifies failure rather than success. The former might take statistics showing an increase in the number of ton/miles per head of the population carried by a country's transport system as proof of economic progress, while to the latter—the Buddhist economist—the same statistics would indicate a highly undesirable deterioration in the *pattern* of consumption.

Natural Resources

Another striking difference between modern economics and Buddhist economics arises over the use of natural resources. Bertrand de Jouvenel, the eminent French political philosopher, has characterized "Western man" in words which may be taken as a fair description of the modern economist:

> He tends to count nothing as an expenditure, other than human effort; he does not seem to mind how much mineral matter he wastes and, far worse, how much living matter he destroys. He does not seem to realize at all that human life is a dependent part of an ecosystem of many different forms of life. As the world is ruled from towns where men are cut off from any form of life other than human, the feeling of belonging to an ecosystem is not revived. This results in a harsh and improvident treatment of things upon which we ultimately depend, such as water and trees.[7]

The teaching of the Buddha, on the other hand, enjoins a reverent and nonviolent attitude not only to all sentient beings but also, with great emphasis, to trees. Every follower of the Buddha ought to plant a tree every few years and look after it until it is safely established, and the Buddhist economist can demonstrate without difficulty that the universal observance of this rule would result in a high rate of genuine economic development independent of any foreign aid. Much of the economic decay of Southeast Asia (as of many other parts of the world) is undoubtedly due to a heedless and shameful neglect of trees.

Modern economics does not distinguish between renewable and nonrenewable materials, as its very method is to equalize and quantify everything by means of a money price. Thus, taking various alternative fuels, like coal, oil, wood, or water power: the only difference between them recognized by modern economics is relative cost per equivalent unit. The cheapest is automatically the one to be preferred, as to do otherwise would be irrational and "uneconomic." From a Buddhist point of view, of course, this will not do; the essential difference between nonrenewable fuels like coal and oil on the one hand and renewable fuels like wood and waterpower on the other cannot be simply overlooked. Nonrenewable goods must be used only if they are indispensable, and then only with the greatest care and the most meticulous concern for conservation. To use them heedlessly or extravagantly is an act of violence, and while complete nonviolence may not be attainable on this earth, there is nonetheless an ineluctable duty on man to aim at the idea of nonviolence in all he does.

Just as a modern European economist would not consider it a great economic achievement if all European art treasures were sold to America at attractive prices, so the Buddhist economist would insist that a population basing its economic life on nonrenewable fuels is living parasitically, on capital instead of income. Such a way of life could have no permanence and could therefore be justified only as a purely temporary expedient. As the world's resources of nonrenewable fuels—coal, oil, and natural gas—are exceedingly unevenly distributed over the globe and undoubtedly limited in quantity, it is clear that their exploitation at an ever increasing rate is an act of violence against nature which must almost inevitably lead to violence between men.

The Middle Way

This fact alone might give food for thought even to those people in Buddhist countries who care nothing for the religious and spiritual values of their heritage and ardently desire to embrace the materialism of modern economics at the fastest possible speed. Before they dismiss Buddhist economics as nothing better than a nostalgic dream, they might wish to consider whether the path of economic development outlined by modern economics is likely to lead them to places where they really want to be. Towards the end of his courageous book *The Challenge of Man's Future,* Professor Harrison Brown of the California Institute of Technology gives the following appraisal:

> Thus we see that, just as industrial society is fundamentally unstable and subject to reversion to agrarian existence, so within it the conditions which offer individual freedom are unstable in their ability to avoid the conditions which impose rigid organization and totalitarian control. Indeed, when we examine all of the foreseeable difficulties which threaten the survival of industrial civilization, it is difficult to see how the achievement of stability and the maintenance of individual liberty can be made compatible.[8]

Even if this were dismissed as a long-term view—and in the long term, as Keynes said, we are all dead—there is the immediate question of whether modernization, as currently practised without regard to religious and spiritual values, is actually producing agreeable results. As far as the masses are concerned, the results appear to be disastrous—a collapse of the rural economy, a rising tide of unemployment in town and country, and the growth of a city proletariat without nourishment for either body or soul.

It is in the light of both immediate experience and long-term prospects that the study of Buddhist economics could be recommended even to those who believe that economic growth is more important than any spiritual or religious values. For it is not a question of choosing between "modern growth" and "traditional stagnation." It is a question of finding the right path of development, the "Middle Way" between materialist heedlessness and traditionalist immobility, in short, of finding "Right Livelihood."

That this can be done is not in doubt. But it requires much more than blind imitation of the materialist way of life of the so-called advanced countries.[9] It requires, above all, the conscious and systematic development of a "Middle Way in technology," as I have called it.[10] A technology more productive and powerful than the decayed technology of the ancient East, but at the same time nonviolent and immensely cheaper and simpler than the labor-saving technology of the modern West.

Notes

1. *Pyidawtha, The New Burma* (Economic and Social Board, Government of the Union of Burma, 1954), p. 10.

2. Ibid., p. 8.

3. Ibid., p. 128.

4. Ananda K. Coomaraswamy, *Art and Swadeshi* (Madras: Ganesh and Co.), p. 30.

5. J. C. Kumarappa, *Economy of Performance*, 4th ed. (Rajghat, Kashi: Sarva-Seva-Sangh-Publication, 1958), p. 117.

6. J. K. Galbraith, *The Affluent Society* (Harmondsworth: Penguin, 1962), pp. 272–273.

7. Richard B. Gregg, *A Philosophy of Indian Economic Development* (Ahmedabad: Navajivan Publishing House, 1958), pp. 140–141.

8. Harrison Brown, *The Challenge of Man's Future* (New York: Viking Press, 1954), p. 255.

9. E. F. Schumacher, "Rural Industries," in *India at Midpassage* (London: Overseas Development Institute, 1964).

10. E. F. Schumacher, "Industrialization through Intermediate Technology," in *Minerals and Industries*, vol. 1, no. 4 (Calcutta, 1964). Vijay Chebbi and George McRobie, *Dynamics of District Development* (Hyderabad: SIET Institute, 1964).

10 The Purpose of Wealth: A Historical Perspective

Gerald Alonzo Smith

The practice of medicine may require the prescription of an addictive stimulant for the sake of good health. The amount of the stimulant is finite and limited by the end. When, however, one takes a stimulant for its own sake, the desire for it becomes infinite since it is no longer limited by a final goal but is an end in itself. The same is true of the output of the economic process which, rather than being used for the sake of achieving the final goal of life, tends to become the final goal itself. Since output is then not limited by any final goal, the desire for it becomes infinite. We get hooked on economic growth. To paraphrase Descartes, such a lifestyle would be based on the philosophical foundation: I make and I buy, therefore I am.

In such a philosophical perspective one's reason becomes subject to the desires of the acquisitive side of one's nature rather than being the dominant partner in the orientation and direction of one's activities. To act irrationally comes to mean only that, given one's desires, one commits some action that is inconsistent with such desires. It makes no difference what one's desires are, because they are seen to be beyond the reach of reason. As long as he used the most efficient tools, the completely mad Captain Ahab was entirely rational in his search for the white whale. No less an economist than Frank Knight has remarked on such a view: "Living intelligently includes more than the intelligent use of means in realizing ends; it is fully as important to select the ends intelligently, for intelligent action directed toward wrong ends only makes evil greater and more certain."[1] More recently, Tibor Scitovsky has written about such activity,

---

Published by permission of the author. Financial support for this research was provided by the Rockefeller Brothers Fund.

"This may well be an example of the higher irrationality of behavior governed by narrowly rational calculation."[2]

Most economists, however, have refused to follow Knight and Scitovsky in a discussion of how one's economic behavior affects the achievement of one's final end. Indeed, the representative modern economist, having been weaned upon a quantitative and arithmomorphic methodology, probably does not even recognize that there is a relationship between the final end of man and his economic activity. Although some economists are more perceptive and recognize that there is a problem, they refuse to discuss the issue because they are quite aware of the pitfalls to which such normative judgments upon the legitimacy of the producer's and consumer's activity lead. Since most economists have never had the time, inclination, or training to formulate an authentic value system based upon a rational reflection of the final end of human nature, it is no wonder that they do not want to enter into that complex and delicate question. It is not hard to understand why such economists readily and mostly unconsciously tend to follow in the well-worn footsteps of their nineteenth-century predecessors who had found their answer to this question in the ethics of utilitarian philosophy. For them happiness implied good, and since every individual was the best judge of his own happiness, the individual's choice of economic goods had to be taken as given (that is, beyond analysis). No wonder the nineteenth-century economist felt an enthusiasm akin to religious fervor toward his economic theory. It helped unlock the door to the greatest happiness of the greatest number simply by making sure that the individual consistently followed out the dictates of his self-interested acquisitive nature.

We who live in the twentieth century are not so optimistic. We realize more clearly the consequences of such utilitarianism and individualism. We know the economists with their tremendous influence in helping to decide where to allocate resources can *not* avoid making a normative judgment on this important issue of the final end of man and economic behavior. The very decision to ignore the question of the hierarchy of natural needs by treating all effective demand as equal and given is in itself a value judgment and, by definition, makes of these demands ends in themselves. The easy way out is not always the right way. Such an analysis (or lack thereof) would not be too harmful, however, if it left only the economics profession in error. But the economics profession provides society with the image of economic society, and this image, in turn, notably affects the economic behavior of society.[3] As Warren Samuels has noted, "Economists should and do participate in the social valuational process, despite disclaimers to the contrary."[4] If the economics profession accepts as appropriate the image that the end of an economic system is to fulfill consumers' effective demands and beyond that nothing useful can be said about the legitimacy of demand either in the aggregate or in the particular, then such a judgment will penetrate in a very subtle manner throughout society and will, in turn, reinforce the economists in their original judgment.

There have, however, been economists during the modern era who have questioned this crucial valuational premise. They have refused to let go of the conclusions reached slowly and with great difficulty about the final goal of one's life by the Greco-Judeo-Christian civilization and to uncritically accept in their place the conclusions proposed by the ethics of utilitarianism and the epistemology of positivism modified by individualism.[5] J. C. L. Simonde de Sismondi in *Nouveaux principes d'économie politique,* John Ruskin in *Unto This Last,* J. A. Hobson in *Wealth and Life: A Study in Value,* and R. H. Tawney in *The Acquisitive Society* have each in his own way rejected the conventional economic wisdom that views increased production and consumption as an end in itself. Each of these humanistic economists looked upon increases in economic activity as a means rather than an end. Each investigated whether the increases in production and consumption experienced during their lifetime benefited man in achieving his final end, which, following the Greco-Judeo-Christian tradition, they defined as life in *all* its dimensions, especially in the higher immaterial dimensions.

J. C. L. Simonde de Sismondi (1773–1842)

Sismondi was the first economist of modern times to question the notion that growth in economic productivity was an end in itself or the same as growth in the public good. It is necessary to remember that Sismondi was writing in an era of transition from the craft system to the factory system and that his criticism was directed against the excesses of that transition period. Sismondi was well equipped for his role as a critic of the excesses of industrialism as he was one of the few economists of his generation who had the historical knowledge and acumen to observe the transitory nature of his era.

We are, and this point cannot be sufficiently stressed, in an altogether new state of society, of which we have absolutely no experience. We tend to separate completely all sorts of ownership from all sorts of work, to break all connections between man and master, to deprive the former of all associations in the profits of the latter.[6]

In a very early age of industrialism Sismondi attempted to orient the economics profession from the rigid abstractions of economic man that ultimately came to be its hallmark. He wanted economics to describe and analyze a changing economic scene but to hold fast to the ancient truths about man. Instead, economics would declare that its concepts and abstractions were permanent and that it was the nature of man that was unknowable and fleeting. Sismondi began his analysis of the economy by rejecting the view that the role of the economist was to maximize wealth in itself. For his view of the importance of wealth Sismondi turned back to the Greeks and especially to Aristotle for his inspiration. "But at least they [the Greeks] never lost sight of the fact that wealth had no other worth than what it contributed to the national happiness; and precisely because their treatment was less abstract, their point of view was oftentimes more just than ours."[7] In both his *Nouveaux principes* and his later *Etudes sur l'économie politique,* Sismondi is insistent throughout on the distinction he obtained from Aristotle between the science of "chremastics," which treats of the accumulation of monetary wealth or items of exchange value for their own sake, and "political economy," which treats of the role that economic production and consumption should play in achieving the final goal of society.

The chremastic science, or the study of the means of increasing wealth, in setting aside the purpose of this wealth, is a false science.[8]

When one takes the increase of economic goods as the end of society, one necessarily sacrifices the end for the means. One obtains more of production, but such production is paid for dearly by the misery of the masses.[9]

Since Sismondi disagreed with the conventional premise that increase in wealth was the final goal of the newly emerging science of economics, it is not surprising that he differed in his resolution to the economic problems facing the society of his era. Simply put, the solution of the conventional economists for the massive social misery that was only too apparent during those early crises of the Industrial Revolution was to expand production. "It [the misery resulting from the economic crisis of 1818–1821] is not a consequence of production being too much increased. Increase it more."[10] According to the classical economics, the general growth of economic productivity presented only minor inconveniences and was overall a good thing in every respect. Did not Say's Law show that supply created its own demand and, therefore, if supply was increasing, demand must also be increasing? Sismondi rejects this line of reasoning as being too abstract:

The science in their hands is so speculative, that it seems to be detached from all practice. It was believed at first that in extricating the theory from all the accessory circumstances, one ought to render it clearer and easier to seize, but the opposite is attained. The new English economists are quite obscure and can be understood only with great effort because our mind is opposed to admitting the abstractions demanded of us. This repugnance is in itself a warning that we are turning away from the truth when, in moral science where everything is connected, we endeavor to isolate a principle and to see nothing but that principle.[11]

Sismondi, on the other hand, develops his analysis by comparing the former economic society in which the majority worked for themselves as craftsmen and tradesmen with the current industrial society in which most laborers worked for others. (It was Sismondi who coined the word *proletariat.*) Since the craftsman's reward was the fruits of his own labor, and the amount of this reward was determined by the natural order of things, he would stop producing when he had reached the point beyond which he would prefer leisure and the fruits of his past labor to the extra income to be had from further labor.

For the laborer who works for himself there is a point reached in the accumulation of wealth beyond which it would appear as folly to accumulate still more, since such a laborer would not be able to increase his consumption in a proportional amount. But the needs of the laborer who works in an industrial society appear to be infinite. . . . No matter how many riches he has massed, there is no point at which he will say: "This is enough."

Moreover this is a serious error into which have fallen most of the modern economists, that they think that the act of consumption is unlimited and always ready to devour an infinite quantity of production. They do not cease from encouraging the nations to produce, to invent new machines, to improve their work so that the quantity of production achieved in the year will always surpass that of the preceding year: they are very distressed when they see the number of improductive workers multiply, they would point out the idle for the indignant public, and in a nation where the power of the worker has been increased by a hundredfold, they want that everyone should work in order to live.[12]

How does it happen that the industrial laborer works beyond that point which he would in a more natural system? Sismondi answers:

If all *les pompons de la richesse* were offered to the manual worker as a recompense for his assiduous travail of twelve and fourteen hours a day, as many do today, there is not one of these workers who would not choose less luxury and more of leisure, less of frivolous ornaments and more of liberty. Such should be the choice of the entire society, if only there was more equality in our society. Every craftsman who profits the total amount of his own industry, when he compares the almost imperceptible pleasure that he would receive from a slightly finer suit of clothes with the additional work that such a suit of clothes entails, would not wish to pay this price. Luxury is not possible except when it is paid for by the work of others. Assiduous and constant labor is able to be procured, not for the sake of frivolities, but only to gain the necessities of life.[13]

For Sismondi overproduction occurs when workers strain to produce more than they would in a system in which they received a larger share of the fruits of the productive process. Because the owners reaped where the laborers sowed, the decision to expand production was made by those who profited from production rather than by those who bore the real cost of labor required by expanded production. Sismondi wrote about the England of the early days of the Industrial Revolution when its laissez-faire economic ideology and factory system engendered a transparent exploitation of the workers unknown to previous centuries. It is important to note that in this critical period of transition Sismondi was unwilling to glorify economic production for its own sake after the manner of the orthodox economists because he did not share their absolute faith in the salvific efficacy of Say's Law, which declared that since supply created its own demand, an increase in production was identical to an increased demand for such production. Sismondi instead asked a more fundamental question:

What, then, is the object of human society? Is it to dazzle the eye with an immense production of useful and elegant things; to daunt the senses with the control which man exercises over nature, and with the precision or the speed with which a human work is executed by lifeless beings? Is it to cover the sea with vessels and the earth with railways which distribute in all directions the products of an ever increasing industry? . . . If such is the case, we have undoubtedly made immense progress as compared with our ancestors; we are rich in inventions, rich in activities, rich in scientific powers, rich in merchandise everywhere; for every nation has produced not only for itself but also for its neighbors. But, if the aim which society ought to accept, in encouraging labor and protecting the fruits of the labor of man, fruits which we call wealth—if these fruits, which consist of oral and intellectual goods as well as material goods, should be the means of improvement as well as of enjoyment, are we sure that we are approaching our goal?[14]

Sismondi's observation of the industrial system reminded him of the story of Gandalin.

In the time of enchantment, Gandalin, who lodged a sorcerer in his home, noticed that every morning the sorcerer would take a broom handle and, saying a few magic words on it, make out of it a water-carrier, who at once would get for him as many pails of water as he desired. One morning Gandalin hid himself behind a door and listened with all his might to overhear the magic words which the sorcerer pronounced for his enchantment. He, however, did not hear what the sorcerer said next to undo it. As soon as the sorcerer went away, Gandalin repeated the experiment; he took the broom handle, pronounced the mysterious words and the broom water-carrier went forward to the river and returned with water, and then again went forward and came back with it, thus again and again; Gandalin's reservoir was already full and the water flooded the room. "It's enough!" cried he, "Stop!" But the machine-man neither saw nor heard; insensible and indefatigable, he would have brought all the water from the river. Gandalin, in his despair, took an axe and hit his carrier with repeated blows. Then he saw the fragments of the broom, upon falling on the ground, immediately get up and reassume the magic form and run to the river. Instead of the carrier, he had now four, eight, sixteen; the more that he struck down the machine-men, the more machine-men got up to do his work in spite of him. The entire river would have passed into his home, if the sorcerer had not fortunately come back and destroyed his enchantment.[15]

Sismondi then comments that

the water, however is a good thing. Water, just as much as work, just as much as capital, is necessary for life. But one is able to have too much, even of the best things of life. . . . Each new application of science and the useful arts, similar to the axe of Gandalin knocking down the machine-men which the magic words had created, only to find soon two, four, eight, sixteen in its place; so the productivity continues to increase with a rapidity without measure. Has not the moment come, or at least is not the moment able to come, when one should say: This is too much?

According to the theory which is professed today in all the schools of political economy, this moment has not yet come, and it is never going to come.[16]

This story of Gandalin epitomizes Sismondi's view of his society—a society that was increasing economic production with a rapidity without measure, but for what? Sismondi's historical studies had taught him that there was more to a superior civilization than just increased material production. Such increased productivity, if brought about because of an unjust economic system, could well do more harm than good. As Sismondi grew older, he grew more pessimistic about his society that would not reduce its frenetic activity and orient its economic production and consumption by some final goal. On September 19, 1834, he wrote in his private journal:

> I read in the *Westminster Review* a striking article on civilization, in which the author points out many of the bad effects of the present system, which hitherto I have been almost the only one to remark. There is much ability in this article, but it inspires one with a melancholy feeling, because the evils are so serious and one does not see the remedies; the *too much* of everything is the evil of the day.[17]

As we leave this perceptive observer of the transition age to modern industrialism and turn to a later age of greater production and abundance, we should not forget that it was Sismondi who first criticized the economic theorists for making the increase of production a national goal. In his time of massive exploitation of laboring men, women, and, let us not forget, children, he questioned whether the cost in human suffering was too great for the frivolous items being produced.

John Ruskin (1819–1900)

Though as late as 1848 John Stuart Mill questioned whether all of the inventions yet produced had "lightened the day's toil of any human being," he nevertheless thought that "they have not yet begun to effect those great changes in human destiny, which it is in their nature and futurity to accomplish."[18] At the same time that Mill was writing these sentiments, another writer was beginning to investigate what effect these great changes in physical inventions and their expanding productivity were having on human destiny.

John Ruskin was born in 1819 into a moderately wealthy London merchant family. Though his family had experienced the power and pleasure that commercial success brought in its train, his puritanic and artistic parents made sure that he never judged the accumulation of wealth to be the main goal of his life. Ruskin was trained as an artist and art critic, and his reputation grew immensely with the successive volumes in his series *Modern Painters*. It was always, however, Ruskin's ambition to bring the beauties and inspiration of the intellectual and artistic world to the representative British worker. When the average British laborer failed to respond to his efforts, Ruskin set himself to the task of finding the cause of such blindness. His investigations into the life and society of the British laboring class rather quickly led him to believe that something was wrong with an economy that produced so much quantity of so little quality, yet brutalized so many people in doing it. Like other English critics of the industrial society, such as Coleridge, Cobbett, Carlyle, Dickens, Arnold, Morris, and many more, Ruskin soon denounced the commercial society of his time for its worship of Mammon, its "gospel of greed,"[19] and the conventional political economy that intellectually supported such a system. As the historian Asa Briggs has noted, "indeed, they [the poets] had probed far more deeply than the political economists into the inner meanings of the processes of change, had taken the world of nature as well as the world of men into the reckoning."[20] Yet of all these humanistic critics, only Ruskin attempted to challenge the economic theorists on their own ground by undertaking the task of thoroughly analyzing and exposing the errors of conventional political economy.

Like Sismondi, Ruskin began his analysis by distinguishing between the true science of political economy and that science which merely attempted to maximize economic productivity.

> The real science of political economy, which has yet to be distinguished from the bastard science, as medicine from witchcraft, and astronomy from astrology, is that which teaches nations to desire and labor for the things that lead to life, and which teaches them to scorn and destroy the things that lead to destruction.[21]

For Ruskin, "the ideal of human life is a union of Spartan simplicity of manners with Athenian sensibility and imagination."[22] Thus for his definition of wealth it was natural for Ruskin to turn back to the Greek Xenophon for his inspiration. Ruskin claimed that the *Economist* of Xenophon "contains a flawless definition of wealth, and an explanation of its dependence for efficiency on the merits and faculties of its possessors;—a definition which cannot be bettered; and which must be the foundation of all true Political Economy among nations, as Euclid is to all time the basis of Geometry."[23]

Ruskin is referring to the first chapter of Xenophon's *Economist*, "The Management of Property, That Is, Whatever Is of Use to a Man, but Is of No Value to Such as Are Slaves to Their Passions," where Xenophon is intent on showing that some economic possessions aid man in living and thus are true wealth or property, and some possessions, on the contrary, contribute to the destruction of man's nature. These latter possessions cannot be considered as true wealth, but must be considered as the opposite of wealth or property—*illth* was Ruskin's label for them.

Then the very same things are property to a man who knows how to use them, and not property to one who does not. For instance, a flute is property to a man who can play on it fairly; but to one who is wholly unskilled in its use it is no more property than mere useless stones would be—unless indeed he sold it.

So it is clear to us that a flute in the hands of a man who does not know how to use it, is not property to him, unless he sells it. So long as he keeps it, it is not property. And indeed, Socrates, we shall thus have reasoned consistently, since we before decided that a man's property must be something that benefits him. If the man does not sell the flute, it is not property, for it is of no use; but if he sell it, it becomes property.

To this Socrates answered, Yes, if he know how to sell it. But if he, again, were to sell it to a man who does not know how to use it, it would not be property even when sold, according to what you say.

Your words, Socrates, seem to imply that not even money would be property unless a man knew how to use it.

Well, you seem to agree with me that a man's property is only what benefits him. Suppose a man were to make this use of his money, to buy, say, a mistress, by whose influence his body would be worse, his soul worse, his household worse, how could we then say that his money was any benefit to him?

We could not—unless, indeed, we are to count as property henbane, the herb that drives mad those who eat it.[24]

This is an important passage for Ruskin and one that he would return to more often than to any other for his inspiration when pursuing problems in political economy. One can see a glimpse of it in the following often-quoted declaration: "And possession is in use only, which for each man is sternly limited; so that such things, and so much of them as he can use, are, indeed, well for him, or Wealth; and more of them, or any other things are ill for him, or illth."[25]

Thus the concept of wealth includes more than just the measurement of one's actual possessions; it includes, secondly, the capability of utilizing them in an appropriate and vital manner.

> "Having" is not an absolute, but a graduated, power; and consists not only in the quantity or nature of the thing possessed, but also (and in a greater degree) in its suitableness to the person possessing it and in his vital power to use it. . . . Wealth, therefore, is "the possession of the valuable by the valiant."[26]

It is clear from this brief discussion of Ruskin's view of wealth that he could not consider the mere accumulation of wealth to be the final goal of either the individual or the nation. He contrasts the erroneous with the correct perception of wealth in the following passage:

> There will be always a number of men who would fain set themselves to the accumulation of wealth as the sole object to their life. Necessarily, that class of men is an uneducated class, inferior in intellect, and more or less cowardly. It is physically impossible for a well-educated, intellectual, or brave man to make money the chief object of his thoughts; just as it is for him to make his dinner the principal object of them. All healthy people like their dinners, but their dinner is not the main object of their lives. So all healthily-minded people like making money—ought to like it, and to enjoy the sensation of winning it; but the main object of their life is not money; it is something better than money.[27]

Ruskin taught that moderate wealth should be the goal. "A nation which desires true wealth, desires it moderately, and can therefore distribute it with kindness, and possess it with pleasure; but one which desires false wealth, desires it immoderately, and can neither dispense it with justice, nor enjoy it in peace."[28] Ruskin urged the individual to recognize that "the law of life is that a man should fix the sum he desires to make annually, as the food he desires to eat daily; and stay when he has reached the limit, refusing increase of business, and leaving it to others, so obtaining due freedom of time for better thoughts."[29] Hence his prescriptions for the running of a state: "I strongly suspect that in a well-organized state, the possession of wealth ought to incapacitate for public office,"[30] and "one of the most important conditions of a healthy system of social economy, would be the restraint of the properties and incomes of the upper classes within certain fixed limits."[31]

This call for moderation and restraint did not fall on fertile ground in Victorian England. James Sherburne has pointed out that "Ruskin's final call for restraint was, perhaps, the most incomprehensible to his Victorian contemporaries. It lies in the sensitive area of social advancement or 'getting-on.' Ruskin denies the 'gospel of whatever we've got, to get more' as vehemently as he does that of 'wherever we are, to go somewhere else.'"[32] The customary reaction was that expressed in a lead article by the *Manchester Examiner and Times* on October 2, 1860: "He [Ruskin] is not worth our powder and shot, yet, if we do not crush him, his wild words will touch the springs of action in some hearts, and ere we are aware a moral floodgate may fly open and drown us all."[33] For better or worse, the *Manchester Examiner and Times* and, one might add, the conventional political economists, were able to keep shut the moral floodgate that Ruskin's thought represented and thus to keep the Victorian economic theory on the dry road of amorality. Though one economist predicted in 1888 that future economic theory would be built with Ruskinian bricks rather than with Ricardian straw,[34] this prediction has simply not come true.

Yet Ruskin's wild words have touched deeply some minds and hearts. Such diverse individuals as the heretical English economist John A. Hobson,[35] the artist and craftsman Eric Gill,[36] the biologist Patrick Geddes,[37] the physical chemist Frederick Soddy,[38] the economic historian R. H. Tawney,[39] Richard T. Ely, a founder of the American Economic Association,[40] the English novelist and distributist G. K. Chesterton,[41] the French novelist Marcel Proust,[42] and Indian pacifist and political leader Mohandas Gandhi[43] have all paid homage to Ruskin and his ideas.

John A. Hobson (1858–1940)

Although John A. Hobson often claimed that he was Ruskin's disciple, and in many of his works indicated that he was merely attempting to fill in some of the gaps in Ruskin's "magnificent plunge" into economic theory which brought "whole civilizations to a grand assise,"[44] it is nonetheless true that John Hobson added to and modified as much as he kept intact from Ruskin's thought. Ewald Grether has described the relationship of Ruskin and Hobson thus: "It is clear that it was neither a faith nor a creed that descended from Ruskin to Hobson, but primarily an attitude."[45] This inherited attitude was that of subjecting standard or conventional economic theory to the test of a humane assessment. Hobson, more than Ruskin, admitted that there was a place for the orthodox economic theory that took the narrow and more quantitative vision of simply allocating scarce resources efficiently among the perceived desires of individuals. Nonetheless, Hobson's plea for a "wider human assessment" of the output of the economic system marks him clearly as a Ruskinian. For both Ruskin and Hobson the discipline of economics had to be moderated by a social ethic and brought under the umbrella of a broader science—the art or science of human welfare.

Hobson began his analysis by pointing to an inconsistency in orthodox economic thought:

> Though everybody agreed that consumption was the final goal, this goal, as such, was nobody's concern. When goods passed through the hands of farmers, manufacturers, and traders, into the hands of consumers, they seemed to pass out of the economic system into a destructive process that took place in privacy and obscurity. . . . [And though] consumption remained the formal end of economic processes, production was the real end.[46]

Such an ostrich attitude toward the problems of evaluating the worth of final consumption could only lead to further error, implied Hobson.

> Only so far as current tastes and appetites are reliable indices of human utility, only so far as we can identify the desired with the desirable, is the evolution of customary standards of life a sound human art. But it is needless to cite the ample evidence of the errors and wastes that are represented in every human standard of consumption.[47]

In order to obviate such errors Hobson attempted to dispel some of the "privacy and obscurity" that surrounded the consumption of economic products; as he perceived the task, "some further adjustment is needed to assess the desired in terms of the desirable."[48]

Hobson first rejected the approach taken by standard economic textbooks when dealing with consumer behavior. Hobson saw behind the façade of effective demand.

> But a study primarily directed to the ascertainment and measurement of elasticity of demand, does not yet accord the disinterested valuations of consumptive processes required by a theory in which consumption is the "sole end." For consumption here only enters the economic field as a factor in markets and the determination of prices, not as the means of realizing the purpose to which the whole economic system is directed.[49]

In order to realize the purpose toward which the whole economic system is directed, Hobson had to determine what it was that was desirable, or, in other words, what was the Ultimate End by which the economic system could be orientated and measured. Hobson's favorite phrase for such an ideal was "organic welfare," about which he once added, "Though in form a mere synonym for good life, it is by usage both more restricted and more precise."[50] In another study he was concerned to show that organic welfare had both a materialistic component and a nonmaterial or artistic, spiritual component. "The organic conception of *mens sana in corpore sano* still stands as the first principle of human welfare. . . . It finds its justification in the truth so strongly enforced by Aristotle that we must first have a livelihood and then practise virtue."[51]

What contributes to a *mens sana in corpore sano?* More specifically, what contribution does the economic process make to a *mens sana in corpore sano?* As Hobson notes, we are immediately "confronted by the question how far the actual economic conduct, with its accompanying desires and gratifications, can be taken as a safe index of the desirable or organic welfare in its true sense." His response is that "we cannot assume a full identity of the income of an individual or a community, expressed in terms of current satisfactions, with that income expressed in terms of human welfare." This is so because "the total process of consumption-production may contain large elements of human waste or error, in that the tastes, desires, and satisfactions which actively stimulate this wealth creation may not conform to the desirable."[52] Later Hobson is more explicit in his condemnation of using the satisfaction of current consumers' effective demand as the Ultimate End of economics. "We cannot admit as the objective of economic activities either the yield of material goods which these activities produce, or the 'psychic income' which they yield as assessed in terms of current deservedness or satisfaction, without reference to their intrinsic desirability." Hobson then echoes Ruskin by immediately adding: "A material or a psychic income may contain 'illth' as an alloy to its wealth."[53]

The notion of excess production appears next in Hobson's analysis. His declaration that "mechanical production can easily outrun organic consumption" reminds one of both Ruskin's concept of "acceptant capacity" and Sismondi's strictures of the political economists "who treat consumption as an unlimited force, always ready to devour an unlimited production."[54]

After surveying the results of actual consumption patterns and the economist's analysis of such consumer behavior, Hobson concludes that "it cannot be said that any adequate study either of the evolution of actual standards of consumption, or of 'desirable' standards, has yet been made." He also indicated the reason for such failure: "Though much attention has been given to the economy of expenditure in equalizing 'marginal utilities,' it has not been clearly recognized that the several margins are themselves determined by processes of utilitarian calculations based on balances of organic requirements." In Hobson's mind, this failure to recognize the determination of the "several margins" is due, at least partly, to the fact that economists fail "to realize adequately that the organic nature of man necessarily stamps itself on his standard of consumption, and that, therefore, the various items of consumption must be studied as contributions toward the organic whole."[55]

Though Hobson never fully resolved in his own mind the absolute value of the items of consumption, he was unwilling to agree with the "popular thinking that is apt to brush aside the questions with the remark that values are matters of individual tastes, and *quot homines tot sententiae.*"[56] He considered such a position to be false because "we know that there exists a substantial body of agreement as to the main constituents of welfare, and even as to the order of their evaluation."[57]

From his analysis of consumption behavior Hobson discovered that an industrial economy has a built-in bias toward excessive production and consumption of economic goods. "This charge of materialism made against the more advanced industrial communities . . . [is] based on an over-stimulation of certain instincts for physical satisfactions, due to the innovating tendencies of modern capitalism with its elaborated apparatus of selling pressures." This leads to an excess that is due to "a hasty exploitation of newly roused tastes that absorb too much of human nature in economic processes. 'Getting and spending, we lay waste our powers.'"[58]

"Getting and spending, we lay waste our powers": the philosophy of materialism seems the inevitable consequence of an industrial economy in which narrow-minded and tunnel-visioned economists only describe a minute portion of the social canvas. Such economists never take off their blinders, and though they may peek from time to time to the left, they never look above to the Ultimate End.

Hobson's solution to combat this inherent bias toward excessive production is the following:

Human energy, therefore, increasingly demands that half the power of mechanical production shall be applied, not to producing more goods but more leisure, that is to say, to so liberating the producer from the strain and burden of specialized production that he may become a skilled consumer, with leisure and energy enough at his free disposal to assimilate the slower gains of scientific production, instead of being overwhelmed by them, while at the same time bringing his harmonised economic standard of living into proper relations with the non-economic activities and satisfactions of his life. This seems impracticable so long as profiteering rules the economic system. For the profit-maker can only gain his end either by working his machines and his workers to their full capacity, and turning out goods so rapidly that his skilled marketeers must induce the general body of workers to take their share in increased goods, not in increased leisure and other non-economic satisfactions, or by restrictions of output that give a wasteful or excessive leisure.[59]

As a fitting summation of J. A. Hobson's contribution to the analysis of wealth and the economic system that produces such wealth, and, at the same time, an introduction to the thought of Richard Henry Tawney, the noted economic historian and student of the current economic scene who will be the subject of the next section, we can quote a passage from Tawney's quite favorable review of Hobson's *Wealth and Life*.

The essence of humanism, perhaps, is the attitude which judges the externals of life by their effect in assisting or hindering the life of the spirit. It is the conviction that the machinery of existence—property and material wealth, and industrial organization, and the whole fabric and mechanism of social institutions—is to be regarded as means to an end, and that this end is the growth towards perfection of individual human beings. In this sense, Mr. Hobson is the greatest of economic humanists. Undisturbed by the roar of the wheels, he approaches the engine with questions most of us are too clever, or too superficial, to condescend to ask. What is the thing for? In what way do its impressive gyrations minister to the dignity and happiness of mankind?[60]

Richard H. Tawney (1880–1962)

During a period of introspection and reflection Richard H. Tawney made a long entry into his diary on July 12, 1913:

As long as individuals think the attainment of moderate material comfort the chief end of life, so long will governments plead as an excuse for not doing this or that that they cannot afford it. If modern England and America are right in believing that the principal aim of man, what should be taught to children, what should serve as a rough standard of merit, what merits approbation and respect, is the attainment of a moderate—or even immoderate—standard of comfort, and that moral questions arise only after this has been attained; then they must be content to go without religion, literature, art, and learning. These are not hard to find for those who really seek them, or who seek them first. But if they are sought second they are never found at all. . . .

What I mean is that the failure of society to make the changes which are obviously important when regarded in bulk is due to the fact that individually we all have a false philosophy of life. We assume that the greatest misfortune which can befall a man is poverty—and that conduct which leads to the sacrifice of income is unwise, impractical, etc.; in short that a man's life should be judged by its yield of income, and a nation's life by its production of wealth. . . .

But supposing unearned incomes, rents, etc. are pooled, will not the world, with its present philosophy, do anything but gobble them up and look up with an impatient grunt for more? That is the real question. It will not be faced in my lifetime because as long as the working classes believe, and believe rightly, that their mentors rob them, so long will they look on the restoration of the booty as *the* great reform, and will impatiently waive aside more fundamental issues, as a traveller robbed by a highwayman declines to be comforted by being told that money, after all, does not buy happiness. But when their masters are off their backs they will still have to face the fact that you must choose between less and more wealth and less and more civilization. . . .

When three or four hundred years hence mankind looks back on the absurd preoccupation of our age with economic issues with the same wonder as, and juster contempt than, we look back on the theological discussions of the middle ages, the names which they will reverence will be those of men who stood out against the prevalent fallacy that the most important problems were economic problems, and who taught men to conquer poverty by despising riches.[61]

Six years later, after the interruption of World War I and its aftermath, Tawney returned to this question of the final end of the economic system. He published his conclusions in *The Acquisitive Society,* which became one of the most controversial books of the 1920s, calling on the British people to reform their fundamental philosophy of life.

These are times which are not ordinary, and in such times it is not enough to follow the road. It is necessary to know where it leads and, if it leads nowhere, to follow another. The search for another involves reflection, which is uncongenial to the bustling people who describe themselves as practical. . . . But the practical thing for a traveler who is uncertain of his path is not to proceed with the utmost rapidity in the wrong direction: it is to consider how to find the right one.[62]

Tawney pointed out that the road upon which England's industrial and economic leaders were leading her—the philosophical road that viewed economic productivity as its own end—had been tried in the past and had been found wanting.

When they desire to place their economic life on a better foundation, they repeat, like parrots, the word "Productivity," because it is the word that rises first in their minds; regardless of the fact that productivity is the foundation on which it is based already, that increased productivity is the one achievement of the age before the war, as religion was of the Middle Ages or art of classical Athens, and that it is precisely in the century which has seen the greatest increase in productivity since the fall of the Roman Empire that economic discontent has been most acute.[63]

Increased productivity alone will not cause social ills to disappear. Such a response is based upon an illusion.

Hence the idea, which is popular with rich men, that industrial disputes would disappear if only the output of wealth were doubled, and every one were twice as well off, not only is refuted by all practical experience, but is in its very nature founded upon an illusion. For the question is one not of amounts but of proportions; and men will fight to be paid $120 a week, instead of $80, as readily as they will fight to be paid $20 instead of $16.[64]

Leaders whose faith is that "riches are not a means but an end" and who imply "that all economic activity is equally estimable, whether it is subordinated to a social purpose or not,"[65] are "like a man who, when he finds that his shoddy boots wear badly, orders a pair two sizes larger instead of a pair of good leather, or who makes up for putting a bad sixpence in the plate on Sunday by putting in a bad shilling the next."[66]

Tawney pointed out the direction of the correct path by harkening back to a central theme of Ruskin's:

The purpose of industry is obvious. It is to supply man with things which are necessary, useful or beautiful, and thus to bring life to body or spirit. In so far as it is governed by this end, it is among the most important of human activities. In so far as it is diverted from it, it may be harmless, amusing, or even exhilarating to those who carry it on, but it possesses no more social significance than the orderly business of ants and bees, the strutting of peacocks, or the struggles of carnivorous animals over carrion.[67]

The true political economist realizes that "all rights . . . are conditional and derivative. . . . They are derived from the end or purpose of the society in which they exist."[68]

Tawney notes that such a frenetic rush to produce without any guiding ultimate principle creates a situation where "part of the goods which are annually produced, and which are called wealth, is strictly speaking, waste . . . [which] should not have been produced at all."[69] And to those who clamor for increased productivity as the solution to society's ills, Tawney responds: "Would not 'Spend less on private luxuries' be as wise a cry as 'Produce more'?" To do so, however, would be "to admit that there is a principle superior to the mechanical play of economic forces . . . and thus to abandon the view that all riches, however composed, are an end, and that all economic activity is equally justifiable."[70]

Comparing "Prussian militarism" to "English industrialism," Tawney argues that both of these ideologies have killed the souls of men by allowing a subordinate social system to dominate their societies. "When the Press clamors that the one thing needed to make this island an Arcadia is productivity, and more productivity, and yet more productivity, that is Industrialism. It is the confusion of means with ends."[71]

Tawney concludes *The Acquisitive Society* by declaring that what English society needs, therefore, is a purpose, a principle of limitation. Such a principle of limitation would divide "what is worth doing from what is not, and settles the scale upon which what is worth doing ought to be done. . . . Above all, it assigns to economic activity itself its proper place as the servant, not the master, of society."[72]

This is not the place to review the historical portion of Tawney's *Religion and the Rise of Capitalism*,[73] but it is appropriate to our analysis to review the conclusions that he drew from his historical studies. J. D. Chambers has succinctly summarized the importance of Tawney's findings:

> As is well known, Tawney's main preoccupation was with the secularization of traditional Christian values in the sixteenth and seventeenth centuries—the greatest secular event, he considered, in the history of Western civilization. It was the first step, in Tawney's view, on the way to the establishment of an acquisitive society based on competition, individualism, and the divine right of self-aggrandisement on the assumption that what is good for one is, in the long run, good for all.[74]

In the concluding chapter of *Religion and the Rise of Capitalism* Tawney returns to many of the concerns that had troubled him in the opening pages of *The Acquisitive Society.* Quoting Berkeley's aphorism "Whatever the world thinks, he who has not much meditated upon God, the human mind and the *summum bonum* may possibly make a thriving earthworm, but will most indubitably make a sorry patriot and a sorry statesman," he notes that "the most obvious facts are the most easily forgotten. Both the existing economic order, and too many of the projects advanced for reconstructing it, break down through their neglect of the truism that, since even quite common men have souls, no increase in material wealth will compensate them for arrangements which insult their self-respect and impair their freedom."[75] Making economic wealth its own end, Tawney points out, goes against the thrust of much of history.

> The distinction made by the philosophers of classical antiquity between liberal and servile occupations, the medieval insistence that riches exist for man, not man for riches, Ruskin's famous outburst, "there is no wealth but life,". . . are different attempts to emphasize the instrumental character of economic activities by reference to an ideal which is held to express the true nature of man.[76]

As we conclude our review of Tawney's analysis of the function of wealth and the economic system, we are led around to the question of what is the "true nature of man." Though Tawney never defined the nature of man in so many words, late in his life he remarked that man, "as known to history, is a religious animal." And he considered modern industrialism and capitalism not as irreligious but as counterreligious with their "idolatry of riches and the idolatry of power."[77] In his diary he had written some twenty years earlier: "If it be asked what is your criterion: why do you condemn this and approve that? I answer that the standard which we apply is really a transcendental, religious, or mystical one."[78] The important thing for Tawney was not to define precisely the ideal—such an achievement was clearly impossible in any total or definitive sense—but to recognize the need to acknowledge the primary importance of such a standard or principle. As he wrote, ideals of religion, art, and understanding "are not hard to find for those who really seek them, or who seek them first." Indubitably, Tawney sought them first. Perhaps that is why another eminent British economic historian, T. S.

Ashton, was able to write of him that "students who had the good fortune to sit at his feet rose with the sense of having been in touch not only with scholarship, but with wisdom."[79]

Conclusion

Wisdom—what is it? In our Western epistemological tradition the most important key to wisdom, the understanding of reality, has been the concept of causation. As Aristotle put it, "Men do not think that they know a thing till they have grasped the 'why' of it." Aristotle pursued this investigation by showing that there are four conceptually distinct and essential causes for every rational human activity utilizing material resources: the *causa materialis*, the *causa efficiens*, the *causa formalis*, and the *causa finalis*. An example is the building of a house, wherein the wood and other materials are the *causa materialis*, the carpenter's labor and the tools are the *causa efficiens*, the blueprint or plan in the carpenter's mind is the *causa formalis*, and the desire to have a home for shelter and comfort is the *causa finalis*. To fully understand this act of building a home, one would have to investigate all four of these causes. Were the most satisfactory materials used? Were the carpenter and his tools efficient? Did he have a good plan or blueprint? Was his final goal consistent with the needs of his nature? All such questions are relevant to understanding this activity.

This same methodological approach has to be used when a social scientist investigates the economic system with the intention of fully comprehending it. In current methodology, however, in economics as well as in other theoretical sciences, two of the types of causes advanced by Aristotle have been largely omitted: the *causa materialis* and the *causa finalis*.[80] In its search for understanding and its quest for reform, economics has tended to utilize only the *causa efficiens* and the *causa formalis*, and of these two it has emphasized the *causa efficiens* much more than the *causa formalis*. More accurately, the *causa formalis* is often submerged by the *causa efficiens*. As Phyllis Colvin has remarked, "theoretical science tends to meld these two fundamentals of explanation, concentrating far more on the *causa efficiens* than on the *causa formalis*."[81] "Doing" becomes its own justification. We no longer ask "Doing what?"; even less do we explore "Doing what for what?" or analyze the consequences of "Doing what for what and with what?"

To attempt to investigate the economic system of reasonable purposive beings and to neglect the *causa materialis* and *causa finalis* of such a system while merging the *causa formalis* into the *causa efficiens* would have appeared to Aristotle to be shirking one's intellectual responsibility and to ensure that one would end up with the most dangerous kind of knowledge—half-knowledge or, more accurately, quarter-knowledge.

The four economists reviewed in this study were aware of this limitation of much economic analysis, this blindered vision, and they endeavored to broaden the perspective of economists so that they would have to include a critical study of the *causa finalis* of the economic system within their analysis. They realized that only that analysis which can lead to an appropriate *causa finalis* is destined to give satisfaction in fruition, since only the appropriate *causa finalis* gives a fundamental unity to the problem of human behavior. Because of the loss of an appropriate *causa finalis,* there is an unsatisfied craving in modern man. Since he has lost control of his direction, he attempts to regain what he has lost by an inordinate and frantic search for those commodities that will stimulate his senses and thus temporarily quiet his sense of lost direction, or by seeking that power which allows one to control the lives of others. This control over other things and other people becomes a surrogate for having control of one's own life. Lacking the power and ability (and the will) to submit the direction of one's own life to an appropriate final goal, one seeks the false remedy of gaining power over things and persons. Economic production and power become their own ends. In so doing, modern man aggravates, rather than resolves, the sickness of society. It is only by returning to first principles, to a study of man and his appropriate final goal, that this sickness can be healed. As R. G. Hawtrey has noted, "Even in exploring human motives and behavior a cognizance of right ends is the foundation of firm and confident conclusions."[82] If such an exploration of the appropriate *causa finalis* is omitted, then although one may show much scholarship by manipulating vast amounts of data with precision and rigor in the largest of computer models, it may not lead to wisdom. Are our students in touch with scholarship or with wisdom?

Notes

1. Frank Knight, *The Economic Organization* (Harper & Row, 1933), p. 4.

2. Tibor Scitovsky, *The Joyless Economy* (Oxford University Press, 1976), p. 236.

3. See Kenneth Boulding, *The Image: Knowledge in Life and Society* (University of Michigan Press, 1956), for a thorough discussion of the important role that such images play in society.

4. Warren J. Samuels, "Normative Premises in Regulatory Theory," *Journal of Post Keynesian Economics* 1 (1978), 112.

5. For a thorough historical discussion of what factors led to the breakdown of this Greco-Judeo-Christian view of human nature and its consequent replacement by a more individualistic and less theological view, see the three following works of Alasdair MacIntyre: *After Virtue: A Study in Moral Theory* (University of Notre Dame Press, 1984); *Whose Justice? Which Rationality?* (University of Notre Dame Press, 1988); and *Three Rival Versions of Moral Enquiry* (University of Notre Dame Press, 1990).

6. J. C. L. Simonde de Sismondi, *Nouveaux principes d'économie politique ou de la richesse dans ses rapports avec la population*, 2d ed. (Paris, 1827), 2:434. (Translation by the author.)

7. *Nouveaux principes*, 1:20.

8. J. C. L. Simonde de Sismondi, *Etudes sur l'économie politique* (Paris, 1837–1838), 1:4. (Translation by the author.)

9. *Nouveaux principes*, 2:140.

10. J. R. McCulloch, "The Opinions of Messrs. Say, Sismondi, and Malthus, on the Effects of Machinery and Accumulation, Stated and Examined," *Edinburgh Review* 35 (1821), 106–107. See also J. B. Say, A *Treatise of Political Economy* (1834), pp. 139, 140n.

11. *Nouveaux principes*, 1:55–56.

12. *Nouveaux principes*, 1:75–76.

13. *Nouveaux principes*, 1:79.

14. *Etudes sur l'économie politique*, 1:27.

15. *Etudes sur l'économie politique*, 1:60–62.

16. *Etudes sur l'économie politique*, 1:60–62.

17. "Extracts from the Private Journal and Letters of M. de Sismondi," *Political Economy and the Philosophy of Government: A Series of Essays from the Words of M. de Sismondi*, translated and edited by an anonymous author (London, 1847), pp. 450–451; the emphasis is in the original. Sismondi would also have liked a recent striking article by Alan Durning on the same topic in

State of the World 1991, ed. Lester Brown et al., entitled "Asking How Much Is Enough" (W. W. Norton, 1991), pp. 153–170.

18. J. S. Mill, *The Principles of Political Economy* (1915 edition), p. 751.

19. "Our large trading cities bear to me very nearly the aspect of monastic establishments in which the roar of the millwheel and the crane takes the place of other devotional music; and in which the worship of Mammon or Moloch is conducted with a tender reverence and an exact propriety; the merchant rising to his Mammon matins with the self-denial of an anchorite, and expiating the frivolities into which he may be beguiled in the course of the day by late attendance at Mammon vespers." "A Joy for Ever," in *Works of Ruskin*, ed. E. T. Cook and Alexander Wedderburn (G. Allen, 1903–1912), 16:138.

20. Asa Briggs, *The Making of Modern England: 1783–1867* (New York, 1959), p. 401.

21. "Unto This Last," in *Works of Ruskin*, 17:85.

22. "A Joy for Ever," in *Works of Ruskin*, 16:134.

23. "The Economist of Xenophon," in *Works of Ruskin*, 31:27.

24. "The Economist of Xenophon," in *Works of Ruskin*, 31:38–39.

25. "Munera Pulveris," in *Works of Ruskin*, 17:168.

26. "Unto This Last," in *Works of Ruskin*, 17:87–88.

27. *Works of Ruskin*, vol. 18:412.

28. "Munera Pulveris," in *Works of Ruskin*, 17:144.

29. "Munera Pulveris," in *Works of Ruskin*, 17:277–278.

30. *Works of Ruskin*, 17:xlvii.

31. "Time and Tide," in *Works of Ruskin*, 17:322.

32. James C. Sherburne, *John Ruskin or the Ambiguities of Abundance* (Harvard University Press, 1972), p. 277.

33. Quoted in *Works of Ruskin*, 17:xxxi.

34. F. J. Stimson, "Ruskin as Political Economist," *Quarterly Journal of Economics* 2 (1888), 74–109.

35. John A. Hobson, *John Ruskin: Social Reformer* (London, 1898).

36. Eric Gill, "John Ruskin," in It *All Goes Together: Selected Essays* (New York, 1944), pp. 45–47.

37. Patrick Geddes, *John Ruskin: Economist* (Edinburgh, 1884). For Geddes's contribution in this area of economics, see the chapter "Patrick Geddes' Critique of Economics" in *Ecological Economics: Energy, Environment, and Society* by Juan Martinez-Alier (Basil Blackwell, 1987), pp. 89–99.

38. See Frederick Soddy, *Wealth, Virtual Wealth and Debt* (London, 1933), pp. 93ff.

39. R. H. Tawney, "John Ruskin," *Observer* (February 19, 1919). Reprinted in *The Radical Tradition*, ed. Rita Hinden (London, 1966), pp. 42–46.

40. See the introduction by Richard T. Ely in the 1901 edition of *Unto This Last*, edited by Richard T. Ely. The copy of Ruskin's *Munera Pulveris* that was owned by Richard T. Ely (now in the possession of Louisiana State University) shows many jottings in Ely's handwriting.

41. G. K. Chesterton, *Poems of Ruskin with an Essay by the Author* (G. Routledge, 1906).

42. Jean Autret, *L'influence de Ruskin sur la vie, les idées et l'oeuvre de Marcel Proust* (Geneva, 1955).

43. Elizabeth T. McLaughlin, *Ruskin and Gandhi* (Bucknell University Press, 1974).

44. John A. Hobson, *Wealth and Life: A Study in Values* (London, 1929), p. viii.

45. Ewald T. Grether, "John Ruskin–John A. Hobson," in *Essays in Social Economics*, ed. Ewald T. Grether et al. (Freeport, N.Y., 1935), p. 163.

46. *Wealth and Life*, pp. 301–302.

47. *Wealth and Life*, p. 328.

48. *Wealth and Life*, p. vii. For Hobson the words *desired* and *desirable* had important and precise, though quite different, meanings. *Desired* signified what consumers actually wanted. *Desirable* stood for what they *should* want.

49. *Wealth and Life*, pp. 304–305.

50. John A. Hobson, *Work and Wealth: A Human Evaluation* (Macmillan, 1914), p. 12.

51. *Wealth and Life*, p. 47.

52. *Wealth and Life*, pp. xxiii–xxiv.

53. *Wealth and Life*, p. 130.

54. Sismondi, *Nouveaux principes*, 1:76.

55. Hobson, *Wealth and Life*, p. 305. Hobson would undoubtedly appreciate the recent book by Tibor Scitovsky, *The Joyless Economy: An Inquiry into Human Satisfaction and Consumer Dissatisfaction* (see note 2), which can be viewed precisely as a study of the standards of consumption and their contributions to man's "organic whole."

56. "However many individuals, that many ideas."

57. *Wealth and Life*, p. 51.

58. *Wealth and Life*, p. 309.

59. *Wealth and Life,* p. 339.

60. R. H. Tawney, "Book Review of *Wealth and Life* by J. A. Hobson," *The Political Quarterly* 1 (1930), 276–277.

61. *R. H. Tawney's Commonplace Book,* ed. J. M. Winter and D. M. Joslin (Cambridge University Press, 1972), pp. 60–62.

62. R. H . Tawney, *The Acquisitive Society* (New York, 1920) p. 2.

63. *The Acquisitive Society,* p. 62.

64. *The Acquisitive Society,* p. 42.

65. *The Acquisitive Society,* p. 37.

66. *The Acquisitive Society,* p. 5.

67. *The Acquisitive Society,* p. 8.

68. *The Acquisitive Society,* p. 51.

69. *The Acquisitive Society,* pp. 37–38.

70. *The Acquisitive Society,* p. 39.

71. *The Acquisitive Society,* p. 46.

72. *The Acquisitive Society,* pp. 182–183.

73. For such a review, see C. K. Wilber, "The 'New' Economic History Reexamined: R. H. Tawney on the Origins of Capitalism," *American Journal of Economics and Sociology* 33 (1974), 249–258.

74. J. D. Chambers, "The Tawney Tradition," *Economic History Review* 24 (1971), 356.

75. R. H. Tawney, *Religion and the Rise of Capitalism* (Mentor, 1954), p. 233.

76. Tawney, *Religion and the Rise of Capitalism,* p. 233.

77. "A Note on Christianity and the Social Order," in *The Attack and Other Papers* (London, 1953), p. 191.

78. *R. H. Tawney's Commonplace Book,* p. 64.

79. T. S. Ashton, "Richard Henry Tawney (1880–1962)," *Proceedings of the British Academy* 48 (1962), 479.

80. See Phyllis Colvin, "Ontological and Epistemological Commitments and Social Relations in the Sciences," in *The Social Production of Scientific Knowledge,* ed. E. Mendelsohn, P. Weingart, and R. Whitley (Boston, 1977), pp. 103–129. Also see Nicholas Georgescu-Roegen, *The Entropy Law and the Economic Process* (Harvard University Press, 1971), pp. 182–191, for an analysis of why the social sciences have neglected the study of the *causa finalis* and the harm that has resulted from such neglect. As for Georgescu-Roegen's appreciation for the importance of the *causa materialis,* one could argue that the whole point of *The Entropy Law and the Economic Process* was to show the need for

including the *causa materialis* in one's economic analysis. What else is low-entropy matter-energy except the basic *causa materialis* of an economy?

Alfred N. Whitehead has expressed somewhat the same opinion: "For instance, the enterprises produced by the individualistic energy of the European people presuppose physical action directed to final causes. But the science which is employed in their development is based on a philosophy which asserts that physical causation is supreme and which disjoins the physical cause from the end." *Science and the Modern World* (Mentor, 1948), p. 111.

81. Colvin, "Ontological and Epistemological Commitments," p. 109.

82. R. G. Hawtrey, "The Need for Faith," *The Economic Journal* 56 (1946), 364.

11 Ecology, Ethics, and Theology

John Cobb

Western ethics, like Western thought generally, has been radically anthropocentric. But the disruptions of ecological systems that have brought the world to crisis have called our attentions to the need for adopting different patterns of behavior with respect to our environment. Ecological crisis opens the question of whether the moral necessity of behavioral changes follows from the same anthropocentric principles we have had in the past or whether our ethical principles themselves are partly at fault and need alteration. If the latter, on what ground can such alteration be effected? Surely ultimate ethical principles cannot be altered only because we do not like their results.

There are two basic elements in almost any ethical theory, although one or the other is often more implicit than explicit. We require some judgment as to what is good or desirable, and we require some principles of right action. Those who concentrate on the first of these elements, developing a theory of value, often assume that when their work is done the answer to the question of how we ought to act is self-evident. That is, having determined good or desirable values, they suppose that we ought to maximize those values. But this ignores the complex question of exactly what our obligation is if maximization presents conflicts. Should we maximize values in general or our own values? Should we maximize present values or be equally concerned about future values? Are there any acts that we should not perform regardless of the value of their consequences? Is the relation of our present action to past commitments a relevant factor in deciding how we ought now to act? Answering these ques-

Published by permission of the author, with additions written especially for this volume.

tions is not a further development of a theory of value but a treatment of the formal questions that belong to ethics proper.

Value Theory

In a value theory it is customary to distinguish types of values. One important distinction is between intrinsic value and instrumental value. An automobile is of instrumental value to me in that it enables me to get places and thereby to have experiences that I would otherwise be denied. Perhaps it may also be instrumentally valuable to me as an object of aesthetic contemplation or as contributory to my feeling of power. So long as its value is only instrumental it must be measured by its potential for contributing to the intrinsic value of my feeling of beauty and power, and by that alone. Hence intrinsic value is our primary consideration.

Although the distinction between instrumental and intrinsic values is important, it should not be exaggerated or misunderstood. There may be some things whose *only* value is instrumental, but there is nothing whose only value is intrinsic. That is, everything or every event has consequences for other things or events. These consequences can be evaluated as relatively favorable or unfavorable. Hence, though not everything can be evaluated in terms of its intrinsic value, everything does have its instrumental value.

It is my contention that whereas everything has instrumental value, only feeling is the locus of intrinsic value. The existence or nonexistence of something that has no feeling seems to me a matter of indifference unless it somehow contributes to the feeling belonging to something else, either actually or potentially. That is to say, what has no feeling has no intrinsic value and its instrumental value is a function of its contribution to something that does feel. Feeling, on the other hand, is intrinsically valuable, while it also has instrumental value in its relation to other feelings.

This doctrine is not far removed from common sense or from traditional theories of value. Utilitarianism proposed to regard only pleasure (and pain) as valuable (and disvaluable). Pleasure is of course a matter of feeling. Critics of utilitarianism have rightly complained that either the notion of pleasure must be taken very broadly or else it must be recognized that men find other feelings beside pleasure valuable. But in general they agree that the focus of intrinsic value is in the sphere of feeling.

However, it is noteworthy that despite a passing reference to the value of feelings of other sentient beings by John Stuart Mill in Chapter II of *Utilitarianism,* utilitarians limited the feeling that is valuable to the feeling humans have, a restriction most value theorists have tended to leave unquestioned. To be concerned about the feelings of other animals has appeared sentimental, and philosophers are even more eager than most men not to appear sentimental.

The most famous opposition to this restriction came from G. E. Moore. It is striking that in order to oppose it he thought he had to give up the locus of value in feeling. In place of *feeling,* Moore argues that *good,* a nonnatural property, is an objective ingredient in states of affairs. When we contemplate alternative states of affairs, he argues, we recognize one as better than the other, but our judgment is not based on the amount of pleasure present in these alternative states of affairs.

Moore proposes, in Section 50 of his *Principia Ethica,* that we consider two "worlds" in the following way.

> Let us imagine one world exceedingly beautiful. Imagine it as beautiful as you can; put into it whatever on this earth you most admire—mountains, rivers, the sea; trees, and sunsets, stars and moon. Imagine all these combined in the most exquisite proportions, so that no one thing jars against another, but each contributes to increase the beauty of the whole. And then imagine the ugliest world you can possibly conceive. Imagine it simply one heap of filth, containing everything that is most disgusting to us, for whatever reason, and the whole, as far as may be, without one redeeming feature.

In order to separate out the question of the value of these worlds in themselves from their value for a human observer, Moore asks that we suppose that no human being can ever *see* either world. Do we not, he then asks, still believe that it is better for the beautiful world to exist rather than the ugly one?

Moore, however, has not been very convincing. At least he has not convinced me. The beauty and ugliness of which he speaks are relational qualities, that is, they do not exist apart from the way certain formal patterns are apprehended by the human observer. Even though he asks us to imagine them as existing apart from observation or the possibility of observation, we are still visualizing them. But we are asked to suppose they have no effect, actual or potential, on any visualizing activity at all. In that case it simply has no meaning to say that one is beautiful and the other ugly.

My view of value can be sharply juxtaposed to Moore's by introducing into the ugly world something from the filth heap that has not been previously included: a number of worms and insects that men agree are utterly repulsive. Let us suppose that we had purposely excluded such life from the beautiful world. And let us assume further that the ugly world constitutes a suitable environment for our insects and worms, that these are able to secure adequate food and are free from excessive pain. In that case I would contend the ugly world had more value than the beautiful one. For in the ugly world there exist feelings of a level excluded from the beautiful world, namely, the feelings of the worms and insects. Since men are excluded in principle from both worlds, the fact that men would prefer to see and live in the beautiful world is irrelevant. If, on the other hand, we characteristically fill the beautiful world with birds and animals, and restrict residence in the ugly one to insects and worms, I would affirm much greater value to the beautiful world, not because it would appear beautiful to men, which of course it would, but because the feelings of birds and animals are more valuable than those of insects and worms. This I believe is more realistic because there is in fact considerable correlation between what we find beautiful and the sort of environment hospitable to higher forms of life.

Thus, whereas I reject Moore's position, I wish strongly to affirm with him, against the anthropocentric tendencies of most value theory, that values do exist apart from man's knowledge of them. If I contemplate two situations that I suppose no man to be cognizant of or affected by, one in which a dog is thoroughly enjoying life and the other in which it is suffering agony from a broken leg, I have no difficulty in judging between them. Enjoyment is an intrinsic good, and agony is an intrinsic evil, whether or not men know about it.

How far does intrinsic value extend beyond man? The question requires us to consider how far feeling extends beyond human feeling. To suppose that only men have feelings is surely arbitrary and contrary to the clear implications of our evolutionary connection with other forms of animal life. The reasonable issue is only how far to extend the *category of feeling* or *experience*. Do unicellular organisms feel? What about the individual cells in multicellular organisms? Can a sharp line be drawn between cells and subcellular entities? My own view is that no line at all can be drawn, that wherever one deals with actual unitary entities one is dealing with feelings. But of course much that is most important in human feeling depends on such a high evolutionary product as consciousness, and I doubt that there is consciousness where there is no central nervous system. Hence, the feeling we can attribute to lower forms of life, and a fortiori to so-called inanimate entities like molecules and submolecular forms, is very different from human feeling.

The major basis in the West of drawing sharp lines separating what is valuable in itself from what is not has been the doctrine of the psyche or soul. For Aristotle psyche was the principle of life and hence was attributed to plants and animals as well as to man. Even so, he made a distinction between the vegetable, animal, and rational souls, limiting the last to man. Christian theology stressed the uniqueness of the rational soul, associated it with the image of God, which in Genesis is attributed only to man, and viewed it as the object of divine redemption. From a religious and ethical point of view, that made absolute the gulf between human souls and souls at other levels. Indeed, the term *soul* came to refer self-evidently to human soul. The hierarchy of levels of soul gave way to the dualism of ensouled man and soul-less animals.

Evolutionary views of living things should have reintroduced a more hierarchical conception. The human psyche as we now know it must have developed from simpler forms of life similar to those now found among other animals. Because of the differences between men and other animals, and because of the now traditional restriction of usage of the word *soul,* we may continue to limit soul terminologically to human soul if we like. But we must avoid the too apparent implication of a radical and abrupt difference. Functions and subjective experiences analogous to human psychic life are attributable to other complex animals as well.

Utilitarianism affirmed that each human being should count for as much as every other. This is a laudably democratic principle, but it has highly questionable features even when applied to humans. It is not really evident that the advantage of a mongoloid idiot or human vegetable should count equally with that of a healthy child. Also there is no objective way of determining the point in development at which a fertilized ovum should count equally with an adult. And the principle cannot function at all when we recognize that there are subhuman intrinsic values as well. It would be quite arbitrary to count a dog's pleasure or pain as equally important with that of a human being. If we extended such a principle to microorganisms the absurdity would become still more apparent. Hence the extension of intrinsic value beyond man to the subhuman world forces consideration of criteria for appraising values.

Utilitarianism did, of course, have criteria. Pleasure is good and pain is bad. What it assumed was that the state of feeling of any person could be plotted somewhere along the continuum between optimal pleasure and maximal pain. It ignored the fact that among persons there may be significant differences in the degree of feeling present. For example, there may be a very intense experience in which the factors of extreme pleasure and extreme pain are so nearly balanced that the utilitarian calculus would yield a negligible value either way. There may be another experience at very low ebb of feeling in which such minimal feeling as is present is purely pleasurable. This would be assigned a top plus score on the utilitarian calculus. The distortion thus introduced would be immeasurably magnified when the subjects compared were a man and a unicellular organism.

For a satisfactory theory of intrinsic value we require initially a quite different measure than pain and pleasure. It must be a quantitative measure of the experience as such rather than of its pleasurableness. A man enjoys *more* experience than do paramecia. How can this "more" be interpreted? One measure is intensity of feeling. We can distinguish between experiences according to their intensity, and, other things being equal, we can meaningfully assert that more is happening in the more intense experience than in the less intense experience. But intensity may be gained at the expense of breadth and inclusiveness: how much is included in an experience and how different factors harmonize within it are also meaningful in comparing experiences. Of two experiences one may be more intense and the other more inclusive. Between these a judgment of which is "more" of an experience would be difficult to come by. However, we can judge that an experience in which inclusiveness *and* intensity are combined involves more feeling than would an experience consisting of either in isolation. Thus intensity and inclusiveness and their ideal combination can function as norms for ranking the experiences of men in relation to subhuman forms of life as well as in relation to each other.

There is a high degree of correlation between the amount of pleasure present in an experience and the amount of inclusiveness and intensity. The more broadly we conceive pleasure, the more closely we can correlate it with inclusiveness and intensity. Yet the correlation of pleasure with harmony and intensity of experience is not perfect; as for pain, the lack of correlation is much too great to be overlooked. Pain, therefore, must be introduced as a distinct factor in the value appraisal. Physical pain or spiritual anguish may be so intense that the annihilation of one experience and even of the possibility for future experiences may be preferable to the continuation of a relatively intense painful experience.

Suffice it to say that a fully developed value theory would have to relate the negative value of pain and suffering to the positive values of inclusiveness and intensity in some coherent way. For purposes of this paper the problem is only indicated.

Ethics Proper

With this sketchy indication of the locus of value and criteria of evaluation, we come to the question of ethics proper. Are we obligated to act so as to maximize value? If so, whose value? Are there other considerations that weigh upon ethical choices?

One major question is whether there is any meaning at all in sentences that state obligations, but let us simply assume here that statements of ethical obligation do have meaning. Our task is to decide which general statement, from among several alternatives, is correct. Consider the following:

1. So act as to maximize value for yourself in the present.
2. So act as to maximize value for yourself for the rest of your life.
3. So act as to maximize value for all men for the indefinite future.
4. So act as to maximize value in general.

Of these the first would hardly be viewed as an ethical principle. There are those who suppose that in fact this describes human behavior and that the ethical call to consider a wider sphere is useless. But this is an exceedingly doubtful judgment. Most would recognize that we act with some regard to some future consequences of our actions at least to ourselves.

The second principle is the maxim of prudence. This is recognizably ethical in character. Against the tendency to consider only short-run consequences it calls for full consideration of long-term consequences. Yet it, too, is highly questionable. First, it cannot be defended as describing actual human behavior. There is no clear evidence that men do consider the consequences of their actions for themselves in the distant future more seriously than they consider more immediate consequences to their friends and children. If they do not maximize value for themselves for the rest of their lives, on what grounds can we say they *should?* The argument is sometimes made that consideration of long-term consequences is the one *rational* basis for making decisions. But such a view entails many questionable assumptions. It assumes absolute self-identity through time and absolute separation between one self and all other selves. I believe this is psychologically and metaphysically false, and hence I cannot see how the ethics based upon it can be regarded as uniquely rational.

The third principle is the familiar utilitarian one. An ethical action is one that seeks the greatest good for the greatest number of men. This is profoundly plausible and attractive to all who have been shaped by the Judeo-Christian tradition. Indeed it is a restatement of the fundamental teaching that we should love our neighbors as ourselves.

The basic utilitarian assumption is that it is right to increase value, a principle I accept. That does not mean that people do in fact act in the way such a principle requires. But it means that the sense of rightness points toward this kind of action. I may disregard my neighbor's good and seek only my own, but insofar as I realize that my neighbor's good is in fact, objectively, just as important as my own, I recognize a disproportion between my action and what would be objectively appropriate. I see my action as irrational and hence morally wrong.

My present point, however, is not to defend this view of morality in detail but to point out its instability. If the reason I should seek the greatest good of the greatest number of persons is that it is right to increase value, then limiting this action to human value is arbitrary. It could be justified only if subhuman entities had no intrinsic value. Since the denial of intrinsic value to subhuman entities is false, excluding subhuman entities from the influence of the third principle is without justification. Therefore, only the fourth principle is sufficiently encompassing to be stable and acceptable.

The calculation of pleasure and pain and the multiplication by the number of persons affected, called for by utilitarianism, has never been predictable in detail. Although it has provided a rough and useful guide for making decisions, the extension of such calculations to the subhuman world would be impossibly complex. To be at all functional, we require an image of that state of affairs in which some optimum of value obtains.

We are helped toward an image for this state of affairs by the idea of the biotic pyramid, a concept that describes the movement of life from the soil and the microorganisms therein through vegetation, through the herbivores, and to the carnivores and primates. The total amount of value in a pyramid is roughly correlative with the richness of the base, the number of levels, the diversity of forms and total numbers at each level, and the complexity of living forms at the top. These measures correlate closely with each other.

The more valuable biotic pyramids would clearly place man at the top. Hence the biotic pyramid does not provide an antihuman view of value. Nevertheless, up to a considerable, not yet determined point, man, unlike any other species, can increase his numbers at the top of the pyramid by reducing the number of levels in the pyramid and the diversity of life at each level. For the rich biotic community of the American prairie, man substituted the wheat field and thereby fed a much larger human population. Hence there is a tension between the comprehensive biotic model that is inclusive of man and specialized biotic models in which man's needs alone are better satisfied.

We are warned today that the highly specialized biotic communities produced by man are more precarious than we had previously supposed and that for man's own survival it is important that he modify them. However, that is not now the issue before us. *Without* such specializations the total human population would have to be much smaller than it can be *with* such specializations. There is therefore also a tension between what is optimal for man alone and what is optimal when viewed in terms of the biotic pyramid. The problem we face is how to balance these optimalities. If we count only human values, the levels and diversity in the biotic pyramid will be relevant only insofar as they support economic and aesthetic values in human experience, that is, they will be relevant *only* instrumentally. Unless we deny altogether the hierarchical ranking of values, a ranking reflected in the pyramid itself, we cannot discount the great increase of value a larger human population gives so long as population size does not impair the quality of human life. But if we take seriously the fact that all forms of life have value, we cannot ignore the loss of value entailed by man's simplification of the pyramid.

This discussion points to the need for moderating, without renouncing, man's structuring of the world around his own needs. He must of course develop a more realistic view of his actual long-term needs and seek to practice the utilitarian ethic more wisely. When he does so, the value of the biotic pyramid will be more adequately conserved than it is now. But the force of the present argument is that ethical action will require still further moderation of man's actions so as to give greater scope to the biotic pyramid. He must learn to balance his values against the others rather than to judge the others as only instrumental to his. Practically, this points, for example, to the moral obligation to preserve wilderness even beyond the values accruing to man from it.

Thus far the ethical principles considered have been oriented entirely to consequences. There is another tradition in philosophical ethics that is sharply critical of exclusive attention to the consequences of actions, a tradition that interests itself instead in the intrinsic rightness of actions. The most famous spokesman of this tradition is Kant. Kant's position is extreme. He seems to say, incorrectly, I believe, that the advantages accruing from actions are irrelevant to ethical judgment. However, analysis of his thought indicates that there are two important considerations introduced by his approach that are neglected in the utilitarian approach followed above.

First, Kant points out that an ethical action must be in some way generalizable. It is not enough to calculate that the probable consequences of one's acting in a certain way will increase value. One must also ask what the consequences would be if people in general acted in that way. For example, I might calculate that there would be more increase in value by picking and taking home some wildflowers than by leaving them in the woods. But that would not make this action ethically right unless I could also decide that value would be increased in general by others who picked wildflowers and took them home under similar circumstances. What Garrett Hardin pointed out in "The Tragedy of the Commons" graphically illustrates this principle in relation to the ecological crisis. Since the reader can follow Hardin's logic in another part of this volume, I will not develop it here.

Second, Kant shows that the judgment of the rightness of an action must include consideration of its relationship to the past as well as to the future. If I have made a serious promise, I should not simply ignore that fact when I later face a decision impinging on it. Kant is so convinced of this that he makes extreme—and I think false—judgments about the absolute moral necessity of living up to promises. Most of us could agree that there are many circumstances under which I ought not fulfill a promise no matter how solemnly I made it. But that is not to say that the fact of my having made the promise should have no weight in deciding what to do. If that were the case, as Kant notes, the very notion of a promise would be destroyed.

I have selected the notion of promise as an example. Kant stresses truth telling. The very nature of society presupposes some kind of mutual commitment to truthfulness except when there are overriding reasons against it. Society presupposes other commitments as well. The acceptance of a job entails implicit as well as explicit commitments. The legal enforcement of contracts is intended to support their moral weight. Marriage is based on mutual vows. Bringing children into the world is understood to entail obligations to them.

My point is that in general what happened at some time before making a decision sets a context for reflection on what one should do in the decision. This is complicated by the fact that what others have done for us may be as important as what we have ourselves done. For me to ignore the great generosity previously extended me by another man when I make a present decision that affects him would be morally wrong. However, what one's debts *really* are cannot be determined by examining *one's feelings* of gratitude.

There is no simple objective way of determining what commitments one makes in his life, what debts one incurs, and how all of these are to be balanced against each other. Much of the anguish of the ethically sensitive person comes from his realization of the impossibility of living up to all his commitments and repaying all his debts. Further, giving too much consideration to the way in which present action should relate to these "givens" from the past can block fresh and creative action in the present that is oriented to the production of new and greater values in the future.

Theology

At this point I find myself forced across the threshold from ethics proper to theology. How can a man deal with the inescapable experience of guilt that is engendered by ethical sensitivity? Should he desensitize himself? Or can he persuade himself that he is in fact not guilty? Or can he satisfy himself by balancing the scales and then tipping them toward the side of virtue? Or must he constantly defend himself from the self-accusation of guilt that he projects into every criticism directed toward him by others, however gentle? Or can he experience both the reality and the forgiveness of guilt?

These are all important questions, but they are not the ones that can be appropriately treated here. For our present purposes other functions of theology are more pertinent.

First, since there are no objective bases for determining exactly what commitments one has made and how his indebtedness is to be evaluated, how one defines his basic perspective on life and his fundamental self-understanding become crucial. One man may see himself as self-made, owing little or nothing to society and family. Another man may see society chiefly as a corrupt institution and a corrupting force, and he may locate any power that works for good in the virtue of individual men. A third man may see all that is most valuable in his own life as given to him through society and he may deplore his own tendencies toward ingratitude and violation of the rules of society. Evaluation of these several judgments, so important for the functioning of the ethical principles we discussed earlier, belongs to theology.

The most important theological question for our discussion is whether men have any commitments or debts beyond the limits of human society. Now the theological diversity becomes still clearer. Certainly Kant had no perception of such a relation of ethics to the natural environment. And in this respect his position is typical of that of Western ethics in general. But there are exceptions.

It is possible to see one's life neither as self-made nor as the product of human society alone but as a gift of the total evolutionary process. If I view myself primarily in this way, then it is appropriate for my response to be one of gratitude. The fitting ethical action then is service of that to which I find myself so comprehensively indebted. To serve the evolutionary process can be understood to mean furthering its inclusive work. One would then strive in general to contribute to the progress or growth of life in all its diversity of forms, beginning with human life but by no means limiting oneself to it.

This religious objective is thus far stated very vaguely. Its clarification is a theological task. What is the "total evolutionary process"? And how should we understand it? It would, of course, be possible to understand evolution in such a way that commitment to it would have quite opposite effects from those I have listed. For example, if one's vision of evolution is dominated by the notion of "survival of the fittest" then he might rejoice in man's continuing success in stamping out all competitive forms of life. He might also encourage ruthless competition among human individuals and societies so as to accelerate the evolutionary process if evolution were understood in this way.

I suggest that far more basic to the evolutionary process than survival of the fittest is the urge for survival itself. In living things there is an urge for life, for continued life, for more and better life. Theories of evolution describe the results of this pervasive urge, which certainly produces competition as well as cooperation. But theories of evolution presuppose the urge, apart from which there would be nothing to compete for. But I think this urge itself, rather than the formulae that describe aspects of its consequences, is what one may reasonably feel indebted to for his existence.

If this urge works toward more and better life, one must have some criteria for understanding what that means. Here we can return generally to the theory of value with which we began. More and better life is that in which there is enhancement of feeling. Feeling is enhanced when it can be more inclusive and intense and when pain is not excessive. Consciousness marks and enriches higher levels of feeling. A man may reasonably understand the very rich potentiality of feeling that is his as the product of millions of years of evolutionary development in which the urge for more and better life has been at work. One's sense of what is ethically appropriate can be deeply affected by this vision.

Even here, however, the implications can be ambiguous. Since the values of human experience represent the consummate achievement of the evolutionary process, one might still deduce that evolution's lesser achievements are of trivial importance for the furtherance of the evolutionary process. One might suppose that it is appropriate for ethics to be instrumental in furthering the development, through the human species, of superhuman forms of life. The question here is how to appraise the rich variety of the products of the evolutionary process. Is there value in the variety as such, apart from the relatively minor value that most evolutionary products have for human contemplation, exploitation, and study. Does the evolutionary process in some way prize its own products?

The importance of this question can be seen if we consider again the kind of intrinsic value that variety can have. We have seen that inclusiveness enriches the value of experience. Reducing what is available for inclusion thus reduces the potential value of subsequent experience. Thus reduction of the number of species of living things on the earth would mean some reduction also of potential for future value. Yet the great variety of species now existing has but trivial relevance for most human experience, and if the value of variety is based on consideration of relevance alone, it would count but little.

When we say we feel variety has great value, we tend to think of this variety in terms of an inclusive perspective. We conceive of the biota of the planet earth not as life viewed by man but as life viewed by a larger, more inclusive perspective. When we do so, we attribute a value to the whole that is greatly enriched by all the complex contrasts and interrelations of the parts, man being one of those parts.

Is the perspective from which this rich value can be contemplated a real perspective? The *idea* of an inclusive perspective is a real idea, but if it is only an idea in the minds of men, the values a *real* perspective would generate do, in fact, not obtain, and the prizing of variety of life is then mostly sentimental.

If, however, reality is such that there is an inclusive perspective in addition to the limited ones that are human, the value of the variety of life is real. And our callous disregard of the values of the whole for the sake of values of parts is a violation and desecration that has great ethical importance. To believe this is to believe, implicitly if not explicitly, in God.

Theology has yet another importance to an ecological ethics, and in this connection theology and ecological ethics relate to each other much the way theology relates to purely humanistic ethics; but it is worthwhile to consider this additional relation in the context of ecological ethics: ethical theory focuses attention on clearly conscious decision making, telling us how we should balance the factors on which we reflect when making such decisions. Conscious decision making depends on calculations of probable consequences of alternative actions and on the relation of action to behavior shaped by past commitments and obligations.

I do not want to disparage ethical behavior. In comparison with the widespread tendency to thoughtless selfishness, mere conventionalism, and compulsiveness, ethical behavior is of utmost desirability. Yet ethical living or ethical behavior also has its problems and limitations. We can take time to note only two.

First, life is a constant series of subtle decisions, many of them unconscious, and it is therefore easy to exaggerate the importance of what are really rather rare instances calling for reflective decision making. To extend reflective decision making as a norm too far toward trivial decisions would be to make wholesome living impossible.

Second, ethical reflection necessarily operates with the knowledge that is presently available. The decision must be made in terms of the expectation of consequences established by this limited knowledge. This is inevitable. That part of this knowledge which can be easily and articulately expressed also takes precedence over less easily expressible sensitivities and implicit understandings. The subtle lure of as yet unimagined values has little opportunity to play its role. Ethical action is almost always conservative.

The problem can be seen in human history. It can be illustrated in the characteristic tensions between art and morality. Moral principles tend to formulate and enforce practices supportive of fully apprehended goods, i.e., those values recognized and established in the community. The artist is often exploring the fringes of his sensibility in ways that cannot but be destructive of the established order of values.

If we place these considerations in the still wider context of the whole evolutionary process they become still more important. The urge toward continued, increased, and enhanced life has pushed and pulled living things through hundreds of millions of years toward new and unforeseeable forms. Unforeseeable ends cannot come into the calculation of the utilitarian ethicist. Hence, to serve the evolutionary process cannot be simply identical with making ethical decisions calculated to further it. We need to work with the process rather than only to manipulate it.

What then is the alternative? The alternative, I think, would be sensitivity to the urge toward life as it operates both within oneself and in the entire world. The alternative would be attunement of the self to that creative process. And this can lead to a spontaneity that is informed by rational ethics but at the same time transcends rational ethics.

Can Biblical Faith Help?

But what is the relation of all this to the dominant religious traditions of the West—those rooted in the Bible? Lynn White, Jr., has argued that the roots of our ecological crisis lie in the attitudes fostered in western Europe by the reading of the Bible. His point is that Western culture, shaped by the Bible, pictured human beings as separated from all other creatures and dominating them. Westerners felt a divine imperative to master nature both technologically and scientifically. We see now that this mastery was also over our own bodies and by males over females. Instead of committing ourselves to God as the One who evokes and enhances life in us and in all things, we have understood faith in God as the expression of our transcendence of nature and of our rightful imposition of our own purposes upon it.

There is much truth in this picture of Western Christianity. It has widely fostered a hierarchical view in which God is above us and nature below, and it has tended, especially in its nineteenth- and twentieth-century forms, to associate itself with philosophies that are radically anthropocentric. Even though science as a whole has cut its ties to Christianity, the drive for intellectual mastery that it expresses is an inheritance from medieval Christianity. Even today typical Christian ways of thinking are as much a part of the ecological problem as a help toward its solution. It has not been a mistake to look away from modern Christian theology and toward American Indians and ancient Taoists for a vision of the interdependence of humanity and the remainder of nature.

Nevertheless, a fresh look at the Bible shows us that the opposition of nature and history, so prominent in recent Protestant theology, is not present in the Bible itself. The Bible is not anthropocentric but theocentric, and the God who is the center of all things is the creator of all nature, not only of human beings. God declares the plants and animals good quite independently of the creation of human beings. God knows and cares for plants and animals as well as for human beings. In short, God is the Creator of all life.

In the New Testament it is clear that God is not only the remote and transcendent power who externally imposes order upon things. God creates and redeems by entering in the world. This is the incarnational view that is the heart of Christianity. The life in the world expresses the divine presence. Note how this is formulated in the immensely influential prologue to John's gospel:

> When all things began the Word already was. The Word dwelt with God, and what God was, the Word was. The Word, then, was with God at the beginning, and through him all things came to be; no single thing was created without him. *All that came to be was alive with his life.* . . . So the Word became flesh. (New English Bible, John 1:1–4a, 14a; emphasis mine)

If we Christians take seriously the idea that the Word, which is Christ, and which fully became flesh in Jesus, is also found in the life of all living things, we will cease to be part of the problem and will be freed to participate in the healing of a suffering biosphere. Our service to God as Christ will be the commitment to God as the urge and call to life and its enhancement and not an orientation away from the natural world toward a purely human and transcendent sphere. We will love and serve the God who cares about the grasses of the field and the death of a sparrow and not one who treats nature as a mere stage on which the human drama is enacted.

If philosophy, science, and biblical faith are defined in terms of their dominant expressions in the recent past, all are enemies of the future. They have shared in blinding us to our continuity with the natural world and our obligations to it. They have set us on a course that leads to catastrophe for human beings as well as other creatures.

But philosophy and science are not committed to repeat their major past expressions. To do so is not philosophical or scientific. The commitment to truth demands revision in the light of evidence which, while far from new, has new power of conviction.

Likewise, biblical faith is not committed to repeat the forms it has taken in recent centuries. There is nothing in the Bible itself that requires the dualism, the anthropocentrism, and the human arrogance that have blinded us to the inherent value of other creatures. Commitment to Christ as the Way, Truth, and Life (John 14:6) requires instead that we follow the way of openness to truth in the service of life. The cross is not a club with which to destroy other living things but a symbol of willingness even to suffer that others may live. We know now that the others are not only human beings but, as the Hindus and Buddhists say, all sentient beings or, as the Old Testament says, all flesh. A penitent Christianity transformed by an authentic recovery of its own normative sources can contribute depth and vision to the profound reorientation needed in our public life.

12 The Abolition of Man

C. S. Lewis

It came burning hot into my mind, whatever he said and however he flattered, when he got me to his house, he would sell me for a slave.

John Bunyan

"Man's conquest of Nature" is an expression often used to describe the progress of applied science. "Man has Nature whacked" said someone to a friend of mine not long ago. In their context the words had a certain tragic beauty, for the speaker was dying of tuberculosis. "No matter," he said, "I know I'm one of the casualties. Of course there are casualties on the winning as well as on the losing side. But that doesn't alter the fact that it is winning." I have chosen this story as my point of departure in order to make it clear that I do not wish to disparage all that is really beneficial in the process described as "Man's conquest," much less all the real devotion and self-sacrifice that has gone to make it possible. But having done so I must proceed to analyse this conception a little more closely. In what sense is Man the possessor of increasing power over Nature?

Let us consider three typical examples: the aeroplane, the wireless, and the contraceptive. In a civilized community, in peacetime, anyone who can pay for them may use these things. But it cannot strictly be said that when he does so he is exercising his own proper or individual power over Nature. If I pay you to carry me, I am not therefore myself a strong man. Any or all of the three things I have mentioned can be withheld from some men by other men—by those who sell, or those who allow the sale, or those who own the sources of production, or those who make the goods. What we call Man's power is, in reality, a power possessed by some men which they may, or may not, allow other men to profit by. Again, as regards the powers manifested in the aeroplane or the wireless, Man is as much the patient or subject as the possessor, since he is the target both for bombs and for propaganda. And as regards contraceptives, there is a paradoxical, negative sense in which all possible future generations are the patients or subjects of a power wielded by those already alive. By contraception simply, they are denied existence; by contraception used as a means of selective breeding, they are, without their concurring voice, made to be what one generation, for its own reasons, may choose to prefer. From this point of view, what we call Man's power over Nature turns out to be a power exercised by some men over other men with Nature as its instrument.

It is, of course, a commonplace to complain that men have hitherto used badly, and against their fellows, the powers that science has given them. But that is not the point I am trying to make. I am not speaking of particular corruption and abuses which an increase of moral virtue would cure. I am considering what the thing called, "Man's power over Nature" must always and essentially be. No doubt, the picture could be modified by public ownership of raw materials and factories and public control of scientific research. But unless we have a world state this will still mean the power of one nation over others. And even within the world state or the nation it will mean (in principle) the power of majorities over minorities, and (in the concrete) of a government over the people. And all long-term exercises of power, especially in breeding, must mean the power of earlier generations over later ones.

The latter point is not always sufficiently emphasized, because those who write on social matters have not yet learned to imitate the physicists by always including Time among the dimensions. In order to understand fully what Man's power over Nature, and therefore the power of some men over other men, really means, we must picture the race extended in time from the date of its emergence to that of its extinction. Each generation exercises power over its successors: and each, insofar as it modifies the environment bequeathed to it and rebels against tradition, resists and limits the power of its predecessors. This modifies the picture which is sometimes painted of a progressive emancipation from tradition and a progressive control of natural processes resulting in a continual increase of human power. In reality, of course, if any one age really attains, by eugenics and scientific education, the power to make its descendants what it pleases, all men who live after it are the patients of that power. They are weaker, not stronger: for though we may have put wonderful machines in their hands we have preordained how they are to use them. And if, as is almost certain, the age which had thus attained maximum power over posterity were also the age most emancipated from tradition, it would be engaged in reducing the power of its predecessors almost as drastically as that of its successors. And we must also remember that, quite apart from this, the later a generation comes—the nearer it lives to that date at which the species becomes extinct—the less power it will have in the forward direction, because its subjects will be so few. There is therefore no question of a power vested in the race as a whole steadily growing as long as the race survives. The last men, far from being the heirs of power, will be of all men most subject to the dead hand of the great planners and conditioners and will themselves exercise least power upon the future. The real picture is that of one dominant age—let us suppose the hundredth century A.D.—which resists all previous ages most successfully and dominates all subsequent ages most irresistibly, and thus is the real master of the human species. But even within this master generation (itself an infinitesimal minority of the species) the power will be exercised by a minority smaller still. Man's conquest of Nature, if the dreams of some scientific planners are realized, means the rule of a few hundreds of men over billions upon billions of men. There neither is nor can be any simple increase of power on Man's side. Each new power won by man is a power *over* man as well. Each advance leaves him weaker as well as stronger. In every victory, besides being the general who triumphs, he is also the prisoner who follows the triumphal car.

I am not yet considering whether the total result of such ambivalent victories is a good thing or a bad. I am only making clear what Man's conquest of Nature really means and especially that final stage in the conquest, which, perhaps, is not far off. The final stage is come when Man by eugenics, by prenatal conditioning, and by an education and propaganda based on a perfect applied psychology, has obtained full control over himself. *Human* nature will be the last part of Nature to surrender to Man. The battle will then be won. We shall have "taken the thread of life out of the hand of Clotho" and be henceforth free to make our species whatever we wish it to be. The battle will indeed be won. But who, precisely, will have won it?

For the power of Man to make himself what he pleases means, as we have seen, the power of some men to make other men what *they* please. In all ages, no doubt, nurture and instruction have, in some sense, attempted to exercise this power. But the situation to which we must look forward will be novel in two respects. In the first place, the power will be enormously increased. Hitherto the plans of educationalists have achieved very little of what they attempted and indeed, when we read them—how Plato would have every infant "a bastard nursed in a bureau," and Elyot would have the boy see no men before the age of seven and, after that, no women,[1] and how Locke wants children to have leaky shoes and no turn for poetry[2]—we may well thank the beneficent obstinacy of real mothers, real nurses, and (above all) real children for preserving the human race in such sanity as it still possesses. But the man-moulders of the new age will be armed with the powers of an omnicompetent state and an irresistible scientific technique: we shall get at last a race of conditioners who really can cut out all posterity in what shape they please. The second difference is even more important. In older systems both the kind of man the teachers wished to produce and their motives for producing him were prescribed by the *Tao*—a norm to which the teachers themselves were subject and from which they claimed no liberty to depart.[3] They did not cut men to some pattern they had chosen. They handed on what they had received: they initiated the young neophyte into the mystery of humanity which overarched him and them alike. It was but old birds teaching young birds to fly. This will be changed. Values are now mere natural phenomena. Judgments of value are to be produced in the pupil as part of the conditioning. Whatever *Tao* there is will be the product, not the motive, of education. The conditioners have been emancipated from all that. It is one more part of Nature which they have conquered. The ultimate springs of human action are no longer, for them, something given. They have surrendered—like electricity: it is the function of the Conditioners to control, not to obey them. They know how to *produce* conscience and decide what kind of conscience they will produce. They themselves are outside, above. For we are assuming the last stage of Man's struggle with Nature. The final victory has been won. Human nature has been conquered—and, of course, has conquered, in whatever sense those words may now bear.

The Conditioners, then, are to choose what kind of artificial *Tao* they will, for their own good reasons, produce in the Human race. They are the motivators, the creators of motives. But how are they going to be motivated themselves? For a time, perhaps, by survivals, within their own minds, of the old "natural" *Tao*. Thus at first they may look upon themselves as servants and guardians of humanity and conceive that they have a "duty" to do it "good." But it is only by confusion that they can remain in this state. They recognize the concept of duty as the result of certain processes which they can now control. Their victory has consisted precisely in emerging from the state in which they were acted upon by those processes to the state in which they now use them as tools. One of the things they now have to decide is whether they will, or will not, so condition the rest of us that we can go on having the old idea of duty and the old reactions to it. Now can duty help them to decide that? Duty itself is up for trial: it cannot be also the judge. And "good" fares no better. They know quite well how to produce a dozen different conceptions of good in us. The question is which, if any, they should produce. No conception of good can help them to decide. It is absurd to fix on one of the things they are comparing and make it the standard of comparison.

To some it will appear that I am inventing a factitious difficulty for my Conditioners. Other, more simpleminded, critics may ask "Why should you suppose they will be such bad men?" But I am not supposing them to be bad men. They are, rather, not men (in the old sense) at all. They are, if you like, men who have sacrificed their own share in traditional humanity in order to devote themselves to the task of deciding what "Humanity" shall henceforth mean. "Good" and "bad," applied to them, are words without content: for it is from them that the content of these words is henceforward to be derived. Nor is their difficulty factitious. We might suppose that it was possible to say "After all, most of us want more or less the same things—food and drink and sexual intercourse, amusement, art, science, and the longest possible life for individuals and for the species. Let them simply say, This is what we happen to like, and go on to condition men in the way most likely to produce it. Where's the trouble?" But this will not answer. In the first place, it is false that we all really like the same things. But even if we did, what motive is to impel the Conditioners to scorn delights and live laborious days in order that we, and posterity, may have what we like? Their duty? But that is only the *Tao,* which they may decide to impose on us, but which cannot be valid for them. If they accept it, then they are no longer the makers of conscience but still its subjects, and their final conquest over Nature has not really happened. The preservation of the species? But why should the species be preserved? One of the questions before them is whether this feeling for posterity (they know well how it is produced) shall be continued or not. However far they go back, or down, they can find no ground to stand on. Every motive they try to act on becomes at once a *petitio.* It is not that they are bad men. They are not men at all. Stepping outside the *Tao,* they have stepped into the void. Nor are their subjects necessarily unhappy men. They are not men at all: they are artifacts. Man's final conquest has proved to be the abolition of Man.

Yet the Conditioners will act. When I said just now that all motives fail them, I should have said all motives except one. All motives that claim any validity other than that of their felt emotional weight at a given moment have failed them. Everything except the *sic volo, sic jubeo* has been explained away. But what never claimed objectivity cannot be destroyed by subjectivism. The impulse to scratch when I itch or to pull to pieces when I am inquisitive is immune from the solvent which is fatal to my justice, or honour, or care for posterity. When all that says "it is good" has been debunked, what says "I want" remains. It cannot be exploded or "seen through" because it never had any pretensions. The Conditioners, therefore, must come to be motivated simply by their own pleasure. I am not here speaking of the corrupting influence of power nor expressing the fear that under it our Conditioners will degenerate. The very words *corrupt* and *degenerate* imply a doctrine of value and are therefore meaningless in this context. My point is that those who stand outside all judgments of value cannot have any ground for preferring one of their own impulses to another except the emotional strength of that impulse. We may legitimately hope that among the impulses which arise in minds thus emptied of all "rational" or "spiritual" motives, some will be benevolent. I am very doubtful myself whether the benevolent impulses, stripped of that preference and encouragement which the *Tao* teaches us to give them and left to their merely natural strength and frequency as psychological events, will have much influence. I am very doubtful whether history shows us one example of a man who, having stepped outside traditional morality and attained power, has used that power benevolently. I am inclined to think that the Conditioners will hate the conditioned. Though regarding as an illusion the artificial conscience which they produce in us their subjects, they will yet perceive that it creates in us an illusion of meaning for our lives which compares favourably with the futility of their own: and they will envy us as eunuchs envy men. But I do not insist on this, for it is mere conjecture. What is not conjecture is that our hope even of a "conditioned" happiness rests on what is ordinarily called "chance"—the chance that benevolent impulses may on the whole predominate in our Conditioners. For without the judgment "Benevolence is good"—that is, without reentering the *Tao*—they can have no ground for promoting or stabilizing their benevolent impulses rather than any others. By the logic of their position they must just take their impulses as they come, from chance. And Chance here means Nature. It is from heredity, digestion, the weather, and the association of ideas, that the motives of the Conditioners will spring. Their extreme rationalism, by "seeing through" all "rational" motives, leaves them creatures of wholly irrational behaviour. If you will not obey the *Tao,* or else commit suicide, obedience to impulse (and therefore, in the long run, to mere "nature") is the only course left open.

At the moment, then, of Man's victory over Nature, we find the whole human race subjected to some individual men, and those individuals subjected to that in themselves which is purely "natural"—to their irrational impulses. Nature, untrammelled by values, rules the Conditioners and, through them, all humanity. Man's conquest of Nature turns out, in the moment of its consummation, to be Nature's conquest of Man. Every victory we seemed to win has led us, step by step, to this conclusion. All Nature's apparent reverses have been but tactical withdrawals. We thought we were beating her back when she was luring us on. What looked to us like hands held up in surrender was really the opening of arms to enfold us forever. If the fully planned and conditioned world (with its *Tao* a mere product of the planning) comes into existence, Nature will be troubled no more by the restive species that rose in revolt against her so many millions of years ago, will be vexed no longer by its chatter of truth and mercy and beauty and happiness. *Ferum victorem cepit:* and if the eugenics are efficient enough there will be no second revolt, but all snug beneath the Conditioners, and the Conditioners beneath her, till the moon falls or the sun grows cold.

My point may be clearer to some if it is put in a different form. Nature is a word of varying meanings, which can best be understood if we consider its various opposites. The Natural is the opposite of the Artificial, the Civil, the Human, the Spiritual, and the Supernatural. The Artificial does not now concern us. If we take the rest of the list of opposites, however, I think we can get a rough idea of what men have meant by Nature and what it is they oppose to her. Nature seems to be the spatial and temporal, as distinct from what is less fully so or not so at all. She seems to be the world of quantity, as against the world of quality: of objects as against consciousness: of the bound, as against the wholly or partially autonomous: of that which knows no values as against that which both has and perceives value: of efficient causes (or, in some modern systems, of no causality at all) as against final causes. Now I take it that when we understand a thing analytically and then dominate and use it for our own convenience we reduce it to the level of "Nature" in the sense that we suspend our judgments of value about it, ignore its final cause (if any), and treat it in terms of quantity. This repression of elements in what would otherwise be our total reaction to it is sometimes very noticeable and even painful: something has to be overcome before we can cut up a dead man or a live animal in a dissecting room. These objects resist the movement of the mind whereby we thrust them into the world of mere Nature. But in other instances too, a similar price is exacted for our analytical knowledge and manipulative power, even if we have ceased to count it. We do not look at trees either as Dryads or as beautiful objects while we cut them into beams: the first man who did so may have felt the price keenly, and the bleeding trees in Virgil and Spenser may be far-off echoes of that primeval sense of impiety. The stars lost their divinity as astronomy developed, and the Dying God has no place in chemical agriculture. To many, no doubt, this process is simply the gradual discovery that the real world is different from what we expected and the old opposition to Galileo or to "body-snatchers" is simply obscurantism. But that is not the whole story. It is not the greatest of modern scientists who feel most sure that the object, stripped of its qualitative properties and reduced to mere quantity, is wholly real. Little scientists, and little unscientific followers of science, may think so. The great minds know very well that the object, so treated, is an artificial abstraction, that something of its reality has been lost.

From this point of view the conquest of Nature appears in a new light. We reduce things to mere Nature in order *that we* may "conquer" them. We are always conquering Nature, because "Nature" is the name for what we have, to some extent, conquered. The price of conquest is to treat a thing as mere Nature. Every conquest over Nature increases her domain. The stars do not become Nature till we can weigh and measure them: the soul does not become Nature till we can psychoanalyze her. The wresting of powers from Nature is also the surrendering of things *to* Nature. As long as this process stops short of the final stage we may well hold that the gain outweighs the loss. But as soon as we take the final step of reducing our own species to the level of mere Nature, the whole process is stultified, for this time the being who stood to gain and the being who has been sacrificed are one and the same. This is one of the many instances where to carry a principle to what seems its logical conclusion produces absurdity. It is like the famous Irishman who found that a certain kind of stove reduced his fuel bill by half and thence concluded that two stoves of the same kind would enable him to warm his house with no fuel at all. It is the magician's bargain: give up our soul, get power in return. But once our souls, that is, ourselves, have been given up, the power thus conferred will not belong to us. We shall in fact be the slaves and puppets of that to which we have given our souls. It is in Man's power to treat himself as a mere "natural object" and his own judgments of value as raw material for scientific manipulation to alter at will. The objection to his doing so does not lie in the fact that his point of view (like one's first day in a dissecting room) is painful and shocking till we grow used to it. The pain and the shock are at most a warning and a symptom. The real objection is that if man chooses to treat himself as raw material, raw material he will be; not raw material to be manipulated, as he fondly imagined, by himself, but by mere appetite, that is, mere Nature, in the person of his dehumanized Conditioners.

We have been trying, like Lear, to have it both ways: to lay down our human prerogative and yet at the same time to retain it. It is impossible. Either we are rational spirit obliged forever to obey the absolute values of the *Tao*, or else we are mere nature to be kneaded and cut into new shapes for the pleasures of masters who must, by hypothesis, have no motive but their own "natural" impulses. Only the *Tao* provides a common human law of action which can overarch rulers and ruled alike. A dogmatic belief in objective value is necessary to the very idea of a rule which is not tyranny or an obedience which is not slavery.

I am not here thinking solely, perhaps not even chiefly, of those who are our public enemies at the moment. The process which, if not checked, will abolish Man, goes on apace among Communists and Democrats no less than among Fascists. The methods may (at first) differ in brutality. But many a mild-eyed scientist in pince-nez, many a popular dramatist, many an amateur philosopher in our midst, means in the long run just the same as the Nazi rulers of Germany. Traditional values are to be "debunked" and mankind to be cut out into some fresh shape at the will (which must, by hypothesis, be an arbitrary will) of some few lucky people in one lucky generation which has learned how to do it. The belief that we can invent "ideologies" at pleasure, and the consequent treatment of mankind as mere υλη, specimens, preparations, begins to affect our very language. Once we killed bad men: now we liquidate unsocial elements. Virtue has become *integration* and diligence *dynamism*, and boys likely to be worthy of a commission "potential officer material." Most wonderful of all, the virtues of thrift and temperance, and even of ordinary intelligence, are *sales resistance.*

The true significance of what is going on has been concealed by the use of the abstraction Man. Not that the word Man is necessarily a pure abstraction. In the *Tao* itself, as long as we remain within it, we find the concrete reality in which to participate is to be truly human: the real common will and common reason of humanity, alive, and growing like a tree, and branching out, as the situation varies, into ever new beauties and dignities of application. While we speak from within the *Tao* we can speak of Man having power over himself in a sense truly analogous to an individual's self-control. But the moment we step outside and regard the *Tao* as a mere subjective product, this possibility has disappeared. What is now common to all men is a mere abstract universal, an H.C.F., and Man's conquest of himself means simply the rule of the Conditioners over the conditioned human material, the world of post-humanity which, some knowingly and some unknowingly, nearly all men in all nations are at present labouring to produce.

Nothing I can say will prevent some people from describing this lecture as an attack on science. I deny the charge, of course: and real Natural Philosophers (there are some now alive) will perceive that in defending value I defend *inter alia* the value of knowledge, which must die like every other when its roots in the *Tao* are cut. But I can go further than that, I even suggest that from Science herself the cure might come. I have described as a "magician's bargain" that process whereby man surrenders object after object, and finally himself, to Nature in return for power. And I meant what I said. The fact that the scientist has succeeded where the magician failed has put such a wide contrast between them in popular thought that the real story of the birth of Science is misunderstood. You will even find people who write about the sixteenth century as if Magic were a medieval survival and Science the new thing that came to sweep it away. Those who have studied the period know better. There was very little magic in the Middle Ages: the sixteenth and seventeenth centuries are the high noon of magic. The serious magical endeavour and the serious scientific endeavour are twins: one was sickly and died, the other strong and throve. But they were twins. They were born of the same impulse. I allow that some (certainly not all) of the early scientists were actuated by a pure love of knowledge. But if we consider the temper of that age as a whole we can discern the impulse of which I speak. There is something which unites magic and applied science while separating both from the "wisdom" of earlier ages. For the wise men of old the cardinal problem had been how to conform the soul to reality, and the solution had been knowledge, self-discipline, and virtue. For magic and applied science alike the problem is how to subdue reality to the wishes of men: the solution is a technique; and both, in the practice of this technique, are ready to do things hitherto regarded as disgusting and impious—such as digging up and mutilating the dead. If we compare the chief trumpeter of the new era (Bacon) with Marlowe's Faustus, the similarity is striking. You will read in some critics that Faustus has a thirst for knowledge. In reality, he hardly mentions it. It is not truth he wants from his devils, but gold and guns and girls. "All things that move between the quiet poles shall be at his command" and "a sound magician is a mighty god."[4] In the same spirit Bacon condemns those who value knowledge as an end in itself: this, for him, is to use as a mistress for pleasure what ought to be a spouse for fruit.[5] The true object is to extend Man's power to the performance of all things possible. He rejects magic because it does not work,[6] but his goal is that of the magician. In Paracelsus the characters of magician and scientist are combined. No doubt those who really founded modern science were usually those whose love of truth exceeded their love of power; in every mixed movement the efficacy comes from the good elements not from the bad. But the presence of the bad elements is not irrelevant to the direction the efficacy takes. It might be going too far to say that the modern scientific movement was tainted from its birth: but I think it would be true to say that it was born in an unhealthy neighborhood and at an inauspicious hour. Its triumphs may have been too rapid and purchased at too high a price; reconsideration, and something like repentance, may be required.

Is it, then, possible to imagine a new Natural Philosophy, continually conscious that the "natural object" produced by analysis and abstraction is not reality but only a view, and always correcting the abstraction? I hardly know what I am asking for. I heard rumours that Goethe's approach to nature deserves fuller consideration—that even Dr. Steiner may have seen something that orthodox researchers have missed. The regenerate science which I have in mind would not do even to minerals and vegetables what modern science threatens to do to man himself. When it explained it would not explain away. When it spoke of the parts it would remember the whole. While studying the *It* it would not lose what Martin Buber calls the *Thou*-situation. The analogy between the *Tao* of Man and the instincts of an animal species would mean for it new light cast on the unknown thing, Instinct, by the only known reality of conscience and not a reduction of conscience to the category of Instinct. Its followers would not be free with the words *only* and *merely*. In a word, it would conquer Nature without being at the same time conquered by her and buy knowledge at a lower cost than that of life.

Perhaps I am asking impossibilities. Perhaps, in the nature of things, analytical understanding must always be a basilisk which kills what it sees and only sees by killing. But if the scientists themselves cannot arrest this process before it reaches the common Reason and kills that too, then someone else must arrest it. What I most fear is the reply that I am "only one more" obscurantist, that this barrier, like all previous barriers set up against the advance of science, can be safely passed. Such a reply springs from the fatal serialism of the modern imagination—the image of infinite unilinear progression which so haunts our minds. Because we have to use numbers so much we tend to think of every process as if it must be like the numeral series, where every step, to all eternity, is the same kind of step as the one before. I implore you to remember the Irishman and his two stoves. There are progressions in which the last step is *sui generis*—incommensurable with the others—and in which to go the whole way is to undo all the labor of your previous journey. To reduce the *Tao* to a mere natural product is a step of that kind. Up to that point, the kind of explanation which explains things away may give us something, though at a heavy cost. But you cannot go on "explaining away" forever: you will find that you have explained explanation itself away. You cannot go on "seeing through" things forever. The whole point of seeing through something is to see something through it. It is good that the window should be transparent, because the street or garden beyond it is opaque. How if you saw through the garden too? It is no use trying to "see through" first principles. If you see through everything, then everything is transparent. But a wholly transparent world is an invisible world. To "see through" all things is the same as not to see.

Notes

1. Sir Thomas Elyot, *The Boke Named the Governour* (1531), I, iv: "All men except physitions only shulde be excluded and kepte out of the norisery." I, vi: "After that a childe is come to seuen yeres of age . . . the most sure counsaile is to withdrawe him from all company of women."

2. John Locke, *Some Thoughts concerning Education* (1693), § 7: "I will also advise his *Feet to be wash'd* every Day in cold Water, and to have his Shoes so thin that they might leak and *let in Water*, whenever he comes near it." § 174: "If he have a poetick vein, 'tis to me the strangest thing in the World that the Father should desire or suffer it to be cherished or improved. Methinks the Parents should labour to have it stifled and suppressed as much as may be." Yet Locke is one of our most sensible writers on education.

3. [Editor's note: *Tao* means "The Way," the path of virtuous conduct in Confucianism, the ultimate principle of the universe in Taoism. As close synonyms for his usage of *Tao*, Lewis elsewhere suggests "natural law or traditional morality or the first principles of practical reason."]

4. Christopher Marlowe, *Dr. Faustus* (1588), 77–90.

5. Francis Bacon, *Advancement of Learning* (1605), Bk. I (p. 60 in Ellis and Spedding, 1905; p. 35 in Everyman Ed.).

6. Francis Bacon, *Filum Labyrinthi* (1953), i.

III Economics: Interaction of Ends and Means

It is very arguable that the science of political economy as studied in its first period after the death of Adam Smith (1790), did more harm than good. . . . It riveted on men a certain set of abstractions which were disastrous in their influence on modern mentality. It dehumanized industry. . . . It fixes attention on a definite group of abstractions, neglects everything else, and elicits every scrap of information and theory which is relevant to what it has retained. This method is triumphant, provided that the abstractions are judicious. But, however triumphant, the triumph is within limits. The neglect of these limits leads to disastrous oversights.

Alfred North Whitehead, *Science and the Modern World*, 1925

Introduction

Herman E. Daly and
Kenneth N. Townsend

After the fashion of Ruskin and Hobson, whose work was analyzed by Gerald Alonzo Smith in part II, the first article in this part, "On Economics as a Life Science," resists the trends of a narrowly defined, growth-oriented political economy, albeit by focusing on the bottom half of the ends-means spectrum—on what A. J. Lotka called the "biophysical foundations of economics" or what Hobson called the "groundwork for conscious valuation." The close analogies between biology and economics are discussed both in their short-term steady-state aspects and in their long-term evolutionary aspects. Expanding the use of the input-output model, the essay considers the ecological impact of the human economy and the economical impact of the nonhuman ecology.

The following article, "Sustainable Growth: An Impossibility Theorem," considers further implications of the biophysical limitations to economic processes, this time addressing the panacea of "sustainable growth," which had been offered by the United Nations' Brundtland Commission as a proper goal for continued economic activity in a world of dwindling resources. The terms "sustainable development" and "sustainable growth" have caught on, lately, as one and the same enlightened response to the ecological consequences of exponential growth of populations and economies. The article suggests that, in the broadest sense, even steady-state economies are not sustainable over an indefinite future—given that the economy is an open subsystem of the earth's ecosystem, economic growth is not sustainable. By distinguishing between economic growth, which implies the physical accretion or assimilation of materials within the economy, and economic development, which deals with expanded and realized potentialities of human achievement without material growth, the possibility for continued economic development without growth is considered.

Economists have distinguished between market-oriented capitalist economies and socialist command economies on the basis of the differences in equality and efficiency that these systems have historically produced. Inasmuch as both systems depend upon growth as a measure of economic success, however, both are subject to the impossibility theorem. The comparatively recent free exchange of ideas and information between East and West has revealed acutely advanced states of ecological deterioration in socialist economies, brought on by the coupling of a traditional focus on economic growth with the economic inefficiencies of command economies. In effect, the economies of Eastern Europe, the former Soviet Union, and China have shown signs of imminent ecological collapse sooner than the growth-oriented economies of the West. "Steady-State Economics and the Command Economy" argues the need for the economies of the East to move to a steady state, cautioning against the temptation of complacency that might accompany the infusion of efficient-market principles that current economic trends are bringing to the region.

Kenneth Boulding's classic article on "The Economics of the Coming Spaceship Earth" develops the important vision of the economy as an open system that maintains some structure in the midst of an entropic flow or throughput. As Boulding writes, "The essential measure of the success of the economy is not production and consumption at all, but the nature, extent, quality, and complexity of the total capital stock"—in other words, of the structure maintained by the throughput. For this book Boulding has added "Spaceship Earth Revisited," some further thoughts on space colonization, a theme taken up again in the book's postscript.

Over time the focus of environmental economics has changed from preoccupation with the consequences of dwindling reserves of energy that are necessary for fueling economic activity, and that so concerned economists of the stature of Jevons, to an awareness of the "end of pipeline" pollution problems that are occasioned by growth in material and energy flows. The prospect of economically viable, commercial fusion energy bids fair to stretch the usefulness of terrestrial stocks of energy well into the future, all other things being equal. Growth, however, pushes the economy toward an inevitable pollution limit, at which point the environment ceases to be able to recover from the shock of economic activity. The article by

T. H. Tietenberg, "Using Economic Incentives to Maintain Our Environment," summarizes trends in economics in the development of economic incentives to limit pollution.

In "The Steady-State Economy: Toward a Political Economy of Biophysical Equilibrium and Moral Growth," an attempt is made to pull together the case for a steady-state economy and suggest some policies. Three institutional reforms are presented as necessary for a steady-state economy: (1) limits to inequality in the form of a minimum income combined with a maximum income and wealth; (2) a system of transferrable birth quotas for limiting population; (3) a system of depletion quotas, auctioned by the government, in order to limit total throughput according to ecological and ethical criteria while allowing market allocation of the limited total among competing firms and individuals. The last institution should be compared with severance taxes and energy tax plus rebate proposals. The depletion quota plan has the long-run advantage of controlling throughput more rigorously and of forcing the recognition of a biophysical budget constraint. The severance tax has the short-run advantage of requiring less administrative change and of being politically more acceptable. Perhaps the best strategy would be to go first for the severance tax, and then, on the basis of experience with it, debate the advisability of moving to a depletion quota system at a later time.

Further issues, replies to critics, and second thoughts are considered in the postscript.

13 On Economics as a Life Science

Herman E. Daly

There is no wealth but life.
John Ruskin

All flesh is grass.
Isaiah 40:6

I Introduction

The purpose of this essay is to bring together some of the more salient similarities between biology and economics and to argue that, far from being superficial, these analogies are profoundly rooted in the fact that the ultimate subject matter of biology and economics is *one, viz., the life process*. Most of biology concentrates on the "within skin" life process, the exception being ecology, which focuses on the "outside skin" life process (Bates, 1960, pp. 12–13). Economics is the part of ecology that studies the outside-skin life process insofar as it is dominated by *commodities* and their interrelations. In what follows, the traditional economic (outside-skin) and the traditional biological (within-skin) views of the total life process will be considered, both in their *steady-state* aspect and in their *evolutionary* aspect. Finally an approach to a more general "general equilibrium" model will be suggested by considering the human economy from an ecological perspective.

Reprinted by permission from *Journal of Political Economy* 76, no. 3 (May/June 1968), pp. 392–406.

II Biological Analogies in Economics

Analogy is so fundamental to our way of thinking that the ability to recognize analogies is generally considered one of the criteria of intelligence. While there is a vast difference between analogy on the one hand and logical proof and empirical verification on the other, it by no means follows that the former belongs only to poetry and not to science. Analogy is the essence of the inductive side of science. Furthermore, the dominant mode of thought in economics today is the "analytical simile" (Georgescu-Roegen, 1966, pp. 114–124), the mathematical or geometric model based on a Pythagorean analogy between fuzzy, dialectical reality and well-defined, analytic number. The fruitfulness of this analogy for all science is obvious—but it is an analogy nonetheless, with its roots in the same insight that inspired the mystical Pythagorean brotherhood. That economists have also found biological analogies useful is only slightly less obvious. The circular flow of blood and the circular flow of money, the many parallel phenomena of specialization, exchange, interdependence, homeostasis, and evolution are well known. In the opposite direction, economic analogies in biology are also common, as witnessed by Malthus's influence on Darwin and by the very etymology of the word "ecology." Finally, an ultimately central place for biological analogies in economics has been claimed by no less an authority than Alfred Marshall in his famous statement, "The Mecca of the economist lies in economic biology rather than in economic dynamics" (Marshall, 1920, preface, p. 14), and in his further statement that "in the later stages of economics, when we are approaching nearly to the conditions of life, biological analogies are to be preferred to mechanical" (Marshall, 1925, p. 317). Among current economic theorists it would appear that only the works of Kenneth Boulding (1950, 1958, 1966) and Nicholas Georgescu-Roegen (1966) (both freely drawn upon here) reveal a disposition to take Marshall seriously on this point.

Perhaps the intellectual genealogy of the ideas to be developed in this paper can be more specifically indicated by a pair of quotations from two seminal thinkers of the early part of this century—one a biologist (A. J. Lotka) and the other an economist (J. A. Hobson).

Lotka (1956) informs us that

underlying our economic manifestations are biological phenomena which we share in common with other species; and . . . the laying bare and clearly

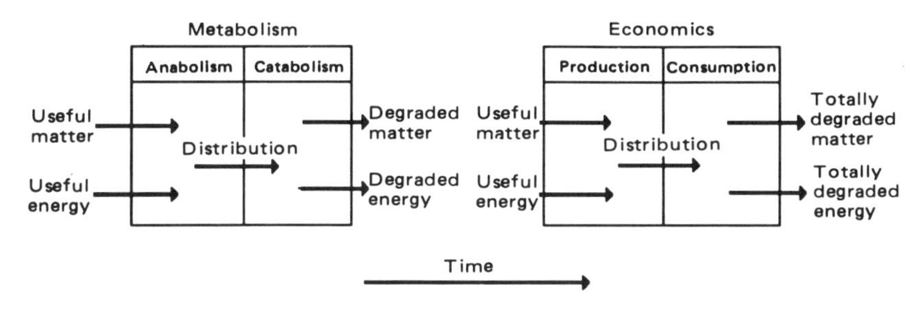

Figure 13.1
The metabolic analogy.

formulating of the relations thus involved—in other words the analysis of the biophysical foundations of economics—is one of the problems coming within the program of physical biology.

Just what these "biophysical foundations" are, and how they support the economic superstructure, is in large part the subject of section V.

From Hobson (1929) we learn that

> all serviceable organic activities consume tissue and expend energy, the biological costs of the services they render. Though this economy may not correspond in close quantitative fashion to a pleasure and pain economy or to any conscious valuation, it must be taken as the groundwork for that conscious valuation. For most economic purposes we are well-advised to prefer the organic test to any other test of welfare, bearing in mind that many organic costs do not register themselves easily or adequately in terms of conscious pain or disutility, while organic gains are not always interpretable in conscious enjoyment.

The "groundwork for conscious valuation" and the "organic test of welfare are ideas with close counterparts in section III, to which we now turn.

III The Steady-State Analogy

The close similarity of the basic within-skin life process of metabolism (anabolism and catabolism) with the outside-skin life process of economics (production and consumption) is evident from figure 13.1.

In either process the only *material* output is *waste*. The purpose (value produced) of the metabolic process is the maintenance of life. The purpose (value produced) of the economic process is the maintenance and enjoyment of life. An accounting balance equation of the life process in value terms would state that the value of life enjoyment plus the value of material waste (zero) equals the sum of the values of all the matter and energy upon which the total life process is based. The total value of life (our subjective estimate thereof) is *imputed* to the total quantity of things necessary for its enjoyable maintenance.[1] The Austrian economists have taught us that this imputation also determines the *relative* values (prices) of individual things according to the principle of diminishing marginal utility, which for Böhm-Bawerk was "the key-stone of all economic theory" (1891, p. 149). Since commodities are priced according to their diminishing marginal utilities, the sum of all goods in the economy valued at their marginal utilities (or prices) would be very small relative to the total utility of all goods (total life value), which is probably infinite.[2] The infinite difference between the finite sum of prices of all goods and the infinite sum of total utility of all goods is an infinite "global consumers' surplus." Hence, insofar as economics concentrates on value in exchange (marginal utility) to the exclusion of value in use (total utility)—to that extent it is concerning itself with only an infinitesimal portion of total life value. This is not meant to minimize the importance of exchange values, since it is precisely by considering margins that we maximize totals. The point is that, while margins are reliable means for *maximizing* totals, they are very treacherous means for *evaluating* totals, as any student who has pondered the diamonds-water paradox must realize. Any sort of economic numerology that, with one-eyed devotion to Pythagoras, insists on glossing over this treachery deserves a thorough dunking in the satirical acid of Jonathan Swift's *A Modest Proposal.*[3] Perhaps Hobson's "organic test of welfare" is simply the idea that it is better to make imprecise statements about unmeasurable but relevant magnitudes (use value, total utility) than to make more precise statements about the measurable but irrelevant magnitude (for evaluating total welfare) of exchange value. Economists shy away from thinking too much about total utility mainly because it is unmeasurable and dependent on value judgments—both embarrassing for a "positive science." But perhaps, as Joan Robinson suggests (1962, p. 54), this aversion to total utility also stems from its tendency to make one question "an economic system in which so much of the good juice of utility is allowed to evaporate out of commodities by distributing them unequally"; furthermore "this egalitarian element in the doctrine was sterilized mainly by slipping from utility to physical output as the object to be maximized." But as we have seen, the ultimate *physical* output of the economic process is waste, and there is no sense in maximizing that!

There is also a balance equation of the life process in physical units, based on the law of conservation of matter-energy. But more significant than the physical balance, from an economic viewpoint, is the one-way, noncircular, irreversible nature of the flow of matter-energy through all divisions of the life process. Since useful (low-entropy) matter-energy is apparently finite, the total life process could be brought to a halt by what Boulding has called "the entropy trap." Thus one of the ultimate natural sources of scarcity, and hence of economic activity, is the second law of thermodynamics (Georgescu-Roegen, 1966, pp. 66–82). Indeed, if one were perversely to insist on a real-cost theory of value, it would seem that entropy, rather than labor or energy, should be the source of value. Even in the subjective theory of value, however, entropy, the common denominator of all forms of scarcity, determines the locations of the margins and hence enters into the determination of marginal utilities and exchange values.

Erwin Schrödinger (1945) has described life as a system in steady-state thermodynamic disequilibrium that maintains its constant distance from equilibrium (death) by feeding on low entropy from its environment—that is, by exchanging high-entropy outputs for low-entropy inputs. *The same statement would hold verbatim as a physical description of the economic process.* A corollary of this statement is that an organism cannot live in a medium of its own waste products. With this principle in mind, one can better appreciate the significance of the following recent observation by J. J. Spengler (1966) in his presidential address to the American Economic Association: "Witness here in America the endless dumping of trash (four pounds per person per day). . . . Indeed, some hold, J. K. Galbraith had better labeled ours an effluent society than an affluent one." This four pounds per person per day does not disappear—it becomes a part of the physical environment in which we must live. Great stress has been put on the reciprocal nature of the relation of fitness between organism and environment by L. J. Henderson (1958). If the organism fits the environment, then it is also the case that the environment is fit for the organism. Henderson argues that there must have been some not-yet-understood process of physical evolution prior to the emergence of life in order for the environment to attain the rather exacting preconditions for supporting life. Thus man's newly acquired ability to degrade his material environment at the rate of four pounds per person per day is likely to be even more dangerous than commonly realized, in view of our ignorance of ecological relations.

How do the economic and metabolic processes fit together? Clearly metabolism is partly contained within the economic subprocess of consumption. Many of the material inputs into metabolism are economic products, and some outputs of metabolism are generally not totally degraded and thus can be further consumed—for example, manure fertilizer and carbon dioxide. But the ultimate physical output of the economic process is totally degraded matter-energy, in Marx's term, "devil's dust." Continuing in Chinese-box fashion, the total economic process is itself a subprocess on the consuming side of the total ecological life process, the producing side of the latter consisting mainly of photosynthesis carried on by green plants, which draw their inputs from the physical environment of air, soil, water, and sunlight.

Both the within-skin and outside-skin life processes have a permanently maintained physical basis that undergoes continual replacement over relatively short time periods (steady-state aspect) and that is capable of qualitative change and reorganization over long periods (evolutionary aspect). In other words "capital" represents "exosomatic organs" and biological organs represent "endosomatic capital." In each case, we observe both short-term depreciation and replacement and long-term technological change. Physical capital is essentially matter that is capable of trapping energy and channeling it to human purposes. Hence, in a very real sense the entire physical environment is capital, since it is only through the agency of air, soil, and water that plant life is able to capture the solar energy upon which the whole hierarchy of life (and value) depends. Should not these elements receive the same care we bestow upon our other machines? And is not any theory of value that leaves them out rather like a theory of icebergs that fails to consider the submerged 90 percent?

IV The Evolutionary Analogy

The material basis of the life process grows when the rate of production (anabolism) exceeds the rate of consumption (catabolism).

Growth merges into development as alteration in the rates of increase of different parts gives rise to new proportions, new qualitative relations, and new technologies. Although development is not well understood by either science, the subtle influence of size on organization has led both biologists and economists to the concept of a proper or optimum scale for a given organizational plan. That Marx, who emphasized this dialectic interplay of quantity and quality, also tended to view economics as a part of natural history is evident in the following quotation (1967, 1:372):

> Darwin has interested us in the history of Nature's Technology, that is, in the formation of the organs of plants and animals, which organs serve as instruments of production for sustaining life. Does not the history of the productive organs of man, of organs that are the material basis of all social organization, deserve equal attention?

The same idea has been expressed by Lotka (1956, p. 208), viz., "Man's industrial activities are merely a highly specialized and greatly developed form of the general biological struggle for existence," and further in a passage (p. 369) that would have pleased Marx:

> The most singular feature of the artificial extensions of our natural body is that they are shared in common by a number of individuals. When the sick man consults the physician, who, we will say, makes a microscopic examination, for example, the patient is virtually hiring a pair of high power eyes. When you drop a nickel into a telephone box, you are hiring the use of an ear to listen to your friend's voice five or ten miles distant. When the workingman accepts a wage of forty dollars for his weekly labor, he is in fact paying to his employers an undetermined amount for the privilege of using his machines as artificial members to manufacture marketable wares.
>
> The modern development of artificial aids to our organs and faculties has exerted two opposing influences.
>
> On the one hand, it has in a most real way bound men together into one body: so very real and material is the bond that society might aptly be described as one huge multiple Siamese twin.
>
> On the other hand, since the control over certain portions of this common body is unevenly distributed among the separate individuals, certain of them may be said in a measure to own parts of the bodies of others, holding them a species of refined slavery, and though neither of the two parties concerned may be clearly conscious of the fact, it is often resented in a more or less vague way by the one less favored.

In biological evolution genes transmit the "knowledge" of organic forms over time, and gene mutations introduce occasional modifications, resulting in the success of the forms best suited to the environment. In economic evolution, culture transmits knowledge over time, and new ideas produce mutant organizations from which competition again determines the fittest. Indeed, Teilhard de Chardin (1959) argues that "cultural evolution" is simply a new evolutionary mechanism that superseded the old mechanism in importance.

A natural history of economic evolution might be built around the theme of "economic surplus" and its progressive growth and cultivation. The original surplus was produced by plants, since they capture more solar energy than that necessary for their own maintenance. Animal life depends on this surplus, and perhaps man's greatest discovery was that he could cultivate and expand that upon which his existence depended, thus "exploiting niggardly nature."[4]

As soon as this primary activity became efficient enough to produce a surplus above the maintenance needs of those engaged in primary production, it became possible to evolve secondary economic activities, etc. Although economic activity moves far away from direct contact with nature, the "biophysical foundations of economics" remain ever present in the background, and it is to these foundations that we now direct our attention.

V The Human Economy in Ecological Perspective

Although the life process is essentially one, it seems that for many analytical purposes the most convenient boundary by which to divide the process is the natural boundary of skin. The outside-skin life process is the subject of ecology, but ecologists abstract from the human economy and study only natural interdependences, while economists abstract from nature and consider only interdependences among commodities and man. But what discipline systematically studies the interdependences that clearly exist between the natural and human parts of the outside-skin life process? Marston Bates, a biologist, addresses himself to this point in the following quotation (1960, p. 247):

> Then we come to man and his place in the system of life. We could have left man out, playing the ecological game of "let's pretend man doesn't exist." But this seems as unfair as the corresponding game of the economists, "let's pretend that nature doesn't exist." The economy of nature and ecology of man are inseparable and attempts to separate them are more than misleading, they are dangerous. Man's destiny is tied to nature's destiny and

the arrogance of the engineering mind does not change this. Man may be a very peculiar animal, but he is still a part of the system of nature.

Any attempt to isolate a segment of reality is always somewhat misleading, but not for that reason less necessary. Our purposes dictate the manner in which we abstract from reality, and as economists well know, many useful purposes can be served by partial analysis—that is, studying one industry in abstraction from its matrix of interconnections with the rest of the economy. While this is a useful procedure for studying the peanut industry, no economist would want to study the automobile industry under such limitations. Too many important feedbacks from the rest of the economy would be left out. Until recently the economy of man was "peanuts" in the total economy of nature. Now it is more like the automobile industry, and to continue *ceteris paribus* treatment of nature (even in general-equilibrium analysis) is indeed dangerous to our purpose if that purpose is to say something about how human wants can best be served.

A rather dramatic example of this kind of danger has been indicated by Dr. Edward Teller (1965), who pointed out that since the Industrial Revolution the tremendous consumption of carbon fuels has resulted in an increased concentration of carbon dioxide in the atmosphere. Since this gas increases the heat retention of the atmosphere, thus raising the average temperature, it may well be that the ultimate effect of the Industrial Revolution will be the melting of the polar ice cap and the inundation of large parts of the world. The more concrete case of the unintentional destruction wrought on the environment by chemical insecticides has been forcefully documented by Rachel Carson (1962). Also, we know that the entire chain of life depends heavily on bacteria—for example, nitrogen fixation and decomposition of dead organisms. Is it not possible that some export from the human economy (for example, detergents) could prove lethal to certain of these organisms? Conversely, might not some human exports be highly beneficial to the propagation of particular disease-causing bacteria? And one need only mention the problem of radioactive fallout. At a less dramatic but increasingly serious level, we have ubiquitous instances of air and water pollution plaguing the world's cities, not to mention the problems of deforestation, soil erosion, and noise.

Such phenomena have long been recognized (grudgingly) in economic theory under the heading of *externalities*—that is, interrelations

Table 13.1
An Expanded Input-Output Table

From	To	
	Human	Nonhuman
Human	(2)	(1)
Nonhuman	(3)	(4)

whose connecting links are external to the economists' abstract world of commodities but very much internal to the world in which we live, move, and have our being. Perhaps "nonmarket interdependence" is a more descriptive term.

It would be easy to liken this concept to a *deus ex machina* lowered into the scene by our theoretical playwrights to save an awkward plot, but it is by no means easy to suggest a better treatment. A better treatment is called for, however, since externalities are spending more time on center stage and less time in the wings than previously. Or, changing the metaphor, to continue theoretical development via continued ad hoc introduction of externalities is reminiscent of adding epicycles and in the long run will lead only to Ptolemaic complications in economic theory. Our economic cosmos is not one of uniform circular motion of commodities among men but one of elliptical orbits through interdependent ecological sectors.

How does one integrate the world of commodities into the larger economy of nature? Perhaps this is a problem in which economics can provide a useful analogy. Leontief's input-output model has proved useful in dealing with phenomena of interdependence, and it may offer the most promising analytical framework within which to consider the above question.[5] Just as the annual flow of gross national product, or final commodities, requires a supporting matrix of flows of *intermediate* commodities, so does the annual flow of all economic commodities (final and intermediate) require a supporting matrix of flows of physical things that carry no price tag but nonetheless are necessary complements to the flows of those things that do carry price tags.

In its simplest input-output representation the total economy can be divided into its human and nonhuman sectors, as in table 13.1.

Cell or quadrant (2) is the domain of traditional economics, that is, the study of inputs and outputs to and from various subsectors within the human-to-human box. Cell (4) represents the traditional area of concern of ecology, the inputs to and outputs from subsectors in the nonhuman-to-nonhuman box. Cells (1) and (3), respectively, contain the flows of inputs from human subsectors to nonhuman subsectors and from nonhuman subsectors to human subsectors. All of the items exchanged in (2) are *economic commodities*, by which we mean that they have positive prices. All items of exchange in cells (1), (3), and (4) may by contrast be labeled *ecological commodities*, which consist of free goods (zero price) and "bads" (negative price). The negative price on bads is not generally observed, since there usually exists the alternative of exporting the bad to the nonhuman economy, which cannot pay the negative price (that is, charge us a positive price for the service of taking the "bad" off our hands, as would be the case if it were transferred to another sector of the human economy). Ecological commodities that are bads are bad in relation to man, not necessarily to the nonhuman world. The difficulty, however, is that these more than gratuitous exports from the human economy in cell (1) are simultaneously inputs to the nonhuman economy and as such strongly influence the outputs from the nonhuman back to the human sector—that is, cell (1) is connected to cell (3) via cell (4), and cell (3) directly influences human welfare.[6] These relationships will perhaps be more evident in table 13.2, which is an expansion of table 13.1, with the four quadrants corresponding to the quadrants of table 13.1. Note that in both tables the basic vision is still a "world of commodities," although a bigger world that now includes both economic commodities (the q_{ij} in quadrant (2)) and ecological commodities (the q_{ij} in quadrants (1), (3), and (4)). The q_{ij} in quadrants (1), (3), and (4) are the "biophysical foundations of economics."

In table 13.2, quadrant (2) is the simplest form of the usual Leontief input-output table, with two transforming sectors (agriculture and industry) and one primary sector (households). Agriculture consists of living transformers of matter-energy, and industry consists of nonliving transformers of matter-energy. The nonhuman economy has likewise been divided into the "transforming sectors" of animal, plant, and bacteria (living sectors) and of atmosphere, hydrosphere, and lithosphere (nonliving sectors). In addition, in row 10 we have a primary-service sector providing the ultimate source of low-entropy matter-energy, the sun, and, in column (10), the great thermodynamic sink into which finally consumed high-entropy matter-energy

Table 13.2
An Expanded Input-Output Table—Detail

Output from	Input to: Agri-culture (1)	Industry (2)	Households (final con-sumption) (3)	Animal (4)	Plant (5)	Bacteria (6)	Atmo-sphere (7)	Hydro-sphere (8)	Litho-sphere (9)	Sink (final con-sumption) (10)	Total
		Quadrant (2)					Quadrant (1)				
1. Agriculture	...	q_{12}	...	...	...	...	q_{17}	...	...	...	Q_1
1. Industry	q_{21}	(q_{22})	q_{23}	...	...	...	q_{27}	...	...	...	Q_2
3. Households (primary services)	...	q_{32}	...	...	...	...	q_{37}	...	...	...	Q_3
		Quadrant (3)					Quadrant (4)				
4. Animal	...	...	...	...	...	...	q_{47}	...	...	...	
5. Plant	...	...	...	...	...	...	q_{57}	...	...	...	
6. Bacteria	...	...	...	...	...	...	q_{67}	...	...	...	
7. Atmo-sphere	q_{71}	q_{72}	q_{73}	q_{74}	q_{75}	q_{76}	(q_{77})	q_{78}	q_{79}	$q_{7,10}$	Q_7
8. Hydro-sphere	...	...	...	...	...	...	q_{87}	...	...	...	
9. Litho-sphere	...	...	...	...	...	...	q_{97}	...	...	...	
10. Sun (primary services)	...	...	...	...	...	...	$q_{10,7}$	...	...	...	

goes, forever degraded as devil's dust.[7] The annual flow of low entropy consists of direct solar energy currently received, plus a running down of the stock of low entropy that came from the sun in the distant past. The table records the passage of low-entropy matter-energy through its life-supporting input-output transformations into high-entropy waste. These transformations are not all known or understood, but certainly the scope they offer for nonmarket interdependence far exceeds the standard examples of externalities in the literature, "somewhat bucolic in nature, having to do with bees, orchards and woods" (Scitovsky, 1954).

Table 13.2 has thus far been considered only as a descriptive catalogue for economically filing vast amounts of information about the exchanges of economic and ecological commodities making up the total economy of life. Any realistic table would probably have to have at least one hundred sectors, and the resulting ten thousand cells would be pigeonholes for storing measured data about the ten thousand most important exchanges in the total economy of life. Would it be possible to convert the table from a descriptive and heuristic device to a statistical tool, a matrix of technical coefficients useful for planning and prediction—that is, could one do with the whole table what Leontief has done with quadrant (2)?

Each row of table 13.2 can be stated as a physical balance equation, thus:

$$\sum_{j=1}^{n} q_{ij} = Q_i;\ i = 1, \ldots, n$$

where i = row and j = column.

Technical coefficients could be defined as $a_{ij} = q_{ij}/Q_j$.

The a_{ij} in quadrant (2) are the usual technical coefficients of the Leontief system, and the a_{ij} in the remaining quadrants are natural technical coefficients. For example, if i is water and j is alfalfa, then a_{ij} would be nine hundred, since it takes nine hundred pounds of water to produce one pound of dried alfalfa (Storer, 1954, p. 96). Assuming all a_{ij} are known, and noting that $q_{ij} = a_{ij}Q_j$, we have the following n equations in n unknowns:[8]

$$\sum_{j=1}^{n} a_{ij}Q_j = Q_i;\ i = 1, \ldots, n.$$

These equations are formally identical to Leontief's quantity table, in which we can sum across rows but not down columns. The assumptions by which Leontief breathes usefulness into this formalism are discussed below and are shown to present no greater theoretical problems for the whole table 13.2 than for quadrant (2). To begin with, Leontief's basic assumption of constant (slowly changing) technology over time seems to be much closer to the facts for table 13.2, since in the nonhuman economy technical change (evolution) is much slower than in the human economy. Linearity or constant-costs assumptions (a_{ij} constant with respect to Q_j) would seem to be at least equally appropriate as a first approximation. Perhaps this assumption, too, is closer to reality for table 13.2, since biological populations grow by adding identical units—hence input-output relations of biological populations are more likely to be proportional to scale (linear) than are such relations for populations of firms (that is, industries) in which new members are never such close replicas of old members. The assumption of single production processes with no joint products appears, at first sight, to be less true for nature than for the human economy. However, this is not at all clear, especially if we include bads and free goods as outputs in our traditional production functions. In general, aggregation and classification criteria used in input-output models (similarity of input structure and fixity of proportions among outputs) would remain applicable in the larger table. Certainly no single classification would give a complete representation of the exquisitely tangled web of physical life relations—but then the usual input-output model is also a very incomplete picture of economic relations. Different classifications can be used to serve different limited purposes.

Although there appear to be no theoretical problems in extending the input-output model in this way, there is the obvious practical difficulty that most of the q_{jj} and a_{ij} in quadrants (1), (3), and (4) have never been measured. Nevertheless they all seem to be measurable or at least subject to indirect calculation. Probably the major reason this information has not been acquired is that we have not had many theoretical pigeonholes into which it would fit. Also, the model does not really require a Laplacian knowledge of the universe, as it may appear to from the presentation. Application can be confined to a given spatial or conceptual region, with an export row and an import column summarizing relations with the "rest of the world." In any case, application appears rather less utopian than "cost-benefit analysis," which on the slender reed of exchange value calculations attempts to "maximize the present value of all benefits less all costs, subject to specified restraints" (Prest and Turvey, 1965, p. 4). In fact, something like table 13.2 would be necessary for identifying "all" costs and benefits in the organic sense of Hobson. The construction of such a table would require the cooperation of many disciplines—which may be a point in its favor.[9]

In conclusion, to summarize and support the point of view taken here, I can do no better than to remind the reader of the introductory aphorisms from Ruskin and Isaiah and to quote Lotka (1956, p. 183) one last time:

For the drama of life is like a puppet show in which stage, scenery, actors and all are made of the same stuff. The players indeed, "have their exits and their entrances," but the exit is by way of translation into the substance of the stage; and each entrance is a transformation scene. So stage and players are bound together in the close partnership of an intimate comedy; and if we would catch the spirit of the piece our attention must not all be absorbed in the characters alone, but must be extended also to the scene, of which they are born, on which they play their part, and with which, in a little while, they merge again.

Notes

1. Value is not permanently imputed to the (nonmaterial) technology within which matter and energy are used, unless that technology is made artificially scarce by patents. Following Schumpeter we can say that a new technology, while it is temporarily scarce by virtue of its novelty, will earn a temporary profit but will not receive a permanent imputed share of total value produced.

2. To say that "total life value" is infinite is not to say that it is ultimate—"For whosoever will save his life shall lose it: and whosoever will lose his life for My sake shall find it. For what is a man profited, if he shall gain the whole world and lose his own soul? Or what shall a man give in exchange for his soul?" (Matt. 16:25, 26). On the commonsense infinitude of total utility, see Böhm-Bawerk (1891), book III, pp. 147–153.

3. In which, using exchange value calculations, Swift logically demonstrates the "economic desirability" of eating children!

4. And, Marx would argue, man even discovered that he could "cultivate and extract" an analogous surplus from other men in the factory "hothouse."

5. The Leontief input-output model derives from a line of thought beginning with François Quesnay's "tableau économique," which was described by Mirabeau as "the great discovery that glorifies our century and will yield posterity its fruits." (For an exposition see Leontief, 1966.) It is more than coincidental that we should find the input-output model relevant to econom-

ics considered as a life science, since Quesnay (a physician) and the physiocrats emphasized the supremacy of nature and the biological analogy.

6. If the reader will pardon the liberties taken with Luke 11:24–26 we may say that sometimes a bad cast out of cell (2) wanders through the waterless places of cells (1) and (4) seeking rest. And finding none it gathers seven new bads, which then descend upon the well-garnished human household through the back door of cell (3). And the last state of that household is worse than the first.

7. Cf. Lotka's (1956, chap. xxiv) concept of the "world engine."

8. If we separate out household consumption as having no meaningful "technical" coefficients, then we would have n equations in $2n$ unknowns (n of the Q_j and n of the q_{ik}, where k is the household sector). Arbitrarily setting any n of these magnitudes determines the remaining n unknowns. This corresponds to the "open" Leontief model. The assumption of technical coefficients for the household sector would give the "closed" model.

9. [Editors' note: For a critical review of this and other extensions of the input-output model to cover environmental sectors, see Peter A. Victor, *Pollution: Economy and Environment* (Toronto: University of Toronto Press, 1972).]

References

Bates, Marston. 1960. *The Forest and the Sea*. New York: Random House.

Böhm-Bawerk, E. 1891 *The Positive Theory of Capital*. Translated by William Smart. New York: Stechert.

Boulding, K. E. 1950. *A Reconstruction of Economics*. New York: Wiley.

Boulding, K. E. 1958. *The Skills of the Economist*. Toronto: Clarke, Irwin.

Boulding, K. E. 1966. "Economics and Ecology." In F. Frazer Darling and John P. Milton, eds., *Future Environments of North America*. New York: Natural History Press.

Carson, Rachel. 1962. *Silent Spring*. Boston: Houghton Mifflin.

Georgescu-Roegen, Nicholas. 1966. *Analytical Economics*. Cambridge, Mass.: Harvard University Press.

Henderson, L. J. 1958. *The Fitness of the Environment*. Boston: Beacon. (Originally published 1913.)

Hobson, J. A. 1929. *Economics and Ethics*. Boston: Heath.

Leontief, Wassily. 1966. *Input-Output Economics*. New York: Oxford University Press.

Lotka, A. J. 1956. *Elements of Mathematical Biology*. New York: Dover. Previously published under the title *Elements of Physical Biology*.

Marshall, Alfred. 1920. *Principles of Economics*. London: Macmillan.

Marshall, Afred. 1925. *Memorials of Alfred Marshall*. Edited by A. C. Pigou. London: Macmillan.

Marx, Karl. 1967. *Capital*. Edited by Friedrich Engels. New York: International Publishers. Reproduction of the English edition of 1887.

Prest, A. R., and R. Turvey. "Cost Benefit Analysis: A Survey." *Economic Journal* (December).

Robinson, Joan. 1962. *Economic Philosophy*. London: Watts.

Schrödinger, Erwin. 1945. *What Is Life?* New York: Macmillan.

Scitovsky, Tibor. 1954. "Two Concepts of External Economies." *Journal of Political Economy* (April).

Spengler, J. J. 1966. "The Economist and the Population Question." *American Economic Review* (March).

Storer, John H. 1954. *The Web of Life*. New York: Devin-Adair.

Teilhard de Chardin, Pierre. 1959. *The Phenomenon of Man*. New York: Harper & Row.

Teller, Edward. 1965. Public address at Louisiana State University, Baton Rouge.

14 Sustainable Growth: An Impossibility Theorem

Herman E. Daly

Impossibility statements are the very foundation of science. It is impossible to: travel faster than the speed of light; create or destroy matter-energy; build a perpetual motion machine, etc. By respecting impossibility theorems we avoid wasting resources on projects that are bound to fail. Therefore economists should be very interested in impossibility theorems, especially the one to be demonstrated here, namely that it is impossible for the world economy to grow its way out of poverty and environmental degradation. In other words, sustainable growth is impossible.

In its physical dimensions the economy is an open subsystem of the earth ecosystem, which is finite, nongrowing, and materially closed. As the economic subsystem grows it incorporates an ever greater proportion of the total ecosystem into itself and must reach a limit at 100 percent, if not before. Therefore its growth is not sustainable. The term "sustainable growth" when applied to the economy is a bad oxymoron—self-contradictory as prose, and unevocative as poetry.

Challenging the Economic Oxymoron

Economists will complain that growth in GNP is a mixture of quantitative and qualitative increase and therefore not strictly subject to physical laws. They have a point. Precisely because quantitative and qualitative change are very different it is best to keep them separate and call them by the different names already provided in the dictionary. *To grow* means "to increase naturally in size by the addition

Reprinted from *Development* 1990, nos. 3/4, pp. 45–47, the journal of the Society for International Development, by permission of the publisher.

of material through assimilation or accretion." *To develop* means "to expand or realize the potentialities of; to bring gradually to a fuller, greater, or better state." When something grows it gets bigger. When something develops it gets different. The earth ecosystem develops (evolves), but does not grow. Its subsystem, the economy, must eventually stop growing, but can continue to develop. The term "sustainable development" therefore makes sense for the economy, but only if it is understood as "development without growth"—i.e., qualitative improvement of a physical economic base that is maintained in a steady state by a throughput of matter-energy that is within the regenerative and assimilative capacities of the ecosystem. Currently the term "sustainable development" is used as a synonym for the oxymoronic "sustainable growth." It must be saved from this perdition.

Politically it is very difficult to admit that growth, with its almost religious connotations of ultimate goodness, must be limited. But it is precisely the nonsustainability of growth that gives urgency to the concept of sustainable development. The earth will not tolerate the doubling of even one grain of wheat 64 times, yet in the past two centuries we have developed a culture dependent on exponential growth for its economic stability (Hubbert, 1976). Sustainable development is a cultural adaptation made by society as it becomes aware of the emerging necessity of nongrowth. Even "green growth" is not sustainable. There is a limit to the population of trees the earth can support, just as there is a limit to the populations of humans and of automobiles. To delude ourselves into believing that growth is still possible and desirable if only we label it "sustainable" or color it "green" will just delay the inevitable transition and make it more painful.

Limits to Growth?

If the economy cannot grow forever then by how much can it grow? Can it grow by enough to give everyone in the world today a standard of per capita resource use equal to that of the average American? That would turn out to be a factor of seven,[1] a figure that is neatly bracketed by the Brundtland Commission's call (Brundtland et al., 1987) for the expansion of the world economy by a factor of five to ten. The problem is that even expansion by a factor of four is impossible if Vitousek et al. (1986, pp. 368–373) are correct in their calculation that the human economy currently preempts one-fourth of the global net primary product of photosynthesis (NPP). We cannot go beyond 100 percent, and it is unlikely that we will increase NPP since the historical tendency up to now is for economic growth to reduce global photosynthesis. Since land-based ecosystems are the more relevant, and we preempt 40 percent of land-based NPP, even the factor of four is an overestimate. Also, reaching 100 percent is unrealistic since we are incapable of bringing under direct human management all the species that make up the ecosystems upon which we depend. Furthermore it is ridiculous to urge the preservation of biodiversity without being willing to halt the economic growth that requires human takeover of places in the sun occupied by other species.

If growth up to the factor of five to ten recommended by the Brundtland Commission is impossible, then what about just sustaining the present scale—i.e., zero net growth? Every day we read about stress-induced feedbacks from the ecosystem to the economy, such as greenhouse buildup, ozone layer depletion, acid rain, etc., which constitute evidence that even the present scale is unsustainable. How then can people keep on talking about "sustainable growth" when: (a) the present scale of the economy shows clear signs of unsustainability, (b) multiplying that scale by a factor of five to ten as recommended by the Brundtland Commission would move us from unsustainability to imminent collapse, and (c) the concept itself is logically self-contradictory in a finite, nongrowing ecosystem? Yet sustainable growth is the buzz word of our time. Occasionally it becomes truly ludicrous, as when writers gravely speak of "sustainable growth in the rate of increase of economic activity." Not only must we grow forever, we must accelerate forever! This is hollow political verbiage, totally disconnected from logical and physical first principles.

Alleviating Poverty, Not Angelizing GNP

The important question is the one that the Brundtland Commission leads up to, but does not really face: How far can we alleviate poverty by development without growth? I suspect that the answer will be a significant amount, but less than half. One reason for this belief is that if the five- to tenfold expansion is really going to be for the sake of the poor, then it will have to consist of things needed by the poor—food, clothing, shelter—not information services. Basic goods have an irreducible physical dimension and their expansion will require growth rather than development, although development via improved efficiency will help. In other words, the reduction in resource content per dollar of GNP observed in some rich countries in recent years cannot be heralded as severing the link between economic expansion and the environment, as some have claimed. Angelized GNP will not feed the poor. Sustainable development must be development without growth—but with population control and wealth redistribution—if it is to be a serious attack on poverty.

In the minds of many people, growth has become synonymous with increase in wealth. They say that we must have growth to be rich enough to afford the cost of cleaning up and curing poverty. That all problems are easier to solve if we are richer is not in dispute. What is at issue is whether growth at the present margin really makes us richer. There is evidence that in the US it now makes us poorer by increasing costs faster than it increases benefits (Daly and Cobb, 1989, appendix). In other words we appear to have grown beyond the optimal scale.

Defining the Optimal Scale

The concept of an optimal scale of the aggregate economy relative to the ecosystem is totally absent from current macroeconomic theory. The aggregate economy is assumed to grow forever. Microeconomics, which is almost entirely devoted to establishing the optimal scale of each microlevel activity by equating costs and benefits at the margin, has neglected to inquire if there is not also an optimal scale for the aggregate of all micro activities. A given scale (the product of population times per capita resource use) constitutes a given throughput of resources and thus a given load on the environment, and can consist of many people each consuming little, or fewer people each consuming correspondingly more.

An economy in sustainable development adapts and improves in knowledge, organization, technical efficiency, and wisdom; and it does this without assimilating or accreting, beyond some point, an ever greater percentage of the matter-energy of the ecosystem into itself, but rather stops at a scale at which the remaining ecosystem (the environment) can continue to function and renew itself year after year. The nongrowing economy is not static—it is being continually maintained and renewed as a steady-state subsystem of the environment.

What policies are implied by the goal of sustainable development, as here defined? Both optimists and pessimists should be able to agree on the following policy for the US (sustainable development should begin with the industrialized countries). Strive to hold throughput constant at present levels (or reduced truly sustainable levels) by taxing resource extraction, especially energy, very heavily. Seek to raise most public revenue from such resource severance taxes, and compensate (achieve revenue neutrality) by reducing the income tax, especially on the lower end of the income distribution, perhaps even financing a negative income tax at the very low end. Optimists who believe that resource efficiency can increase by a factor of ten should welcome this policy, which raises resource prices considerably and would give powerful incentive to just those technological advances in which they have so much faith. Pessimists who lack that technological faith will nevertheless be happy to see restrictions placed on the size of the already unsustainable throughput. The pessimists are protected against their worst fears; the optimists are encouraged to pursue their fondest dreams. If the pessimists are proven wrong and the enormous increase in efficiency actually happens, then they cannot complain. They got what they most wanted, plus an unexpected bonus. The optimists, for their part, can hardly object to a policy that not only allows but gives a strong incentive to the very technical progress on which their optimism is based. If they are proved wrong at least they should be glad that the throughput-induced rate of environmental destruction has been slowed. Also severance taxes are harder to avoid than income taxes and do not reduce incentives to work.

At the project level there are some additional policy guidelines for sustainable development. Renewable resources should be exploited in a manner such that:

(1) harvesting rates do not exceed regeneration rates; and

(2) waste emissions do not exceed the renewable assimilative capacity of the local environment.

Balancing Nonrenewable and Renewable Resources

Nonrenewable resources should be depleted at a rate equal to the rate of creation of renewable substitutes. Projects based on exploitation of nonrenewable resources should be paired with projects that develop renewable substitutes. The net rents from the nonrenewable extraction should be separated into an income component and a capital liquidation component. The capital component would be invested each year in building up a renewable substitute. The separation is made such that by the time the nonrenewable is exhausted, the substitute renewable asset will have been built up by investment and natural growth to the point where its sustainable yield is equal to the income component. The income component will have thereby become perpetual, thus justifying the name "income," which is by definition the maximum available for consumption while maintaining capital intact. It has been shown (El Serafy, 1989, pp. 10–18) how this division of rents into capital and income depends upon: (1) the discount rate (rate of growth of the renewable substitute); and (2) the life expectancy of the nonrenewable resource (reserves divided by annual depletion). The faster the biological growth of the renewable substitute and the longer the life expectancy of the nonrenewable, the greater will be the income component and the less the capital set-aside. "Substitute" here should be interpreted broadly to include any systemic adaptation that allows the economy to adjust the depletion of the nonrenewable resource in a way that maintains future income at a given level (e.g., recycling in the case of minerals). Rates of return for the paired projects should be calculated on the basis of their income component only.

However, before these operational steps toward sustainable development can get a fair hearing, we must first take the conceptual and political step of abandoning the thought-stopping slogan of "sustainable growth."

Note

1. Consider the following back-of-the-envelope calculation, based on the crude estimate that the US currently uses 1/3 of annual world resource flows (derived from National Commission on Materials Policy, 1973). Let *R* be current world resource consumption. Then *R*/3 is current US resource consumption, and *R*/3 divided by 250 million is present per capita US resource consumption. Current world per capita resource consumption would be *R* divided by 5.3 billion. For future world per capita resource consumption to equal present US per capita consumption, assuming constant population, *R* must increase by some multiple, call it *M*. Then *M* times *R* divided by 5.3 billion must equal *R*/3 divided by 250 million. Solving for *M* gives 7. World

resource flows must increase sevenfold if all people are to consume resources at the present US average.

But even the sevenfold increase is a gross underestimate of the increase in environmental impact, for two reasons. First, because the calculation is in terms of current flows only with no allowance for the increase in accumulated stocks of capital goods necessary to process and transform the greater flow of resources into final products. Some notion of the magnitude of the extra stocks needed comes from Harrison Brown's estimate that the "standing crop" of industrial metals already embodied in the existing stock of artifacts in the ten richest nations would require more than 60 years' production of these metals at 1970 rates. Second, because the sevenfold increase of net, usable minerals and energy will require a much greater increase in gross resource flows, since we must mine ever less accessible deposits and lower grade ores. It is the gross flow that provokes environmental impact.

References

Brundtland, G. H., et al. 1987. *Our Common Fulure: Report of the World Commission on Environment and Development*. Oxford: Oxford University Press.

Daly, H. E., and J. B. Cobb, Jr. 1989. *For the Common Good: Redirecting the Economy toward Community, the Environment and a Sustainable Future*. Boston: Beacon Press.

El Serafy, S. 1989. "The Proper Calculation of Income from Depletable Natural Resources." In Y. J. Ahmad, S. El Serafy, and E. Lutz, eds., *Environmental Accounting for Sustainable Development*, a UNEP–World Bank Symposium. Washington, D.C.: The World Bank.

Hubbert, M. King. 1976. "Exponential Growth as a Transient Phenomenon in Human History." In Margaret A. Storm, ed., *Societal Issues: Scientific Viewpoints*. New York: American Institute of Physics. (Reprinted in this volume.)

National Commission on Materials Policy. 1973. *Material Needs and the Environment Today and Tomorrow*. Washington, D.C.: US Government Printing Office.

Vitousek, Peter M., Paul R. Ehrlich, Anne H. Ehrlich, and Pamela A. Matson. 1986. "Human Appropriation of the Products of Photosynthesis." *BioScience* 34. (6 May).

15

Steady-State Economies and the Command Economy

Kenneth N. Townsend

As the earth is [man's] original larder, so too it is his original tool house.

Karl Marx, *Capital,* volume 1

From China to Eastern Europe and the Commonwealth of Independent States (CIS), political and economic changes of the past few years suggest that the era of the experiment with the centrally planned, command economy is drawing to a close. What began as an attempt to secure human existence by limiting unbridled individual accumulation of wealth has fallen short of a true resolution of the conflict between man and nature, which had been predicted by Marx. The command economy was to have been the chief testament to humankind's ability to oppose itself to the forces of nature in a manner that secured true freedoms for society. Through the construction of a rationally ordered environment, humankind would improve on the less reliable bounty of nature.

The legacy of central planning, which through five-year plans provided the world with added impetus for aggregate economic growth, will include an impressive cleanup bill for the advanced state of environmental deterioration to be found throughout much of the East. This pollution will daunt efforts to achieve true development for years to come. Under communistic systems of economic development, investment in heavy industry produced environmental deterioration to the extent that morbidity and mortality rates in many areas are on the rise. For every picture from Eastern Europe of a family heading for the West in its antiquated, highly polluting Trabant, there exists a subtle, more harrowing scene of a deceptively

Published for the first time in this volume.

green landscape contaminated with heavy metals, radioactive fallout, and incredibly high levels of deposits of the oxides of sulfur and nitrogen from the smokestacks of industry. In many command economies water is so polluted that it is often unfit even for industrial cooling applications, and air is so foul that school children must periodically be removed from their homes to more hospitable climes to cleanse their lungs of airborne contaminants. Destruction of the environment points to a critical failure of communism.

The Purpose of Economic Growth in Command Economies

Economic growth affords people an improvement of the material standard of living. Yet economic growth makes necessary the increased utilization of a nation's materials and energy resource endowment, as well as an increase in the level of waste emissions into the environment. The environment provides asset value in the human economy, both as a source of low-entropy material and energy inputs and as a sink for high-entropy economic waste. The twin requirements of a sustainable economic development are depletion of physical resources at rates that do not unduly jeopardize the productive capacity of the environment, and emission of waste at rates that do not exceed the absorptive capacity of the environment.

A common theme behind much of the emphasis on economic growth, both in the East and the West, has been the desire by humankind to rationally order the natural environment so as to yield a more secure and sustained bounty than is provided in a state of nature. Nowhere has this urge been more evident than in the centrally planned command economy based upon the tenets of Marxist economic theory. Communism, as originally envisioned by Marx and Engels, was a historically inevitable response to the missteps of capitalism in dealing with an all-too-often harsh nature. Although capitalism provided considerable improvement over humankind's natural state, it did so at the expense of separating labor from its immediate object of production, natural resources, as well as from the full value of its produce.[1] Such market-oriented commodification of labor and resources is incompatible with the rationally ordered state envisioned by the authors of the command economy. Whereas capitalism secured the existence of society, it still left people to face the problem of individual existence. With the "seizing of the means of production by society," however,

> the struggle for individual existence disappears. Then for the first time man, in a certain sense, is finally marked off from the rest of the animal kingdom, and emerges from mere animal conditions of existence into really human ones. The whole sphere of the conditions of life which environ man, and which have hitherto ruled man, now comes under the dominion and control of man, who for the first time becomes the real, conscious lord of Nature, because he has now become master of his own social organisation.[2]

That man would turn out to be an irresponsible lord and master of nature in socialism as well as capitalism did not occur to Engels.

For millennia, the scope and impact on the environment of the human economy was small, relative to the impact of geological and geochemical forces. Today, after generations of exponential growth of population and economies, the impact of the human economy on the environment rivals that of powerful forces of seismic and climatic activities. The pressures of exponential growth of the human population, as well as its stock of exosomatic artifices, now jeopardizes the environmental source and sink.

Historical Roots of the Steady-State Economy in the East

One of the first scientists to recognize the capacity of the human economy's impact to alter the biosphere was Russian mineralogist Vladimir I. Vernadsky. Vernadsky, whose monumental work, *The Biosphere,* was published in 1926, was a pioneer in the fields of mineralogy and geochemistry. As early as the end of the nineteenth century Vernadsky displayed a keen understanding of the inevitability that growth in the scale of human activity would have dangerous consequences for the biosphere and, ultimately, for humankind itself. Like Marx, Vernadsky believed that people collectively could transform the biosphere from a spontaneous organization into a rational, managed organization that would meet not only their material but also their spiritual and aesthetic demands. Vernadsky's term for the rational, managed biosphere was *noosphere,* which comes from the Greek *noos* meaning mind or intellect.[3]

In one sense, Vernadsky's noosphere parallels the development of much of the thought concerning humankind's relationship with the environment that has characterized ecological writings from the former Soviet Union, Eastern Europe, and China. Whereas in the West an ideal environment is typically thought to imply a pristine, natural environment, in the East the concept of a natural environment includes rational, ordered structures imposed on the character of nature by human effort.

Taken at another level, however, Vernadsky's noosphere seems to imply something contrary to the thrust of Marxism-Leninism, which views environmental assets as spontaneous gifts of nature. In a biosphere whose capacity to sustain life is bounded by the finite availability of low-entropy materials and energy, as well as by a finite capacity to absorb high-entropy wastes, the attainment of a rational noosphere suggests intrinsic limits to the process of economic growth. In describing the increased pressures on the biosphere brought about by economic growth, Vernadsky in the East anticipated the development of steady-state economics in the West.

The transformation of the biosphere into Vernadsky's noosphere requires that human activity be organized on a scale that is compatible with constraints imposed by the character of the biosphere. In spite of his prominence during a formative period in the development of command economies in the East, Vernadsky seems to have had little impact in influencing the course of eastern economic development. In the years since his death in 1945, however, a few scholars in the East have kept his vision alive and have built upon it. Echoing Vernadsky, Soviet economist V. Alekseev in 1973 wrote,

> the scale of the transformational activity of people has grown immeasurably and for the first time in history has become comparable with the scale of activity of the natural forces on our planet: with geological, chemical and biological processes. . . . At the present time, the quantity of heat produced by people in the process of their economic activities comprises approximately 0.01% of the quantity of the energy reaching the earth from the sun. A further increase in this quantity (and it is doubling approximately every 10 years), which is equivalent to an increase in the solar constant, i.e., in the amount of heat received from the sun, in approximately 70–80 years may require calling a halt to increases in energy production on earth in order to avoid overheating the planet's surface.[4]

What Vernadsky, Alekseev, and other eastern scholars have discovered is the principle of biophysical constraints on the ultimate extent of economic growth. At roughly the same time as Alekseev, scholars in the West were articulating similar principles. In the United States, Kenneth Boulding,[5] Nicholas Georgescu-Roegen,[6] and Herman E. Daly[7] popularized the concept that humankind must strike a balance between economic growth and environmental deterioration.

As Vernadsky understood years ago, before people generally began to consider such things, the attainment of a rational, ordered noosphere requires that the human economy be harmonious with the physical environment. In command economies this has long been understood to imply a condition in which nature and humankind change in response to the varying circumstances of each. A natural environment does not have to signify a pristine environment, unaffected by the human presence.

In the context of command economies, large-scale projects have been undertaken to alter the environment to make it more suitable to human needs. The conversion of a biologically diverse, complex forest into a monocrop field is perhaps considered natural. Attempts in the former Soviet Union to divert or even reverse the course of rivers serve as other cases in point. The economy of the CIS is inconvenienced geographically by the fact that its rivers generally flow south-to-north, away from the arable lands in the southern part of the country. Consequently, plans have been made to reverse the flow of certain rivers so as to irrigate fields in the south. Decisions to permit the gradual drainage of the Aral Sea, weighing the benefits of irrigation against the benefits of preserving the sea for other uses, serve as other examples of modifications of the environment that have not contributed positively to formation of the noosphere. In the Aral Sea region only one-third of the sea's water volume remains, too salty for life. Salt and sand from the dry sea bottom blown into the air cause respiratory problems for residents, most of whom are environmental refugees in spite of some employment in canneries now supplied with fish from a thousand miles away.[8] What must be remembered, however, is that, as Vernadsky and Alekseev have written, the scale of the human economy today is of such a magnitude that modifications of the environment like river diversion projects may affect the biosphere in a highly adverse way.

Environmental Protection in a Command Economy

In theory, unnecessary pollution of the environment should not have occurred under communism. Marxist-Leninist doctrine held that problems arising from environmental disruption were manifestations of a capitalistic organization of factors of production. In a market economy the institution of private property encourages, but does not guarantee, effective management of an asset such as land.

For example, in a market economy an owner of a parcel of land may value it only as a site for dumping waste, which is a use of land that typically does not generate significant Ricardian rents. If society deems the parcel of land more dear in its potential for other use than does the individual who owns it, this fact will be reflected in the parcel's price. This price may induce the individual to sell rather than to hold the land as a dump. The individual has no similar interest in free assets, such as a common pasture or the environment. When an individual dumps garbage on a commons, that person bears only a fraction of the costs of polluting the land yet receives all benefits of not storing garbage on his own property. Costs are thus externalized to society, producing a misallocation of resources. This is the principle of externalities, which was first described by British economist A. C. Pigou[9] and which has been elegantly explained by biologist Garrett Hardin as the "tragedy of the commons."[10]

Theoretically, in a centrally planned economy the externalities problem should disappear. True tragedies of the commons result when individuals privately make decisions concerning the disposition of commonly held resources. With the state serving both as owner and decision-making agent in the case of resources, there should be no possibility for externalizing costs of economic activity onto other parties. Thus in centrally planned economies pollution was frequently treated as a direct consequence of capitalism that would be eliminated through a socialistic pattern of resource control. N. Feitel'man and I. Smagarinskii note:

> The fundamental principles of natural resource utilization under socialist society were formulated by K. Marx, who wrote: ". . . Collective man and associated producers rationally regulate their exchange with nature and place this exchange under their general control instead of allowing it to control them like a blind force: they make this exchange with the least expenditure of effort and under conditions that are worthiest and most appropriate to their human nature."[11]

Marshall Goldman has described this problem succinctly: "Because, in a socialist society, the state owns all the means of production, sooner or later the state must bear all the social costs. If each factory were held accountable for both the direct and social costs of its operations, much of the pollution would be treated within the confines of the plant before it could be pushed onto the population as a whole."[12]

As Goldman pointed out twenty years ago, the absence of private ownership of resources meant that plant managers did not have to pay for land, which resulted in overutilization of this resource. Nor did plant managers have to pay for use of the environment as an ecological sink. This has resulted in higher levels of pollution. It was expedient for a plant manager, eager to exceed his production quota, to dump waste directly into the immediate environment, knowing that ultimately the operation could be relocated on other land, which was essentially free. If the enterprise was fined for polluting, finance for payment of the fine could generally be petitioned from the government. The state as bearer and enforcer of social cost remained purely theoretical. It was single-minded in its pursuit of growth and did not wish to slow itself down with anything as mundane as environmental cost, especially when natural resources are in Marxist theory supposed to be free. Thus, as Gorizontov and Prokudin explain,

> Certain ecological difficulties occasionally arise in socialist countries owing to the fact that some ministries, agencies, enterprises, or other organizations do little to make more effective use of natural resources and to improve the protection of the environment. . . . The advantages created by the socialist social system are not yet sufficiently utilized to secure the better utilization and conservation of natural resources."[13]

Owing to sheer size as well as tremendous resource endowment, it is understandable that for so long natural resources and access to the environment as a waste repository were considered essentially free goods in countries like the former Soviet Union and China. To be sure, the economic growth afforded Soviet society during its period of industrialization, and now being achieved in China with its rapid industrialization, has brought about substantial improvements in the material standard of living. Now, however, the scale and complexity of eastern economies necessitate effective rationing of resources and of access to the environment.

Gorizontov and Prokudin recommend economic incentives to regulate the use of scarce resources and the environment. They write:

> The introduction of cost-accounting principles and of a system of economic incentives will promote the better use of natural resources. This requires the substantiation and institution of a rational system of rent payments that will encourage enterprises to make more complete use of natural resources and production wastes. It requires a flexible incentive system that will offer rewards for the more complete utilization of natural resources and the preservation of all components of the environment. It requires a system of sanctions and fines for damage inflicted on the environment by economic activity.[14]

The choice of rationing mechanisms to achieve the desired level of regulation of resource depletion and environmental pollution is, in some sense, a matter of choice that depends upon the legal and political framework of the socialist country. Alekseev has pointed out the problem inherent in the use of pricing of environmental assets as the rationing mechanism. He correctly concludes that it is possible that such a system would not always be accurate.[15] That criticism noted, it must nevertheless be asserted that without effective rationing of environmental assets Vernadsky's fear that the human economy will wreck the biosphere may soon be a reality.

Surveying the Damage

It is a common mistake to assume that command economies present a monolithic face to the world. They are as varied in their makeup as are the market economies. Some countries, such as former East Germany and Czechoslovakia, are highly industrialized, while others, such as Yugoslavia, China, and Poland, retain an agricultural base. In China, population has grown to perilous proportions, while in Poland, Czechoslovakia, and former East Germany population grows slowly or is in decline. Nevertheless, it is possible to characterize the centrally produced economic plans of the CIS, Eastern Europe, and China as having invariably emphasized, albeit to varying degrees, the development of heavy industry and the collectivization of agriculture in promoting growth. This tendency is especially apparent in the five-year plans of the Eastern European and Soviet economies. Polish economist Aleksander Müller writes:

> The regularities governing the industrialization process in the Soviet Union and in the East European countries reveal the application of one and the same uniformized strategy resulting from the imposition upon the latter countries of the Soviet model, whose essential characteristics were fixed by J. Stalin in 1926. That model was based on the [assumption that] socialist industrialization must begin with the construction of heavy industry, whose development determines the possibilities of the technical reconstruction and modernization of the entire economy.[16]

The impact of Stalinization upon the development of command economies has been to produce highly concentrated pockets of industrialization, in which large-scale plant design coupled with energy inefficiency produces intense pollution of the immediate environment. Mining, refining, and manufacturing industries throughout the Eastern bloc and China generate some of the highest levels of both energy use and pollution per unit of GNP produced in the world. The CIS remains the least energy-efficient industrial economy. China remains even less energy-efficient than the CIS.[17] Poland is particularly inefficient in its patterns of energy consumption. In the United States, for example, consumption of energy per constant 1980 $US of GNP was 20,645 kilojoules in 1987. In China the figure was more than double, at 42,962 kilojoules. In Poland the estimate was more than four times the U.S. level, at 88,829 kilojoules.[18]

A review of environmental conditions in certain command economies and former command economies reveals a pattern of advanced ecological deterioration. Many of these countries now work feverishly to develop environmental protection policies that will remedy the effects of the Stalinist period of growth. However, as recent information reaching the West has attested, soil, water, and air pollution under communism reveals a very irresponsible dominance over the forces of nature. What follows is a brief survey of the ecological damage that has accumulated in certain European and Asian command economies.

Poland

Prior to Word War II, Poland, Czechoslovakia, and Germany were highly industrialized economies in Central Europe. During the period under Stalin, these countries were depended upon to produce the heavy industry necessary for economic independence in the Warsaw Pact countries. In Poland this translated into growth in the cement, steel, metals, shipbuilding, and chemicals industries. Extractive industry, such as large-scale open-pit strip mining of coal and minerals, as well as large production facilities for iron, steel, and other commercially important metals are the norm.

This pattern of development led to environmental deterioration of serious proportions. Mining, smelting, and forging operations have left air, soil, and water in Poland heavily contaminated. Growth in agriculture has contributed greatly to phosphorous and nitrogen pollution. Increases in the use of automobiles has added to the burden of air pollution in cities. Finally, extensive use of energy-intensive, large-scale patterns of industrial production has left Poland with unenviable per capita energy use to per capita GNP ratios.[19]

Poland derives its energy requirements from hydrocarbon fuels. Some 97 percent of Poland's energy comes from the burning of coal, with 58 percent of that figure coming from the use of hard coal and 42 percent from the use of soft brown coal.[20] This reliance upon hydrocarbon fuels, coupled with high ratios of energy used per unit of GNP produced, has resulted in enormous quantities of hydrocarbon emissions into the atmosphere annually. By the late 1980s Poland ranked fifteenth in the production of greenhouse gases, such as carbon dioxide, methane, and chlorofluorocarbons,[21] although it ranked only 45th in per capita GNP.[22]

Another consequence of heavy hydrocarbon fuel use has been the deposition of sulfur dioxide. By the early 1980s deposits of sulfur dioxide reached 2,825 thousand metric tons per year.[23] This, combined with deposits of the oxides of nitrogen and suspended particulates in the air, subjected Poles to abnormally difficult breathing conditions.

Air pollution is worst in the southern half of Poland, principally in the Upper Silesian Industrial Region and the area surrounding Krakow. Former party secretary in Upper Silesia Zdislaws Grudzien reports: "Of all gases and dust emitted in Poland, 27.5 percent fall on the Katowice mining region, which makes up only 2.1 percent of the total surface area of the country." By 1984, 1,470 kilograms of dust, 800 kilograms more than the healthful limit, fell on every square kilometer of the Katowice region. In Katowice, there is smog 183 days of the year.[24] In the Chorzow and Plock areas, SO_2 deposition exceeds 1,000 metric tons per square kilometer.[25] The principal culprit in this type of pollution is the use of high-sulfur lignite for fuel. (Poland, the fourth largest coal producer in the world, chooses to export much of its low-sulfur coal to generate foreign exchange.)

Krakow, situated in a valley and bounded by the Nowa Huta steel works to the southeast and the now-closed Skawina Huta aluminum works to the southwest, has been particularly hard hit. Stefan Jarzebski, Minister of Environmental Protection and Water Economy, has said that Krakow "may have the worst pollution problem in the world."[26] The Skawina Aluminum Works produced half of the smelted aluminum in Poland at its peak level of production. It also contributed 2,200 tons of fluoride gas into the atmosphere annually. So pernicious is the air pollution problem in Krakow that trains are reportedly limited to a top speed of 25 miles per hour owing to the corrosion of rail tracks, and building façades and statuary must be repaired or replaced outright—as was the case with the statues of the apostles that adorned the front of the property of the Peter and Paul Cathedral in Krakow. Statuary adorning the thirteenth-century Wawel Castle are now formless lumps. Because of the formation of hydrochloric and hydrofluoric acids in the atmosphere, even the sixteenth-century gold roof of the Sigismund Chapel of Wawel Cathedral (where Pope John Paul II served as archbishop of Krakow) had to be replaced—the gold was reduced to soluble chlorides.[27]

A second consequence of the industrial development of Poland under communism has been the heavy contamination of soil, especially in and around mining and industrial regions. Soil samples in Krakow and Upper Silesia reveal extraordinarily high levels of contamination from heavy metals. As many as 35 percent of school-age children surveyed displayed symptoms of lead poisoning. Garden plots in Upper Silesia receive so much heavy metal pollution that garden produce is laden with zinc, copper, cadmium, and lead. Measurements for lead in soil around Krakow reveal between 42 and 8,890 milligrams of lead per kilogram of soil.[28] It is reported that soil samples from the towns of Olkosz and Slakow show the highest levels of lead and cadmium contamination ever recorded in the world.[29]

One popular theory explaining the phenomenon of deforestation from acid rain maintains that acid rain from hydrocarbon emissions makes aluminum and other metals in soil soluble. These metals are thus more easily accumulated within the tissues of trees—especially coniferous forests—in which they are toxic. In Upper Silesia nearly 200,000 hectares of coniferous forests are reported to be dead or dying.[30]

Fresh water as a resource is scarce in Poland, by world standards. In 1990, per capita annual internal renewable water resources averaged 7.69 thousand cubic meters worldwide. In Poland, by comparison, per capita annual internal renewable water resources averaged just 1.29 thousand cubic meters.[31] Not only does Poland have limited access to fresh water but much of its surface waters are heavily contaminated.

Poland's river water is classified as follows: Class I, fit for human consumption; Class II, fit for animal consumption; Class III, fit only for industrial use; Class IV, unfit even for industrial use. By 1980, only 19 percent of Poland's rivers were rated Class I or II, down from 41 percent only three years previously. Also by 1980, fully 48 percent of river water was classified Class IV, too corrosive even for industrial cooling applications. Poland's main river, the Vistula, is so polluted that only 432 of its 1,068 kilometers are even Class III.[32]

Poland, too, is a heavy polluter of the Baltic Sea. Of 309,000 metric tons of nitrogen introduced into the entire Baltic annually, some 91,400 metric tons are introduced from Poland. Approximately 5,700 metric tons of phosphorous runoff, or 20 percent of the total runoff into the entire sea, emanate from Poland. 1.3 million metric tons of organic material, measured as BOD, enter the Baltic through Poland. It is estimated that Poland contributes 30 percent of the total pollution load into the entire Baltic, even though Poland constitutes about 8.3 percent of the coastline adjoining the Baltic.[33]

All this pollution has contributed to increases in morbidity and mortality in Poland. Air pollution in nearly all of Poland's major cities exceeds Poland's own permissible limits by 50 times. Approximately one-quarter of Poland's population lives in areas where average concentrations of sulfur dioxide exceed one milligram per cubic meter of air. Residents of Upper Silesia are reported to have a 15 percent higher incidence of circulatory disease, 30 percent more cancer, and 47 percent more respiratory disease than do Poles generally.[34] Poland's official tourist agency, Orbis, recommends that visitors schedule no more than a three-day stay in Krakow, owing to the health effects of air pollution in that historic city. School children in Upper Silesia are given mandatory two-week vacations to the north, in order to partially clear their lungs of airborne contaminants.[35] In Poland generally, life expectancy is decreasing. Male mortality due to cardiovascular disease is seven to eight times higher than for women, and is increasing.[36] World Health Organization data suggest that in Poland male mortality rates due to circulatory disease are 485.9 per 100,000, versus 298.3 per 100,000 in the United States.[37]

Czechoslovakia

Czechoslovakia, like Poland, has caused a significant ecological problem resulting from the buildup of industry and agriculture during the last forty years. Open-furnace smelting of iron, steel, and aluminum, as well as the centuries-old practice of mining coal, nickel, and zinc, have resulted in high levels of heavy metal deposition in soil. Invariably, some of these metals are taken up by plants that are eaten by people. In northern Bohemia and in Slovakia, regions that are mineral-rich and consequently sites for mining and refining operations, statistics from the Ministry of Health reveal higher levels of morbidity and mortality.[38]

Czechoslovakia comprises 4.58 million hectares of land covered by forests, of which 32,400 hectares are reported to to have been lost.[39] According to some reports, over half a million hectares of arable land have been damaged by pollution from industry, with 400,000 hectares of productive forest land being "seriously damaged."[40] Reportedly, 60 percent of the land in Bohemia is acidified.[41]

Bedric Moldan, former Minister for Environment for the Czech lands, has collected samples of unusual dirt nodules, taken from southern Moravia, that wash out of fields during flash floods. The nodules are as hard as stones. Moldan claims, "Neither soil scientists nor chemists can find any precedent for that degree of degradation of soil."[42]

Fresh water is slightly more plentiful in Czechoslovakia than in Poland, though it is still scarce by world standards. In 1990, compared with worldwide per capita annual internal renewable water resources of 7.69 thousand cubic meters, Czechoslovakia averaged just 1.79 thousand cubic meters.[43] One state publication has reported that in Slovakia half of the total river length is "seriously or very seriously" contaminated.[44]

Like Poland, Czechoslovakia derives its energy requirements from hydrocarbon fuels, principally through burning coal. Although Czechoslovakia is more heavily industrialized than Poland, it fares somewhat better in terms of greenhouse gas production. Still, Czechoslovakia ranks twenty-eighth on the list of countries producing greenhouse gases.[45]

Air pollution is worst in Bohemia in northwestern Czechoslovakia, and in the heavily industrial regions of Slovakia to the east. In these parts of the country high levels of suspended particulates, fluorine gas emissions, sulfur dioxide, and oxides of nitrogen have been reported. The principal source of air pollution in this region is the use of high-sulfur lignite for fuel. As is common throughout Central and Eastern Europe, Czechoslovakia burns great quantities of lignite, both for industrial use and for home heating. This problem is so prevalent in northern Bohemia that children are removed from the area for one month each year as a health measure.[46]

Unlike Poland, which has placed a moratorium on considerations of the use of fission power for the generation of electricity, Czechoslovakia produces power with nuclear energy. Significantly, reactor designs in Czechoslovakia are patterned after the Soviet Union's design employed at Chernobyl, which poses questions concerning the safety of this arrangement.

China

The People's Republic of China, with its rapid economic development engendered by Deng Xiaoping, is experiencing environmental problems very similar to those of the more industrialized countries of Eastern Europe. Alas, development for China exacts a harsher price than for other countries. By virtue of its sheer size, and of the advanced decay of its environment, China's rapid economic development since 1978 appears unsustainable in light of China's extreme level of overpopulation. With more than 1.12 billion people, China is easily the world's most populous country.

China is a rural nation, with some 800 million of its people living outside cities. China, like other command economies in which farming has been collectivized, struggles to feed its population, but succeeds nevertheless. Commencing in 1978, China began to dismantle the communal system of rural farming. The commune was replaced with the individual household responsibility system. With this new system households would contract with the government to produce crops. Coupled with substantial increases in government purchase prices for farm products, this reform resulted in an increase in agricultural production at an average rate of 6.7 percent per annum during the 1978–1985 period.

In addition to rural economic reform, the Chinese began in the early 1980s to remake the economies of cities. Four Economic Special Zones were created, and in 1984 14 cities were designated as "open cities." In theses areas increasingly free economic trade and, in particular, foreign commerce, were permitted to flourish. By mid-decade some 400,000 new enterprises and 163,000 new investment projects had been created.

Predictably, the consequences of the rapid economic development policies were spectacular. According to Nicholas Lardy, during much of the 1980s China's gross national product grew at an annual rate of eight to nine percent per year. China's exports, as a fraction of GNP, were growing even faster than Japan's. Finally, China's industrial sector was the most rapidly growing in the world, prior to the suppressed revolution of 1989, with an annual growth rate of ten percent, according to Lardy.[47]

In spite of economic liberalization in China, the increases in production and consumption of more than a billion people produce often staggering levels of pollution. Between one-third and one-half million hectares of arable land, between the sizes of Rhode Island and Delaware, are lost to industry annually in China.[48] According to the *Beijing Review*, by the year 2000 average annual expansion of deserts will be 666,000 hectares. Vaclav Smil has written that "it is almost certain that outside a few protected and inaccessible areas (mainly in southeastern Tibet), there will be no natural forests left in China by the beginning of the next century."[49]

Seventy percent of China's air pollution comes from the burning of coal. Although Chinese coal tends to be relatively low in sulfur content, its use by the Chinese is so pervasive that in some regions corrosiveness of rain water has become so pronounced that it burns holes in people's garments.[50] Compared with Tokyo, East Asia's largest city, air pollution levels in Beijing are 5 to 17 times higher for particulate matter, 3 to 6 times higher for SO_2, and four times higher for the oxides of nitrogen. Not surprisingly, mortality rates attributable to lung cancer are increasing in China.[51]

Industrial waste gas accounts for 19.5 percent of air pollution in China.[52] According to Qu Geping, chief of China's Ministry for Environmental Protection, about 250,000 of China's 400,000 new enterprises are "serious polluters." According to one survey, polluters discharge 12.04 million tons of smoke and 9.74 million tons of sulfur dioxide annually.[53]

China's most pressing pollution problem, however, is contamination of water, especially in the huge metropolis of Shanghai. Every day in Shanghai, approximately 3–4 million cubic meters of sewage is emitted from factories and homes, which flows mainly into Suzhou Creek and the Huangpu River.[54] Some environmental experts cite the Suzhou as the filthiest stretch of flowing water in the world.[55] According to Lardy, in the summer months, the ratio of water to municipal waste in Shanghai's rivers falls to 1:1 or even below.[56] Cholera and hepatitis epidemics are common in the summer months in this region of China. It has been estimated that as many as one third of all Chinese have active hepatic infections at any given time.

The record of accumulation of soil, air, and water pollution in command economies suggests that the energy-intensive, large-scale pattern of industrial development of eastern economies has resulted in the environmental pressures that had been predicted by Vernadsky. Although Poland, eastern Germany, Czechoslovakia, the CIS, and other command economies have begun the move toward a more decentralized, market orientation, the fund of pollution already accumulated under communism is proving burdensome. Poland, for instance, is interested in joining the European Community, now that it has abandoned communism. The Environmental Protection Ministry estimates, however, that Poland will have to set aside the next four years of GNP just to meet the entry standards for the EC concerning environmental quality. Management of the environment in the East must change radically if Vernadsky's *noosphere* is to be attained.[57]

Achieving a Steady State in Eastern Economies

Marx, Engels, Lenin, and Mao all shared the honorable vision of improving the human condition in advocating a communistic economy. Unlike Vernadsky, however, they shared a preanalytic vision of the world in which the human economy would not likely exhaust the spontaneous gifts of nature. The dismal record of resource efficiency and pollution efficiency in eastern economies suggests that growth constraints at the waste emission margin require drastic changes in patterns of eastern economic development. To some degree, liberalization of command economies, permitting the use of market incentives to produce efficiently and to eliminate needless industrial wastes, can be counted upon to improve pollution rates in the eastern world. Yet capitalist market systems and socialist command systems both remain committed to the same preanalytic vision of infinite sources and sinks. Relative efficiencies of market systems, compared with command economies, enable capitalist economies to get away with this faulty vision longer. Command economies, with their production and pollution inefficiencies, have collapsed sooner. The danger is that the East thinks that once it adopts a market system it can grow forever—that market efficiency will permit an ever-expanding scale.

This problem can be understood from the perspective of analysis of different sources of a society's environmental impact. Environmental impacts, such as pollution, can be calculated in terms of three factors:

$$I \equiv P \times \frac{Y}{P} \times \frac{I}{Y}.$$

Impact (*I*) is the product of population (*P*) times output per capita or "affluence" (*Y*/*P*) times impact per unit of output (*I*/*Y*) or "technology." A steady-state economy requires that impact be less than the environmental capacity. To reduce impact one must reduce one or more of the three factors. An *x* percent reduction in any factor causes an *x* percent reduction in the product *I*, so in an arithmetic sense the three factors are of equal importance. But in any concrete context it makes sense to ask which factor is most likely to permit an *x* percent reduction. As a broad generalization it seems that the South has greatest room for reducing population growth; the North has greatest room for reducing affluence; and the East has greatest room for improving efficiency in impact per unit of output. Since the East is eager to increase "affluence," and since population growth is zero in some countries (and China, the exception, is making strong efforts to limit it), that leaves "technology" as the factor sure to be emphasized.[58]

One approach to producing sustainable economic development is to advocate policies in the East that would result in the creation of steady-state economies. By limiting the growth in the production of population and the stock of physical artifacts, and by limiting resource throughput to low values commensurate with maintenance of physical capital stocks adequate to produce reasonable living standards, eastern economies may achieve a kind of resolution of the conflict between man and nature that had been anticipated by Marx.

In a steady-state economy, as explained by Daly, the goal of achieving macrostability in the economy, by limiting throughput to sustainable levels, is pursued within a framework of microvariability, which is achieved through reliance on policies that offer people maximal choice in resource use decisions. Marketable permits to extract virgin nonrenewable resources, as well as marketable permits to pollute, offer individual choice within a framework of "mutually coerced, mutually agreed upon" aggregate limits on production and pollution.

In the United States, ample experience with the "command and control" approach to pollution abatement points to the inefficiency typical of such attempts to clean up. The US Environmental Protection Agency has been wont not only to set pollution limits for industry, but then to specify the "best available technology" to be used in compliance with those limits. This approach to pollution abatement is too inflexible. It fails to account for the fact that firms face different costs of compliance, owing to the individual peculiarities of production processes.

A better approach to limiting pollution would have the government issue pollution permits to firms, which businesses could subsequently trade. Possession of permits would enable businesses to emit untreated pollutants up to the quantity specified on the permits. Beyond this point, a firm must either treat its emissions or acquire additional permits on the open market. Permitting has been shown to limit pollution in the least costly way, a desirable feature in a world of scarce resources.

Here, Poland leads the way in revamping environmental policy. With its commitment to develop a market economy, Poland has recently announced the establishment of two permitting areas—one in the severely polluted Katowice region of Silesia—in which permits to pollute will be issued and traded. Such policies bid fair to improve the efficiency of environmental clean-up efforts.

Pollution permits would minimize needless pollution of the environment. In countries in which overpopulation is a problem, most notably China, this concept could be extended to the reduction of family sizes as well. Kenneth Boulding was an early advocate of such a system of marketable birth licenses, receiving considerable criticism for his efforts.[59] According to Boulding's plan a nation could set a goal for a stable population size. Birth licenses could be issued in increments that sum to the total number of births which, given the death rate of the nation's population, would yield zero growth of population. Individuals, however, would be free to sell and buy existing licenses on the open market, producing an efficient allocation of the permits to bear children—those willing to pay for licenses could have more children than the initial distribution of licenses to all people would permit. Since people face different costs associated with bearing and raising children, and exhibit different tastes and preferences concerning the desire to raise families, it can be expected that marketable licenses would produce a more agreeable arrangement than would a system such as China's in which the government simply limits family size for all persons to a predetermined level, e.g., one child per family in the case of China.

Permits for the depletion of nonrenewable scarce resources could be expected to work in much the same fashion as birth permits and pollution permits. Efficiency criteria should be comparable as well.

Summary

Exponential economic growth is incompatible with survival in a biosphere that is finite in its capacity to yield material and energy resources and in its capacity to absorb economic wastes. Consequently, it is morally troubling that five-year plans in command economies should endorse economic growth without fail. If the world is observed to have a finite capacity to support human life, economic growth, at some point, becomes a matter of redistributing to present generations the material sustenance required to support future human lives. Although those not yet born have little or no recourse and negotiating power over the present, it must surely occur to those now living that at some point increases in material well-being are scarcely worth the cost of even distant future human misery.

Perhaps more important, ecological deterioration in the eastern world is resulting in the needless impoverishment of current lives that must contend with foul air, contaminated water, and poisoned farmland. Patterns of economic growth in former command economies, and in those that yet cling to the centralized system of production and distribution of goods and services, suggest the desirability of adopting policies most likely to yield sustainable results.

Whatever regulations are ultimately used, it is important to recognize the need to reconsider the efficacy of growth of the economy for its own sake. Instead, attempts should be made to develop the economy, by improving the efficiency with which existing levels of stocks of physical capital yield utility for consumers. Given the finite nature of the earth's crust, as well as the unforgiving rules of thermodynamics that permit no perfect recycling of materials or energy, no growth state is sustainable for a long and prosperous future. Sooner or later, and preferably sooner, the economies of the world must achieve a rational, ordered noosphere in which the hallmark of the economy is development, not growth. To quote academician T. Khachaturov, "A great deal of work remains to be done to halt the deterioration of the environment and to create the best possible conditions for the existence of future generations."[60]

Notes

Some ideas contained in this essay are adapted from Kenneth N. Townsend and Andrew R. Shoemaker, "The Chinese Environment: An Analysis of

Environmental Policy Problems within the People's Republic of China," in *The Ecological Economics of Sustainability: Making Local and Short-Term Goals Consistent with Global and Long-Term Goals*, ed. Robert Costanza, Ben Haskell, Laura Cornwell, Herman Daly, and Twig Johnson (Environment Working Paper no. 32, The World Bank, Washington, D.C., June 1990).

1. Karl Marx, *Economic and Philosophic Manuscripts of 1844*, in *The Marx-Engels Reader*, ed. Robert C. Tucker, 2d ed. (New York: W. W. Norton, 1978), p. 77.

2. Frederick Engels, *Socialism, Utopian and Scientific*, ed. John E. Elliott (Santa Monica, Calif.: Goodyear, 1981), pp. 479–480.

3. A. L. Yanshin, "Ecological Advances in the Soviet Union," *Environment* 30, no. 10 (December 1988), 8.

4. V. Alekseev, "Protection of the Natural Environment (Economic Aspect)," *Problems in Economics* (November 1973), 24–25.

5. K. E. Boulding, "Economics and Ecology," in F. Frazer Darling and John P. Milton, eds., *Future Environments of North America* (New York: Natural History Press, 1966).

6. Nicholas Georgescu-Roegen, *The Entropy Law and the Economic Process* (Cambridge: Harvard University Press, 1971).

7. Herman E. Daly, "The Economics of the Steady State," *American Economic Review* (May 1974).

8. *World Resources 1990–91: A Report by the World Resources Institute in Collaboration with the United Nations Environment Programme and the United Nations Development Programme* (New York: Oxford University Press, for the World Bank, 1990), p. 171.

9. A. C. Pigou, *The Economics of Welfare*, 2d ed. (London: Macmillan, 1950).

10. Garrett Hardin, "The Tragedy of the Commons," *Science* 162 (December 1968), 1243–1248. (Reprinted in this volume.)

11. N. Feitel'man and I. Smagarinskii, "Regulirovanie ratsional'nogo prirodopol'zovaniia," *Voprosy ekonomiki* (1981, no. 11), pp. 36–47; reprinted in *Problems of Economics* (October 1982), 3.

12. Marshall I. Goldman, "Externalities and the Race for Economic Growth in the USSR: Will the Environment Ever Win?," *Journal of Political Economy* (March/April 1972), 314–315.

13. B. Gorizontov and V. Prokudin, "Okhrana okruzhaiushchei sredy v stranakh-chlenakh SEV," *Voprosy ekonomiki* (1978, no. 4), pp. 68–76; reprinted in *Problems of Economics* (December 1978), 25.

14. Gorizontov and Prokudin, "Okhrana okruzhaiushchei," p. 32.

15. Alekseev, "Protection of the Natural Environment," p. 33.

16. Aleksander M. Müller, "Achievements of and Challenges to Political and Economic Democratization in Poland," presented at the American Marketing

Association 1990 Summer Marketing Educator's Conference, Washington, D.C., 1990, p. 4.

17. *World Resources 1990–91*, p. 147.

18. *World Resources 1990–91*, pp. 316–317.

19. Stanley J. Kabala, "Poland: Facing the Hidden Costs of Development," *Environment* (November 1985), p. 8.

20. Kabala, "Poland," p. 10.

21. *World Resources 1990–91*, p. 15.

22. *World Development Report 1989* (New York: Oxford University Press, for the World Bank, 1989), p. 165.

23. Kabala, "Poland."

24. Adam Zwass, *The Economies of Eastern Europe* (Armonk, N.Y.: M. E. Sharpe, 1984), p. 133.

25. Kabala, "Poland," p. 11.

26. Don Hinrichsen, "In Krakow Even the Buildings Dissolve," *International Wildlife* (March/April 1987), 12.

27. Hinrichsen, "In Krakow," p. 14.

28. Hinrichsen, "In Krakow."

29. Kabala, "Poland," p. 12.

30. Kabala, "Poland."

31. *World Resources 1990–91*, p. 331.

32. Kabala, "Poland," p. 9.

33. Kabala, "Poland," p. 10.

34. Kabala, "Poland," p. 12.

35. Jerzy Holzer, "A Demographic and Sociological Picture of Polish Society," speech delivered at Central School for Planning and Statistics, Warsaw, Poland, 23 November 1990.

36. Holzer, "A Demographic and Social Picture."

37. *World Resources 1990–91*, p. 264.

38. Jon Thompson, "East Europe's Dark Dawn," *National Geographic* (June 1991), 53.

39. *World Resources 1990–91*, p. 293.

40. Andrew H. Dawson, *Planning in Eastern Europe* (London: Croom Helm, 1987), p. 125.

41. Sam Bingham, "Czechoslovakian Landscapes," *Audubon* 93 (January 1991), 96.

42. Bingham, "Czechoslovakian Landscapes."

43. *World Resources 1990–91*, p. 331.

44. Dawson, *Planning in Eastern Europe.*

45. *World Resources 1990–91*, p. 15.

46. Frederick Painton, "Where the Sky Stays Dark," *Time* 135 (28 May 1990), 40–42.

47. Nicholas Lardy, address to Third Environmental Protection Delegation from the United States, University of Washington, Seattle, 3 August 1988.

48. Lee Woyen, "Population Pressure and Control" in *Managing the Environment in China*, ed. Qu Geping and Lee Woyen (Tycooly International Publishing Ltd., 1984), p. 132.

49. Vaclav Smil, "China's Environmental Morass," *Current History* 88 (September 1989), 279.

50. Jasper Becker, "Shanghai Succumbs to the Stink Index," *New Scientist*, 2 October 1985, p. 20.

51. Smil, "China's Environmental Morass," p. 287.

52. Wen Boping and Song Diantang, "On the Legislative Control of Air Pollution," in *Environmental Law and Policy in the People's Republic of China*, ed. Lester Ross and Mitchell Silk (New York: Quorum Books, 1987), p. 115.

53. *Beijing Review*, 16–22 May 1988, p. 12.

54. Becker, "Shanghai Succumbs."

55. Seymour Barfield, "China's River Pollution," *Journal of Environmental Health* 49 (March/April 1987), 320.

56. Lardy, address to the Third Environmental Protection Delegation.

57. Lecture by Tomasz Zylicz, Director of Economics, Ministry of Environmental Protection of Poland, Warsaw, 17 July 1991.

58. I am indebted to Herman E. Daly for pointing out the nature of the environmental impact identity to me.

59. Kenneth E. Boulding, *The Meaning of the Twentieth Century* (New York: Harper & Row, 1964), pp. 135–136.

60. T. Khachaturov, "Okhrana prirody v evropeiskikh sotsialisticheskikh stranakh," *Voprosy ekonomiki* (1984, no. 12), pp. 106–116, reprinted in *Problems of Economics* (September 1985), 68.

16 The Economics of the Coming Spaceship Earth

Kenneth E. Boulding

We are now in the middle of a long process of transition in the nature of the image which man has of himself and his environment. Primitive men, and to a large extent also men of the early civilizations, imagined themselves to be living on a virtually illimitable plane. There was almost always somewhere beyond the known limits of human habitation, and over a very large part of the time that man has been on earth, there has been something like a frontier. That is, there was always some place else to go when things got too difficult, either by reason of the deterioration of the natural environment or a deterioration of the social structure in places where people happened to live. The image of the frontier is probably one of the oldest images of mankind, and it is not surprising that we find it hard to get rid of.

Gradually, however, man has been accustoming himself to the notion of the spherical earth and a closed sphere of human activity. A few unusual spirits among the ancient Greeks perceived that the earth was a sphere. It was only with the circumnavigations and the geographical explorations of the fifteenth and sixteenth centuries, however, that the fact that the earth was a sphere became at all widely known and accepted. Even in the nineteenth century, the commonest map was Mercator's projection, which visualizes the earth as an illimitable cylinder, essentially a plane wrapped around the globe, and it was not until the Second World War and the development of the air age that the global nature of the planet really entered the popular imagination. Even now we are very far from

---

From *Environmental Quality in a Growing Economy,* published for Resources for the Future, Inc., by The Johns Hopkins University Press, 1966.

having made the moral, political, and psychological adjustments which are implied in this transition from the illimitable plane to the closed sphere.

Economists in particular, for the most part, have failed to come to grips with the ultimate consequences of the transition from the open to the closed earth. One hesitates to use the terms "open" and "closed" in this connection, as they have been used with so many different shades of meaning. Nevertheless, it is hard to find equivalents. The open system, indeed, has some similarities to the open system of von Bertalanffy,[1] in that it implies that some kind of structure is maintained in the midst of a throughput from inputs to outputs. In a closed system, the outputs of all parts of the system are linked to the inputs of other parts. There are no inputs from outside and no outputs to the outside; indeed, there is no outside at all. Closed systems, in fact, are very rare in human experience, in fact almost by definition unknowable, for if there are genuinely closed systems around us, we have no way of getting information into them or out of them; and hence if they are really closed, we would be quite unaware of their existence. We can only find out about a closed system if we participate in it. Some isolated primitive societies may have approximated to this, but even these had to take inputs from the environment and give outputs to it. All living organisms, including man himself, are open systems. They have to receive inputs in the shape of air, food, water, and give off outputs in the form of effluvia and excrement. Deprivation of input of air, even for a few minutes, is fatal. Deprivation of the ability to obtain any input or to dispose of any output is fatal in a relatively short time. All human societies have likewise been open systems. They receive inputs from the earth, the atmosphere, and the waters, and they give outputs into these reservoirs; they also produce inputs internally in the shape of babies and outputs in the shape of corpses. Given a capacity to draw upon inputs and to get rid of outputs, an open system of this kind can persist indefinitely.

There are some systems—such as the biological phenotype, for instance the human body—which cannot maintain themselves indefinitely by inputs and outputs because of the phenomenon of aging. This process is very little understood. It occurs, evidently, because there are some outputs which cannot be replaced by any known input. There is not the same necessity for aging in organizations and in societies, although an analogous phenomenon may take place.

The structure and composition of an organization or society, however, can be maintained by inputs of fresh personnel from birth and education as the existing personnel ages and eventually dies. Here we have an interesting example of a system which seems to maintain itself by the self-generation of inputs, and in this sense is moving toward closure. The input of people (that is, babies) is also an output of people (that is, parents).

Systems may be open or closed in respect to a number of classes of inputs and outputs. Three important classes are matter, energy, and information. The present world economy is open in regard to all three. We can think of the world economy or "econosphere" as a subset of the "world set," which is the set of all objects of possible discourse in the world. We then think of the state of the econosphere at any one moment as being the total capital stock, that is, the set of all objects, people, organizations, and so on, which are interesting from the point of view of the system of exchange. This total stock of capital is clearly an open system in the sense that it has inputs and outputs, inputs being production which adds to the capital stock, outputs being consumption which subtracts from it. From a material point of view, we see objects passing from the noneconomic into the economic set in the process of production, and we similarly see products passing out of the economic set as their value becomes zero. Thus we see the econosphere as a material process involving the discovery and mining of fossil fuels, ores, etc., and at the other end a process by which the effluents of the system are passed out into noneconomic reservoirs—for instance, the atmosphere and the oceans—which are not appropriated and do not enter into the exchange system.

From the point of view of the energy system, the econosphere involves inputs of available energy in the form, say, of water power, fossil fuels, or sunlight, which are necessary in order to create the material throughput and to move matter from the noneconomic set into the economic set or even out of it again; and energy itself is given off by the system in a less available form, mostly in the form of heat. These inputs of available energy must come either from the sun (the energy supplied by other stars being assumed to be negligible) or it may come from the earth itself, either through its internal heat or through its energy of rotation or other motions, which generate, for instance, the energy of the tides. Agriculture, a few solar machines, and water power use the current available energy income.

In advanced societies this is supplemented very extensively by the use of fossil fuels, which represent, as it were, a capital stock of stored-up sunshine. Because of this capital stock of energy, we have been able to maintain an energy input into the system, particularly over the last two centuries, much larger than we would have been able to do with existing techniques if we had had to rely on the current input of available energy from the sun or the earth itself. This supplementary input, however, is by its very nature exhaustible.

The inputs and outputs of information are more subtle and harder to trace, but also represent an open system, related to, but not wholly dependent on, the transformation of matter and energy. By far the larger amount of information and knowledge is self-generated by the human society, though a certain amount of information comes into the sociosphere in the form of light from the universe outside. The information that comes from the universe has certainly affected man's image of himself and of his environment, as we can easily visualize if we suppose that we lived on a planet with a total cloud-cover that kept out all information from the exterior universe. It is only in very recent times, of course, that the information coming in from the universe has been captured and coded in the form of a complex image of what the universe is like outside the earth; but even in primitive times, man's perception of the heavenly bodies has always profoundly affected his image of earth and of himself. It is the information generated within the planet, however, and particularly that generated by man himself, which forms by far the larger part of the information system. We can think of the stock of knowledge, or as Teilhard de Chardin called it, the "noosphere," and consider this as an open system, losing knowledge through aging and death and gaining it through birth and education and the ordinary experience of life.

From the human point of view, knowledge, or information, is by far the most important of the three systems. Matter only acquires significance and only enters the sociosphere or the econosphere insofar as it becomes an object of human knowledge. We can think of capital, indeed, as frozen knowledge or knowledge imposed on the material world in the form of improbable arrangements. A machine, for instance, originates in the mind of man, and both its construction and its use involve information processes imposed on the material world by man himself. The cumulation of knowledge, that is, the excess of its production over its consumption, is the key to human development of all kinds, especially to economic development. We can see this preeminence of knowledge very clearly in the experiences of countries where the material capital has been destroyed by a war, as in Japan and Germany. The knowledge of the people was not destroyed, and it did not take long, therefore, certainly not more than ten years, for most of the material capital to be reestablished again. In a country such as Indonesia, however, where the knowledge did not exist, the material capital did not come into being either. By "knowledge" here I mean, of course, the whole cognitive structure, which includes valuations and motivations as well as images of the factual world.

The concept of entropy, used in a somewhat loose sense, can be applied to all three of these open systems. In material systems, we can distinguish between entropic processes, which take concentrated materials and diffuse them through the oceans or over the earth's surface or into the atmosphere, and antientropic processes, which take diffuse materials and concentrate them. Material entropy can be taken as a measure of the uniformity of the distribution of elements and, more uncertainly, compounds and other structures on the earth's surface. There is, fortunately, no law of increasing material entropy, as there is in the corresponding case of energy, as it is quite possible to concentrate diffused materials if energy inputs are allowed. Thus the processes for fixation of nitrogen from the air, processes for the extraction of magnesium or other elements from the sea, and processes for the desalinization of sea water are antientropic in the material sense, though the reduction of material entropy has to be paid for by inputs of energy and also inputs of information, or at least a stock of information in the system. In regard to matter, therefore, a closed system is conceivable, that is, a system in which there is neither increase nor decrease in material entropy. In such a system all outputs from consumption would constantly be recycled to become inputs for production, as for instance, nitrogen in the nitrogen cycle of the natural ecosystem.

In the energy system there is, unfortunately, no escape from the grim second law of thermodynamics; and if there were no energy inputs into the earth, any evolutionary or developmental process would be impossible. The large energy inputs which we have obtained from fossil fuels are strictly temporary. Even the most optimistic predictions expect the easily available supply of fossil fuels to be exhausted in a mere matter of centuries at present rates of use. If the rest of the world were to rise to American standards of power consumption, and still more if world population continues to increase, the exhaustion of fossil fuels would be even more rapid. The development of nuclear energy has improved this picture, but not fundamentally altered it, at least in present technologies, for fissionable material is still relatively scarce. If we should achieve the economic use of energy through fusion, of course, a much larger source of energy materials would be available, which would expand the time horizons of supplementary energy input into an open social system by perhaps tens to hundreds of thousands of years. Failing this, however, the time is not very far distant, historically speaking, when man will once more have to retreat to his current energy input from the sun, even though with increased knowledge this could be used much more effectively than in the past. Up to now, certainly, we have not got very far with the technology of using current solar energy, but the possibility of substantial improvements in the future is certainly high. It may be, indeed, that the biological revolution which is just beginning will produce a solution to this problem, as we develop artificial organisms which are capable of much more efficient transformation of solar energy into easily available forms than any that we now have. As Richard Meier has suggested, we may run our machines in the future with methane-producing algae.[2]

The question of whether there is anything corresponding to entropy in the information system is a puzzling one, though of great interest. There are certainly many examples of social systems and cultures which have lost knowledge, especially in transition from one generation to the next, and in which the culture has therefore degenerated. One only has to look at the folk culture of Appalachian migrants to American cities to see a culture which started out as a fairly rich European folk culture in Elizabethan times and which seems to have lost skills, adaptability, folk tales, songs, and almost everything that goes up to make richness and complexity in a culture, in the course of about ten generations. The American Indians on reservations provide another example of such degradation of the information and knowledge system. On the other hand, over a great part of human history, the growth of knowledge in the earth as a whole seems to have been almost continuous, even though there have been times of relatively slow growth and times of rapid growth. As it is knowledge of certain kinds that produces the growth of knowledge in general, we have here a very subtle and complicated system, and it is hard to put one's finger on the particular elements in a culture which make knowledge grow more or less rapidly, or even which make it decline. One of the great puzzles in this connection, for instance, is why the takeoff into science, which represents an "acceleration," or an increase in the rate of growth of knowledge in European society in the sixteenth century, did not take place in China, which at that time (about 1600) was unquestionably ahead of Europe, and one would think even more ready for the breakthrough. This is perhaps the most crucial question in the theory of social development, yet we must confess that it is very little understood. Perhaps the most significant factor in this connection is the existence of "slack" in the culture, which permits a divergence from established patterns and activity which is not merely devoted to reproducing the existing society but is devoted to changing it. China was perhaps too well organized and had too little slack in its society to produce the kind of acceleration which we find in the somewhat poorer and less well organized but more diverse societies of Europe.

The closed earth of the future requires economic principles which are somewhat different from those of the open earth of the past. For the sake of picturesqueness, I am tempted to call the open economy the "cowboy economy," the cowboy being symbolic of the illimitable plains and also associated with reckless, exploitative, romantic, and violent behavior, which is characteristic of open societies. The closed economy of the future might similarly be called the "spaceman" economy, in which the earth has become a single spaceship, without unlimited reservoirs of anything, either for extraction or for pollution, and in which, therefore, man must find his place in a cyclical ecological system which is capable of continuous reproduction of material form even though it cannot escape having inputs of energy. The difference between the two types of economy becomes most apparent in the attitude towards consumption. In the cowboy economy, consumption is regarded as a good thing and production likewise; and the success of the economy is measured by the amount of the throughput from the "factors of production," a part of which, at any rate, is extracted from the reservoirs of raw materials and noneconomic objects, and another part of which is output into the reservoirs of pollution. If there are infinite reservoirs from which material can be obtained and into which effluvia can be deposited, then the throughput is at least a plausible measure of the success of the economy. The Gross National Product is a rough measure of this total throughput. It should be possible, however, to distinguish that part of the GNP which is derived from exhaustible and that which is derived from reproducible resources, as well as that part of consumption which represents effluvia and that which represents input into the productive system again. Nobody, as far as I know, has ever attempted to break down the GNP in this way, although it would be an interesting and extremely important exercise, which is unfortunately beyond the scope of this paper.

By contrast, in the spaceman economy, throughput is by no means a desideratum, and is indeed to be regarded as something to be minimized rather than maximized. The essential measure of the success of the economy is not production and consumption at all, but the nature, extent, quality, and complexity of the total capital stock, including in this the state of the human bodies and minds included in the system. In the spaceman economy, what we are primarily concerned with is stock maintenance, and any technological change which results in the maintenance of a given total stock with a lessened throughout (that is, less production and consumption) is clearly a gain. This idea that both production and consumption are bad things rather than good things is very strange to economists, who have been obsessed with the income-flow concepts to the exclusion, almost, of capital-stock concepts.

There are actually some very tricky and unsolved problems involved in the questions as to whether human welfare or well-being is to be regarded as a stock or a flow. Something of both these elements seems actually to be involved in it, and as far as I know there have been practically no studies directed towards identifying these two dimensions of human satisfaction. Is it, for instance, eating that is a good thing, or is it being well fed? Does economic welfare involve having nice clothes, fine houses, good equipment, and so on, or is it to be measured by the depreciation and the wearing out of these things? I am inclined myself to regard the stock concept as most fundamental, that is, to think of being well fed as more important than eating, and to think even of so-called services as essentially involving the restoration of a depleting psychic capital. Thus I have argued that we go to a concert in order to restore a psychic condition which might be called "just having gone to a concert," which, once established, tends to depreciate. When it depreciates beyond a certain point, we go to another concert in order to restore it. If it depreciates rapidly, we go to a lot of concerts; if it depreciates slowly, we go to a few. On this view, similarly, we eat primarily to restore bodily homeostasis, that is, to maintain a condition of being well fed, and so on. On this view, there is nothing desirable in consumption at all. The less consumption we can maintain a given state with, the better off we are. If we had clothes that did not wear out, houses that did not depreciate, and even if we could maintain our bodily condition without eating, we would clearly be much better off.

It is this last consideration, perhaps, which makes one pause. Would we, for instance, really want an operation that would enable us to restore all our bodily tissues by intravenous feeding while we slept? Is there not, that is to say, a certain virtue in throughput itself, in activity itself, in production and consumption itself, in raising food and in eating it? It would certainly be rash to exclude this possibility. Further interesting problems are raised by the demand for variety. We certainly do not want a constant state to be maintained; we want fluctuations in the state. Otherwise there would be no demand for variety in food, for variety in scene, as in travel, for variety in social contact, and so on. The demand for variety can, of course, be costly, and sometimes it seems to be too costly to be tolerated or at least legitimated, as in the case of marital partners, where the maintenance of a homeostatic state in the family is usually regarded as much more desirable than the variety and excessive throughput of the libertine. There are problems here which the economics profession has neglected with astonishing singlemindedness. My own attempts to call attention to some of them, for instance, in two articles,[3] as far as I can judge, produced no response whatever; and economists continue to think and act as if production, consumption, throughput, and the GNP were the sufficient and adequate measure of economic success.

It may be said, of course, why worry about all this when the spaceman economy is still a good way off (at least beyond the lifetimes of any now living), so let us eat, drink, spend, extract and pollute, and be as merry as we can, and let posterity worry about the spaceship earth. It is always a little hard to find a convincing answer to the man who says, "What has posterity ever done for me?" and the conservationist has always had to fall back on rather vague ethical principles postulating identity of the individual with some human community or society which extends not only back into the past but forward into the future. Unless the individual identifies with some community of this kind, conservation is obviously "irrational."

Why should we not maximize the welfare of this generation at the cost of posterity? "Après nous, le déluge" has been the motto of not insignificant numbers of human societies. The only answer to this, as far as I can see, is to point out that the welfare of the individual depends on the extent to which he can identify himself with others, and that the most satisfactory individual identity is that which identifies not only with a community in space but also with a community extending over time from the past into the future. If this kind of identity is recognized as desirable, then posterity has a voice, even if it does not have a vote; and in a sense, if its voice can influence votes, it has votes too. This whole problem is linked up with the much larger one of the determinants of the morale, legitimacy, and "nerve" of a society, and there is great deal of historical evidence to suggest that a society which loses its identity with posterity and which loses its positive image in the future loses also its capacity to deal with present problems, and soon falls apart.[4]

Even if we concede that posterity is relevant to our present problems, we still face the question of time-discounting and the closely related question of uncertainty-discounting. It is a well-known phenomenon that individuals discount the future, even in their own lives. The very existence of a positive rate of interest may be taken as at least strong supporting evidence of this hypothesis. If we discount our own future, it is certainly not unreasonable to discount posterity's future even more, even if we do give posterity a vote. If we discount this at five percent per annum, posterity's vote or dollar halves every fourteen years as we look into the future, and after even a mere hundred years it is pretty small—only about one-and-a-half cents on the dollar. If we add another five percent for uncertainty, even the vote of our grandchildren reduces almost to insignificance. We can argue, of course, that the ethical thing to do is not to discount the future at all, that time-discounting is mainly the result of myopia and perspective, and hence is an illusion which the moral man should not tolerate. It is a very popular illusion, however, and one that must certainly be taken into consideration in the formulation of policies. It explains, perhaps, why conservationist policies almost have to be sold under some other excuse which seems more urgent, and why, indeed, necessities which are visualized as urgent, such as defense, always seem to hold priority over those which involve the future.

All these considerations add some credence to the point of view which says that we should not worry about the spaceman economy at all, and that we should just go on increasing the GNP and indeed the Gross World Product, or GWP, in the expectation that the problems of the future can be left to the future, that when scarcities arise, whether of raw materials or of pollutable reservoirs, the needs of the then present will determine the solutions of the then present, and there is no use giving ourselves ulcers by worrying about problems that we really do not have to solve. There is even high ethical authority for this point of view in the New Testament, which advocates that we should take no thought for tomorrow and let the dead bury their dead. There has always been something rather refreshing in the view that we should live like the birds, and perhaps posterity is for the birds in more senses than one; so perhaps we should all call it a day and go out and pollute something cheerfully. As an old taker of thought for the morrow, however, I cannot quite accept this solution; and I would argue, furthermore, that tomorrow is not only very close, but in many respects it is already here. The shadow of the future spaceship, indeed, is already falling over our spendthrift merriment. Oddly enough, it seems to be in pollution rather than in exhaustion that the problem is first becoming salient. Los Angeles has run out of air, Lake Erie has become a cesspool, the oceans are getting full of lead and DDT, and the atmosphere may become man's major problem in another generation, at the rate at which we are filling it up with gunk. It is, of course, true that at least on a microscale, things have been worse at times in the past. The cities of today, with all their foul air and polluted waterways, are probably not as bad as the filthy cities of the pretechnical age. Nevertheless, that fouling of the nest which has been typical of man's activity in the past on a local scale now seems to be extending to the whole world society; and one certainly cannot view with equanimity the present rate of pollution of any of the natural reservoirs, whether the atmosphere, the lakes, or even the oceans.

I would argue strongly also that our obsession with production and consumption to the exclusion of the "state" aspects of human welfare distorts the process of technological change in a most undesirable way. We are all familiar, of course, with the wastes involved in planned obsolescence, in competitive advertising, and in poor quality of consumer goods. These problems may not be so important as the "view with alarm" school indicates, and indeed the evidence at many points is conflicting. New materials especially seem to edge towards the side of improved durability, such as, for instance, neolite soles for footwear, nylon socks, wash and wear shirts, and so on. The case of household equipment and automobiles is a little less clear. Housing and building construction generally almost certainly has declined in durability since the Middle Ages, but this decline also reflects a change in tastes towards flexibility and fashion and a need for novelty, so that it is not easy to assess. What is clear is that no serious attempt has been made to assess the impact over the whole of economic life of changes in durability, that is, in the ratio of capital in the widest possible sense to income. I suspect that we have underestimated, even in our spendthrift society, the gains from increased durability, and that this might very well be one of the places where the price system needs correction through government-sponsored research and development. The problems which the spaceship earth is going to present, therefore, are not all in the future by any means, and a strong case can be made for paying much more attention to them in the present than we now do.

It may be complained that the considerations I have been putting forth relate only to the very long run, and they do not much concern our immediate problems. There may be some justice in this criticism, and my main excuse is that other writers have dealt adequately with the more immediate problems of deterioration in the quality of the environment. It is true, for instance, that many of the immediate problems of pollution of the atmosphere or of bodies of water arise because of the failure of the price system, and many of them could be solved by corrective taxation. If people had to pay the losses due to the nuisances which they create, a good deal more resources would go into the prevention of nuisances. These arguments involving external economies and diseconomies are familiar to economists and there is no need to recapitulate them. The law of torts is quite inadequate to provide for the correction of the price system which is required, simply because where damages are widespread and their incidence on any particular person is small, the ordinary remedies of the civil law are quite inadequate and inappropriate. There needs, therefore, to be special legislation to cover these cases, and though such legislation seems hard to get in practice, mainly because of the widespread and small personal incidence of the injuries, the technical problems involved are not insuperable. If we were to adopt in principle a law for tax penalties for social damages, with an apparatus for making assessments under it, a very large proportion of current pollution and deterioration of the environment would be prevented. There are tricky problems of equity involved, particularly where old established nuisances create a kind of "right by purchase" to perpetuate themselves, but these are problems again which a few rather arbitrary decisions can bring to some kind of solution.

The problems which I have been raising in this paper are of larger scale and perhaps much harder to solve than the more practical and immediate problems of the above paragraph. Our success in dealing with the larger problems, however, is not unrelated to the development of skill in the solution of the more immediate and perhaps less difficult problems. One can hope, therefore, that as a succession of mounting crises, especially in pollution, arouse public opinion and mobilize support for the solution of the immediate problems, a learning process will be set in motion which will eventually lead to an appreciation of and perhaps solutions for the larger ones. My neglect of the immediate problems, therefore, is in no way intended to deny their importance, for unless we make at least a beginning on a process for solving the immediate problems we will not have much chance of solving the larger ones. On the other hand, it may also be true that a long-run vision, as it were, of the deep crisis which faces mankind may predispose people to taking more interest in the immediate problems and to devote more effort for their solution. This may sound like a rather modest optimism, but perhaps a modest optimism is better than no optimism at all.

Notes

1. Ludwig von Bertalanffy, *Problems of Life* (New York: Wiley, 1952).

2. Richard L. Meier, *Science and Economic Development* (New York: Wiley, 1956).

3. Kenneth E. Boulding, "The Consumption Concept in Economic Theory," *American Economic Review* 35, no. 2 (May 1945), 1–14; and "Income or Welfare?," *Review of Economic Studies* 17 (1949–1950), 77–86.

4. Fred L. Polak, *The Image of the Future*, vols. I and II, translated by Elise Boulding (New York: Sythoff, Leyden and Oceana, 1961).

17 Spaceship Earth Revisited

Kenneth E. Boulding

I have no claims to priority in the use of the metaphor spaceship earth, though I think I did think it up independently, and it was a metaphor so appropriate to its time that it would have been very surprising if somebody had not thought it up. It is still a very good metaphor, but as with all metaphors, one has to be careful with it. When a metaphor parades as a model, it can sometimes be very dangerous and misleading, particularly as metaphors are so much more convincing than models and are much more apt to change people's images of the world. The spaceship metaphor stresses the earth's smallness, crowdedness, and limited resources; the need for avoiding destructive conflict; and the necessity for a sense of world community with a very heterogeneous crew. On these grounds the metaphor is certainly as good today as it was in the 1960s. One of the paradoxes, indeed, is that we seem to be more theoretically aware of the spaceship earth model than we were in the 1960s, with the Club of Rome reports, the energy crisis, and the United Nations population conference. But this theoretical awareness does not seem to have penetrated down to the level of political consciousness in the life and awareness of the ordinary human being. It seems to be very hard to organize a long-run crisis. Certainly the ordinary American today has very little sense of crisis. Every time he turns a switch, the lights go on; every time he pulls up to a gas station, there is gas for his car. The problems of the 1970s—inflation, the arms race, rising crime rates, battered children, tax revolts, and so on—are far more political and sociological than they are related to the long-run problems of a society nonsustainable in terms of energy, materials, and pollution.

Written especially for the preceding edition of this book.

Furthermore, something has happened in the 1970s that may have an enormous potential significance for the human race or may have none at all: the development of reasonably serious proposals for space colonies. This may sound like science fiction, and there may well be some unknown factor that would make them impossible. The idea is not, however, absurd, simply because of the low energy requirements of moving materials about in empty space. On earth we need energy mainly to overcome gravity and friction.

Solar energy, the great white hope of the virtuous, does not look very much better today than it did in the sixties, at least on the surface of the earth. It is fine for hot-water heating, moderately fine for space heating—but this is only 20 percent of our energy use. It still looks very expensive for electricity in spite of the tremendous advance in photovoltaic cells, and we are certainly no further along than we were in regard to solar-developed fuel. The essential unpredictability of knowledge and technology, of course, means that in another ten years the whole situation may look very different. The probability of our failing to solve the problem of cheap solar energy in usable forms is at least high enough to be worrying. In outer space, on the other hand, the sun never sets. Solar energy is much more available. The possibility of capturing it to transport materials from the moons and asteroids is, at least on paper, not preposterous.

Going into space, however, would be a transition for the evolutionary process as great as that involved in going from sea to land. One of the group of Princeton scientists who have been spearheading the interest in space colonies, in a paper at the American Association for the Advancement of Science in 1978, proposed, perhaps a little tongue-in-cheek, that the carrying capacity of the solar system for human beings might be one sextillion—about a billion times the present population of the earth. This, one presumes, would involve filling a wide band on either side of the orbit of the earth with space colonies using solar energy and mining the asteroids for materials. This is outrageous, but it may be no more outrageous than the whole process of evolution. Perhaps the greatest case for space colonies is that they would reintroduce isolation and variety into the evolutionary process. The most worrying thing about the earth is that there seems to be no way of preventing it from becoming one world. If there is only one world, then if anything goes wrong, everything goes wrong. And by the generalized Murphy's Law, every system has some positive probability, however low, of irretrievable catastrophe. Evolution is able to persist on the earth because of the isolation and variety of its ecosystems. Thus a total catastrophe on Krakatoa was recoverable because it did not affect the more distant ecosystems, which then recolonized the destroyed island. Similarly, the total collapse of the Mayan Empire, an early example indeed of a Club of Rome report, affected Charlemagne or the Emperor of China not at all. They knew nothing of it.

Ultimately, of course, we must face the spaceship earth on earth. Uncertainty, however, is the principal property of the future, and time horizons themselves have an irreducible uncertainty about them. Perhaps the greatest weakness of the metaphor is that the spaceship presumably has a clear destination and a mission to accomplish. It is essentially a planned economy. The evolutionary process, however, is not a significant planned economy any more than an ecosystem. The biological ecosystem is not even a community, in spite of the fact that biologists sometimes call it that, it is the wildest example of free private enterprise and does not even have a mayor. I have argued that evolution is now moving out of the *biogenetic,* that is, DNA, genes and all that, into the *noogenetic,* which is the transmission of learned structures from one generation to the next by a teaching and learning process. This is a transition that begins with the human race. It may be as profound a transition as the development of DNA itself. I get the uneasy feeling sometimes that the human race is the link between the biogenetic and the noogenetic, and we may eventually produce a self-reproducing solid-state intelligence that would be our evolutionary successor. I won't like it, having great race prejudice in favor of the human race, but we have to admit that all species are endangered and that every species is a link, even biologically, between the biogenetic know-how of the gene that preceded it and those that will follow it. It would be presumptuous of us to think that the human race is any more than a link in the great evolutionary process of the universe that moves majestically from the unknown Alpha to the even more unknown Omega.

18 Using Economic Incentives to Maintain Our Environment

T. H. Tietenberg

Environmental regulators and lobbying groups with a special interest in environmental protection in the United States have traditionally looked upon the market system as a powerful and potentially dangerous adversary. That the market has unleashed powerful forces that clearly have acted to degrade the environment has been widely lamented. At the same time, growth proponents have traditionally seen environmental concerns as blocking projects that had the potential to raise living standards significantly.

Conflict and confrontation became the modus operandi for dealing with this clash of objectives. However, the climate for dealing effectively with both concerns has improved dramatically in the last few years. Not only have growth proponents learned that in many cases short-term wealth enhancement projects that degrade the environment are ultimately counterproductive, but environmental groups have come to realize that poverty itself is a major threat to environmental protection.

No longer are economic development and environmental protection seen as an "either-or" proposition. Sustainability has become an important, if still somewhat vaguely defined, criterion for choosing among alternative economic growth paths. But the focus has shifted toward the identification of policies or policy instruments that can promise the alleviation of poverty while protecting the environment.

One approach that is generating much interest is known as an economic incentives approach to environmental regulation. Instead of mandating prescribed actions, such as requiring the installation of a particular piece of control equipment, this approach achieves envi-

Reprinted from *Challenge*, March-April 1990, pp. 42–46, by permission of the author and the publisher.

ronmental objectives by changing the economic incentives of the agents. This can be done with fees or charges, transferable permits, or even liability law. By changing the incentives an individual agent faces, that agent can use his typically superior information to select the best means of meeting his assigned responsibility.

Environmental groups and regulators have come to realize that the power of the market can be harnessed by economic incentive policies for the achievement of environmental goals. The change in attitude was triggered by a recognition that this former adversary, the market, can be turned into a powerful ally. Among their other virtues, approaches relying on economic incentives can reduce the conflict between environmental protection and economic development, can ease the transition to a sustainable (rather than exploitive) relationship between the economy and the environment, and can encourage the development of new, more environmentally benign production processes. Furthermore, economic incentive approaches have been used in the United States and Europe since the mid-1970s, giving us a wealth of experience with how they work in practice.

To be sure, experiences in one country do not automatically transfer to another. The translation of experiences from heavily industrialized nations to less industrialized nations is particularly difficult. Differences in cultural traditions, stages of development, and political superstructure, to name but a few, are formidable barriers to the diffusion of ideas. But we have discovered that at least a portion of what we have learned is relevant.

If the scientists are correct in their assessment of the effect of human activities on the greenhouse effect, the industrialized world has engaged in a development path that in some ways ill prepares it for the future. As the necessity for limiting carbon dioxide intensifies, the types of high-energy, fossil fuel–based production processes that are currently the rule will experience a relative decline in competitiveness as their true social costs become recognized and internalized.

Transforming an existing industrial structure to reposition it for the future will be an expensive proposition. The less industrialized nations have an advantage in that they can design their industrial base in such a way as to position themselves for the environmental realities posed by the future. Anticipating change is almost always cheaper than retrofitting after the fact.

Let me offer an example to illustrate the point: I recently had the opportunity to consult for both California and Sweden on ways of bringing air quality under control. The contrast between the problems these two faced was striking. Sweden has high gasoline prices and a well-developed public transit system. Land use patterns are configured to take advantage of this type of transportation system. Los Angeles on the the other hand is a city that depends almost exclusively on the automobile. Because it is so spread out and the population is so dispersed, a public transit alternative is difficult to implement. Efficient use of mass transit requires the existence of high-density travel corridors, a condition that exists in much of Sweden but not in Los Angeles. The regulators in Los Angeles face a much more difficult problem, not only because of the rather unique geography of their city, but also because existing land use patterns reduce the possible control options significantly. The dispersed development strategy, which was made possible by low gasoline prices, has made achieving the ambient standards much more difficult and much more expensive than it would have been had the same population been encompassed within a more energy-efficient land use pattern from the beginning.

How can economic incentives be used to provide the kinds of signals that will make sustainable development possible? Perhaps the best way to answer this question is to show how this approach has worked in practice.

Saving the New Zealand Fishery

One application involves the use of an economic incentives approach to reduce the stress on an already overexploited renewable resource. Due to their desirability and the traditional open access to the fishery, the populations of several species of fish were being depleted off the coast of New Zealand. While the need to reduce the amount of pressure being put on the populations was rather obvious, how to accomplish that reduction was not at all obvious.

Although it was relatively easy to prevent new fishermen from entering the fisheries, it was harder to figure out how to reduce the pressure from those who had been fishing in the area for years or even decades. Because fishing is characterized by economies of scale, simply reducing everyone's catch proportionately wouldn't make much sense. That would simply place higher costs on everyone and waste a great deal of fishing capacity as all boats sat around idle for a significant proportion of time. A better solution would clearly be to have fewer boats harvesting the stock. That way each boat could be used closer to its full capacity without depleting the populations. Which fishermen should be asked to give up their livelihood and leave the industry?

The economic incentive approach addressed this problem by imposing transferable catch quotas on all fish harvested from the fishery. The revenues derived from the annual fees associated with these quotas were used to buy out fishermen who were willing to forego any future fishing. Essentially each fisherman stated the lowest price that he or she would accept for leaving the industry; the regulators selected those who could be induced to leave at the lowest price, paid the stipulated amount from the revenues, and retired their licenses to fish for the particular species.

It wasn't long before a sufficient number of licenses were retired and the populations protected. Because the program was voluntary, those who left the industry only did so when they had been adequately compensated. Meanwhile, those who paid the fees realized that this small investment would benefit them greatly in the future as the populations recovered.

Economic incentives can not only be used to reduce the conflict between economic development and environmental protection, they can make economic development the vehicle by which greater environmental protection is achieved. In the mid-1970s, several geographic US regions were in violation of air quality standards designed to protect human health.

At that point, the law provided that new industries would not be allowed to move into these areas if they added any more of the pollutant responsible for the standard being violated. Since even those potential entrants adopting the most stringent control technologies would typically add some of the pollutant, this was a serious political blow to mayors eager to expand their employment and tax base. How could they allow economic growth while assuring that air quality would steadily improve to the level dictated by the ambient standard?

Regulators adopted the economic incentive approach known as the "offset policy." Under this policy, firms already established in these polluted areas that chose to voluntarily control their emissions more than required under the prevailing regulations would be allowed to have those excess emission reductions certified as "emission reduction credits." Once certified, their operating permits would be tightened to assure that the reductions were permanent. These emission reduction credits could then be sold to new firms seeking to move into the city, provided that the acquiring firm bought 1.2 emission reduction credits for each 1.0 units of emissions added by the new plant. Air quality improved every time a new firm moved into the area.

Though definitive data are not available, several estimates suggest that some 2,000 to 2,500 offset transactions have taken place. With this policy, the confrontation between economic growth and environmental protection was diffused. New firms were not only allowed to move into polluted cities, but they became one of the main vehicles for improving the quality of the air. Economic growth facilitated, rather than blocked, air quality improvement.

Not long after this episode, it became clear that much tighter controls were needed on older, established sources of pollution in those areas still in violation of the standards. How would the responsibility for additional control be allocated among the sources? Because the emitters were extremely diverse, the menu of technologies was vast, and regulatory staffs were too small to pick out the best technology in each setting, providing a satisfactory answer to this question was no easy task. How were regulators going to solve the problem cost-effectively when they had so little hard information about the choices?

The Bubble Policy

Their solution was to put the market to work for them under what is now known as the "bubble policy." They started by imposing emission standards on the established sources in the most reasonable manner they could, recognizing that in practice many of these standards would inevitably turn out to be unreasonable in the glare of hindsight.

Established sources were then encouraged to create "emission reduction credits" for emission reductions over and above those required by the standards. Once certified, these credits could be stored for subsequent use by the creating source when it wished to expand, or transferred to another source for its use in meeting its standard.

Sources that could control their emissions relatively cheaply generally created and sold these credits, and sources that found themselves confronting very expensive options bought the credits in lieu of installing unreasonably expensive equipment. The result was that the costs of meeting ambient standards were considerably lower than they would have been if the standards had been imposed and no transfers permitted.

This policy has also introduced some extremely beneficial flexibility into the regulatory process. One of the emission reduction credit trades, for example, involved a firm that planned to shut down operations at one plant and build a new one. Under the old regulations, either it would be required to install very expensive pollution control equipment at its old plant, equipment that would be absolutely useless once the new plant was built, or it would simply be let off the hook until the new plant was built, possibly encouraging some foot-dragging in the construction process.

With the bubble policy, the plant was allowed to lease emission reduction credits from another nearby facility. By acquiring these credits the source guaranteed that the region benefited from the cleaner air that it would have received if it had been forced to install the temporary equipment, but at much lower cost and waste of resources. Additionally, since financial outlays were required to acquire these credits for the entire time until the new plant was operating, no incentive for foot-dragging was created.

Another source of flexibility introduced by this approach involves the capability to control previously unregulated sources. This may be a particularly appealing attribute for those countries with currently underdeveloped environmental regulatory systems. For a number of reasons, ranging from the financial difficulties faced by the firm to a failure by regulators to recognize a particular pollution source, some sources are either not regulated or are regulated less than would be socially desirable. What makes this situation particularly appalling is the fact that many of these sources could be controlled at a lower cost than those sources that have to be controlled to a correspondingly higher degree to compensate for their immunity.

The Emissions Trading Program, the umbrella program that contains both the bubble and offset policies described above, provided a solution. Immune sources could voluntarily control their emissions more than required by their standards and sell the credits to other sources. These other sources would gladly purchase the credits because this was significantly cheaper than controlling their own emissions to a higher degree.

In this way those firms that were already subject to a higher degree of control avoided having to ratchet their own controls up ever further to unreasonable levels to compensate for the lack of control in immune firms. In essence the responsibility for identifying additional sources of control had been transferred to the market.

The characteristic of emissions trading that makes the flexibility discussed in the previous paragraph possible is its ability to separate the question of what control is undertaken from the question of who ultimately pays for it. This characteristic is potentially extremely important, not only for domestic but for international environmental policy.

Global Issues

While I have not thought through this aspect completely, let me sow a few seeds that may trigger some thoughts of your own. Two of the "big issues" in global environmental protection are who should bear the costs of protecting resources of global significance and, to the extent international cost sharing is supported, how it can be accomplished. Emissions trading provides one possible vehicle for addressing the latter question.

Suppose, for example, that international agreements are ultimately signed requiring some reduction in the expected growth of carbon dioxide emissions. One proposal for achieving these reductions envisions creating a version of the offset policy for use in holding the emissions of greenhouse gases at their current levels.

Under this proposal, major new sources of greenhouse gases would be required to offset their emissions, assuring that the total emissions would not increase. Offsets could be generated by conservation, recycling, or by retiring older, heavily emitting plants. In the case of carbon dioxide, major new emitters could even invest in tree plantations that would tend to absorb the excess gas. By creating a new market for these offsets, an incentive to invest in offset-creating activities would be stimulated. Meanwhile, the higher prices associated with creating the greenhouse gases, due to the need to acquire offsets, would stimulate the creators to search for ways to reduce their emissions. Transferable offsets would therefore simultaneously encourage both source control and mitigating strategies.

Because the greenhouse gas pollutants are global in nature (the location of the emissions does not matter), global markets in offsets would be possible. This is a tremendously powerful characteristic because larger markets offer the opportunity for larger cost savings. Furthermore, by selling offsets, Third World countries could undertake environmentally sound investments (such as reforestation or protecting forests scheduled for harvest) that were financially underwritten by the developed world.

Another rather different example of how this concept can be applied involves water conservation in arid regions. In some parts of the United States, well-intentioned regulations have had an unintentional side effect: they have discouraged water conservation even when water is scarce. In essence, the regulations don't allow those conserving the water to recover any of the costs associated with conservation by transferring the conserved water to others willing to pay for it, so the incentive to conserve is lost.

On January 17, 1989, a historic agreement was negotiated between a growers association, a major user of irrigation water, and the Metropolitan Water District (MWD) of California, a public agency that supplies water to the Los Angeles area. Under that agreement the MWD will bear the capital and operating costs, as well as the indirect costs (such as reduced hydro power), of a huge program to reduce seepage losses as the water is transported to the growers and to install new water-conserving irrigation techniques.

In return, the growers will get all of the conserved water. Everyone gains. The district gets the water it needs at a reasonable price; the growers get virtually the same amount of irrigation benefits as they got before without being forced to bear large additional expenditures that would be difficult to pass along to consumers.

Judicial Remedies

The courts are beginning to get into the act as well; judicial remedies for environmental problems are beginning to take their place alongside regulatory remedies. Take, for example, the problem of cleaning up toxic waste sites that have already been closed. Current US law allows the government to sue all potentially responsible parties who contributed to the contaminated site (such as waste generators or disposal site operators).

These suits accomplish a double purpose: (1) they assure that the financial responsibility for contaminated sites is borne by those who directly caused the problem, and (2) they encourage those who are currently using those sites to exercise great care, lest they be forced to bear a large financial burden in the event of an incident. The traditional remedy of going to the taxpayers would have resulted in less revenue raised, fewer sites restored, and less adequate incentives for users to exercise care.

Europe has tended to depend more on another type of economic incentive, the effluent or emission charge. This approach places a per unit fee on each unit of pollution discharged. Faced with the responsibility for paying for the damage caused by their pollution, firms recognize it as a normal cost of doing business. This recognition triggers a search for possible ways to reduce the damage, including changing inputs, changing the production process, transforming the residuals to less harmful substances, and recycling by-products. The experience in the Netherlands, a country where the fees are higher than in most other countries, suggests that the effects can be dramatic. These fees have another desirable attribute—they raise revenue. Successful development, particularly sustainable development, requires a symbiotic partnership between the public and private sectors. To function as an equal partner the public sector must be adequately funded. If it fails to raise sufficient revenue, the public sector becomes a drag on the growth process, but if it raises revenue in ways that distort incentives, that too can act as a drag on development. Effluent or emission charges offer the realistic opportunity to raise revenue for the public sector without producing inefficient incentives.

Whereas other types of taxation discourage growth by penalizing legitimate development incentives, emission or effluent charges provide incentives for sustainable development. Some work from the United States suggests that the drag on development avoided by substituting effluent or emission charges for more traditional revenue-raising devices such as capital gains, income, and sales taxes could be significant.

Incentives for forward-looking public action are as important as those for private action. The current national income accounting system provides an example of a perverse economic incentive. Though national income accounts were never intended to function as a device for measuring the welfare of a nation, in practice that is how they are used. National income per capita is a common metric for evaluating how well off a nation's people are. Yet the current construction of those accounts sends the wrong signals. Rather than recognizing the *Exxon Valdez* spill for what it was, namely a decline in the value of the endowment of natural resources in the area, it is recorded as an increase in the national income. The spill boosted GNP!

All the clean-up expenditures served to increase national income, but no account was taken of the consequent depreciation of the natural environment. Under the current system, the accounts make no distinction between growth that is occurring because a country is "cashing in" its natural resource endowment with a consequent irreversible decline in its value, and sustainable growth where the value of the endowment remains. Only when suitable corrections are made to these accounts will governments be judged by the appropriate standards.

The power of economic incentives is certainly not inevitably channeled toward the achievement of sustainable growth. They can be misapplied as well as appropriately applied. Tax subsidies to promote cattle ranching on the fragile soil in the Brazilian rain forest stimulated an unsustainable activity, which has done irreparable damage to an ecologically significant area. They are such powerful devices for channeling market forces that it is particularly important that they be used judiciously.

We live in an age when the call for tighter environmental controls intensifies with each new discovery of yet another injury modern society is inflicting on the planet. But resistance to additional controls is also growing with the recognition that compliance with each new set of controls is more expensive than the last. While economic incentive approaches to environmental control offer no panacea, they frequently do offer a practical way to achieve environmental goals more flexibly and at lower cost than more traditional regulatory approaches. That is a substantial virtue.

19 The Steady-State Economy: Toward a Political Economy of Biophysical Equilibrium and Moral Growth

Herman E. Daly

There is nothing in front but a flat wilderness of standardization either by Bolshevism or Big Business. But it is strange that some of us should have seen sanity, if only in a vision, while the rest go forward chained eternally to enlargement without liberty and progress without hope.

G. K. Chesterton

The Concept of a Steady-State Economy

The steady-state economy (SSE) is defined by four characteristics:

1. A constant population of human bodies.

2. A constant population or stock of artifacts (exosomatic capital or extensions of human bodies).

3. The levels at which the two populations are held constant are sufficient for a good life and sustainable for a long future.

4. The rate of throughput of matter-energy by which the two stocks are maintained is reduced to the lowest feasible level. For the population this means that birth rates are equal to death rates at low levels so that life expectancy is high. For artifacts it means that production equals depreciation at low levels so that artifacts are long lasting, and depletion and pollution are kept low.

Only two things are held constant—the stock of human bodies and the total stock or inventory of artifacts. Technology, information,

---

Reprinted in part from the University of Alabama Distinguished Lecture Series, no. 2, 1971, by permission of the University of Alabama, and in part from *Steady-State Economics: The Economics of Biophysical Equilibrium and Moral Growth*, 1977, by permission of W. H. Freeman and Company, with revisions and additions by the author, 1979, 1992.

wisdom, goodness, genetic characteristics, distribution of wealth and income, product mix, and so on are *not* held constant. In the very long run, of course, nothing can remain constant, so our concept of a SSE must be a medium-run concept in which stocks are constant over decades or generations, not millennia or eons.

Three magnitudes are basic to the concept of a SSE:

1. *Stock* is the total inventory of producers' goods, consumers' goods, and human bodies. It corresponds to Irving Fisher's (1906) definition of capital and may be thought of as the set of all physical things capable of satisfying human wants and subject to ownership.
2. *Service* is the satisfaction experienced when wants are satisfied, or "psychic income" in Fisher's sense. Service is yielded by the stock. The quantity and quality of the stock determine the intensity of service. There is no unit for measuring service, so it may be stretching words a bit to call it a magnitude. Nevertheless, we all experience service or satisfaction and recognize differing intensities of the experience. Service is yielded over a period of time and thus appears to be a flow magnitude. But unlike flows, service cannot be accumulated. It is probably more accurate to think of service as a "psychic flux" (Georgescu-Roegen, 1966, 1971).
3. *Throughput* is the entropic physical flow of matter-energy from nature's sources, through the human economy and back to nature's sinks; it is necessary for maintenance and renewal of the constant stocks (Boulding, 1966; Daly, 1968; Georgescu-Roegen, 1971).

The relationship among these three magnitudes can best be understood in terms of the following simple identity (Daly, 1974):

$$\frac{\text{service}}{\text{throughput}} \equiv \frac{\text{service}}{\text{stock}} \times \frac{\text{stock}}{\text{throughput}}$$

The final benefit of all economic activity is service. The original useful stuff required for yielding service, which cannot be produced by man but only used up, is low-entropy matter-energy—in other words, the throughput. But throughput is not itself capable of directly yielding service. It must first be accumulated into a stock of artifacts; it is the stock that directly yields service. We can ride to town only in a member of the existing stock of automobiles. We cannot ride to town on the annual flow of automotive maintenance expenditures, nor on the flow of newly mined iron ore destined to be embodied in a new chassis, nor on the flow of worn rusting hulks in junkyards. Stocks may be thought of as throughput that has been accumulated and "frozen" in structured forms capable of satisfying human wants. Eventually the frozen structures are "melted" by entropy, and what flowed into the accumulated stocks from nature then flows back to nature in equal quantity, but in entropically degraded quality. Stocks are intermediate magnitudes that belong at the center of analysis and provide a clean separation between the cost flow and the benefit flux. On the one hand, stocks yield service; on the other hand, stocks require throughput for maintenance. Service yielded is benefit; throughput required is cost.

The identification of cost with throughput should not be interpreted as implying a "throughput or entropy theory of value." There are other costs, notably the disutility of labor and the accumulation time required to build up stocks. In the steady state we can forget about accumulation time since stocks are only being maintained, not accumulated. The disutility of labor can be netted out against the services of the stock to obtain net psychic income or net service. In the steady state, then, the value of net service is imputed to the stocks that render the service, which is in turn imputed to the throughput that maintains the stocks. It is in this sense that throughput is identified with cost. The opportunity cost of the throughput that maintains artifact A is the service sacrificed by not using that throughput to maintain more of artifact B. The throughput is a physical cost that is evaluated according to opportunity cost principles. However, the opportunity cost of the throughput must be evaluated not only in terms of alternative artifact services forgone (which the market does), but also in terms of natural ecosystem services forgone as a result of the depletion and pollution caused by the throughput (which escapes market valuation). Depletion reduces the service of availability of the resource to future people who cannot bid in present markets, and pollution reduces the ability of the ecosystem to perform its life support services. The true opportunity cost of an increment in throughput is the greater of the two classes: artifact service sacrificed and ecosystem services sacrificed. Thus throughput should be thought of as a cost-inducing physical flow, rather than identified with cost itself, which by definition must always be a sacrificed benefit, not a physical magnitude. In like manner the stock is a benefit-yielding physical magnitude and should not be identified with benefit or service itself. This is the case even for short-lived artifacts whose physical degradation is an immediate consequence of use, for instance, gasoline in the tank of a car.

We can arrive at the same basic result by following Irving Fisher's reasoning. Fisher (1906) argued that every intermediate transaction involves both a receipt and an expenditure of identical magnitude which cancel out in aggregation of the total income of the community. But once the final user has obtained the asset, there is no further exchange and canceling of accounts among individuals. The service yielded by the asset to the final consumer is the "uncanceled fringe" of psychic income, the final uncanceled benefit left over after all intermediate transactions have canceled out. Subtracting the psychic disservices of labor, Fisher arrived at *net psychic income,* the final net benefit of all economic activity. It is highly interesting that Fisher did not identify any original, uncanceled, real cost against which the final value of net psychic income should be balanced. Here we must supplement Fisher's vision with the more recent visions and analyses of Boulding (1966) and Georgescu-Roegen (1966, 1971) concerning the physical basis of cost. As everyone recognizes, the stock of capital wears out and has to be replaced. This continual maintenance and replacement is an unavoidable cost inflicted by entropy. Fisher treated it as canceling out in the aggregate: house repair is income to the account of the carpenter and an identical outgo to the account of the house. But Fisher did not trace the chain all the way back to any uncanceled fringe at the beginning that would correspond to uncanceled final costs in the same way that net psychic income corresponds to uncanceled final benefits. If we do this, we come to the unpaid contribution from nature: the provision of useful low-entropy matter-energy inputs and the absorption of high-entropy waste matter-energy outputs. These contributions from nature have no costs of production, only a cost of extraction or disposal, which is paid and enters the canceling stream of accounts. But we do not pump any money down into a well as we pump oil out, nor do we dump dollars into the sea along with our chemical and radioactive wastes. If service is an "uncanceled fringe," then so is throughput. In other words, if we consolidate the accounts of all firms and households, everything cancels out except service and throughput.

In the SSE a different behavior mode is adopted with respect to each of the three basic magnitudes. (1) *Stock* is to be *"satisficed"*—maintained at a level that is sufficient for an abundant life for the present generation and ecologically sustainable for a long (but not infinite) future.[1] (2) *Service* is to be *maximized,* given the constant stock. (3) *Throughput* is to be *minimized,* given the constant stock. In terms of the two ratios on the right-hand side of the identity, this means that the ratio (service/stock) is to be maximized by maximizing the numerator with the denominator constant, while the ratio (stock/throughput) is maximized by minimizing the denominator with the numerator constant. These two ratios measure two kinds of efficiency: service efficiency and maintenance efficiency.

Service efficiency (service/stock) depends on allocative efficiency (does the stock consist of artifacts that people most want and are they allocated to the most important uses?), and on distributive efficiency (is the distribution of the stock among alternative people such that the trivial wants of some people do not take precedence over the basic needs of others?). Standard economics has much of value to say about allocative efficiency, but it treats distribution under the heading of social justice rather than efficiency, thus putting it on the sidelines of disciplinary concern. Although neoclassical economists carefully distinguish allocation from distribution in static analysis, they seem not to insist on any analogous distinction between intertemporal allocation (one person allocating over different stages of his lifetime) and intertemporal distribution (distribution between different people, that is, present people and future people). Intertemporal distribution is a question of ethics, not a function of the interest rate. The notion of optimal allocation over time must be confined to a single lifetime unless we are willing to let ethics and distributional issues into the definition of optimum. Neoclassical economics seems inconsistent, or at least ambiguous, on this point.

Maintenance efficiency (stock/throughput) depends on durability (how long an individual artifact lasts) and on replaceability (how easily the artifact can be replaced when it finally does wear out). Maintenance efficiency measures the number of units of time over which a population of artifacts yields its service, while service efficiency measures the intensity of that service per unit of time. Maintenance efficiency is limited by the entropy law (nothing lasts forever; everything wears out). Service efficiency may conceivably increase for a very long time, since the growing "magnitude," service, is nonphysical. There may, however, be physical limits to the capacity of human beings to experience service. But the definition of the SSE is in terms of physical stocks and throughput and is not affected by whether or not service could increase indefinitely.

Conceptually it is easier to think of stock as the operational policy variable to be directly controlled. Practically, however, as will be seen below, it would be easier to control or limit throughput directly and allow the stock to reach the maximum level sustainable by the fixed throughput. This presents no problems.

The above concepts allow us to make an important distinction between growth and development. *Growth* refers to an increase in service that results from an increase in stock and throughput, with the two efficiency ratios constant. *Development* refers to an increase in the efficiency ratios, with stock constant (or alternatively, an increase in service with throughput constant). Using these definitions, we may say that a SSE develops but does not grow, just as the planet earth, of which it is a subsystem, develops without growing.

How do these concepts relate to GNP, the more conventional index of "growth"? GNP makes no distinction among the three basic magnitudes. It simply adds up value estimates of some services (the service of those assets that are rented rather than purchased, including human bodies, and omitting the services of all owned assets not rented during the current year, with the exception of owner-occupied houses), plus the value of the throughput flow (maintenance and replacement expenditures required to maintain the total stock intact), plus the value of current additions to stock (net investment). What sense does it make to add up benefits, costs, and change in inventory? Services of the natural ecosystem are not counted, and, more important, services sacrificed are not subtracted. In fact, defensive attempts to repair the loss of ecosystem services are added to GNP. The concept of a SSE is independent of GNP, and what happens to GNP in the SSE simply does not matter. The best thing to do with GNP is to forget it. The next best thing is to try to replace it with two separate social accounts, one measuring the value of service (benefit) and the other measuring the value of throughput (cost). In this way costs and benefits could be compared, although this aggregate macrolevel comparison is not at all essential, since regardless of how it turns out the behavior modes remain the same with respect to each of the three basic magnitudes. If we really could get operational cost and benefit accounts, then we might optimize the level of stocks by letting it grow to the point where the marginal cost of an addition to stock just equals the marginal benefit. But that is so far beyond our ability to measure that satisficing will for a long time remain a better strategy than optimizing. Aggregate economic indices should be treated with caution, since there are always some kinds of stupid behavior that would raise the index and thus become "justified."

Neither the concept nor the reality of a SSE is new. John Stuart Mill (1881) discussed the concept in his famous chapter on the stationary state. Historically, people have lived for 99 percent of their tenure on earth in conditions very closely approximating a steady state. Economic growth is essentially a phenomenon of the last 200 years, and only in the last 50 years has it become the dominant goal of nations. Growth is an aberration, not the norm. Development can continue without growth and is, in fact, more likely under a SSE than under a growth economy.

Even "cornucopians" like Weinberg and Goeller (1976) evidently consider a SSE to be a precondition for achieving their Age of Substitutability, in which "society will settle into a steady state of substitution and recycling . . . assuming, of course, a stable population." But why postpone the SSE to some hypothetical future age? Why not seek to come to terms with the SSE now, before we use up the remaining easily available resources that could help in making the transition? Why continue to fan the fires of growth up to the point where the flame's appetite is so voracious that even to maintain it in a steady state would require technologies and social institutions that are so demanding and unforgiving as to reduce the quality of life to that of a regimented community of social insects? Freedom is in large measure a function of slack, of the distance between maximum carrying capacity and actual load. A system operating at its carrying capacity has no room for error or for the freedom that permits error.

Social Institutions

The social institutions of control for a SSE are of three kinds: those for maintaining a constant population, those for maintaining a constant stock of physical wealth, and those governing distribution. In all cases the guiding design principle for social institutions is to provide the necessary control with a minimum sacrifice of personal freedom, to provide macrostability while allowing for microvariability, to combine the macrostatic with the microdynamic (Luten, n.d.).

The Distribution Institution

The critical institution is likely to be that of the minimum and maximum limits on income and the maximum limit on wealth. Without some such limits private property and the whole market economy lose their moral basis, and there would be no strong case for extending the market to cover birth quotas and depletion quotas as a means of institutionalizing environmental limits. Exchange relations are mutually beneficial among relative equals. Exchange between the powerful and the powerless is often only nominally voluntary and can easily be a mask for exploitation, especially in the labor market, as Marx has shown.

There is considerable political support for a minimum income, financed by a negative income tax, as an alternative to bureaucratic welfare programs. There is no such support for maximum income or maximum wealth limits. In the growth paradigm there need be no upper limit. But in the steady-state paradigm there must be an upper limit to the total, and consequently an upper limit to per capita income as well. A minimum wealth limit is not feasible, since we can always spend our wealth and could hardly expect to have it restored year after year. The minimum income would be sufficient. But maximum limits on both wealth and income are necessary, since wealth and income are largely interchangeable, and since, beyond some point, the concentration of wealth becomes inconsistent with both a market economy and political democracy. John Stuart Mill (1881) put the issue very well:

> Private property, in every defense made of it, is supposed to mean the guarantee to individuals of the fruits of their own labor and abstinence. The guarantee to them of the fruits of the labor and abstinence of others, transmitted to them without any merit or exertion of their own, is not of the essence of the institution, but a mere incidental consequence, which, when it reaches a certain height, does not promote, but conflicts with, the ends which render private property legitimate.

According to Mill, private property is legitimated as a bastion against exploitation. But this is true only if everyone owns some minimum amount. Otherwise, private property, when some own a great deal of it and others have very little, becomes the very *instrument* of exploitation rather than a guarantee against it. It is implicit in this view that private property is legitimate only if there is some distributist institution (as, for example, the Jubilee year of the Old Testament) that keeps inequality of wealth within justifiable limits. Such an institution is now lacking. The proposed institution of maximum wealth and income plus minimum income limits would remedy this severe defect and make private property legitimate again. It would also go a long way toward legitimating the free market, since most of our blundering interference with the price system (e.g., farm programs, minimum wage, rent controls) has as its goal an equalizing alteration in the distribution of income and wealth. Thus such a distributist policy is based on impeccably respectable premises: private property, the free market, opposition to welfare bureaucracies and centralized control. It also heeds the radicals' call of "power to the people," since it puts the source of power, namely property, in the hands of the many people, rather than in the hands of the few capitalist plutocrats and socialist bureaucrats.

The concept of private property here adopted is the classical view of John Locke, Thomas Jefferson, and the Founding Fathers. It is emphatically not the apologetic doctrine of big business that the term *private property* evokes today. Limits are built into the very notion of property, according to Locke (quoted in McClaughry, 1974, p. 31):

> Whatsoever, then, a man removes out of the state that nature hath provided and left it in, he hath mixed his labor with it, and joined to it something that is his own, and thereby makes it his property. But how far has God given property to us to enjoy? As much as anyone can make use of to any advantage of life before it spoils, so much may he by his labor fix his property in. Whatever is beyond this is more than his share, and belongs to others.

Clearly, Locke had in mind some maximum limit on property, even in the absence of general scarcity. Locke assumed, reasonably in his time, that resources were superabundant. But he insisted that the right to property was limited. Growing resource scarcity reinforces this necessity of limits. Some of the correlates of this view of private property are listed by McClaughry (1974, p. 32):

> Property should be acquired through *personal effort;* it is a reward for diligent industry and fair dealing. An inheritance or windfall may look and feel like property, and even be used as property, but it lacks this essence of reward for personal effort.
>
> Property implies *personal control* and individual responsibility. Where the putative owners are far removed from the men who make the decisions about the use of their wealth, this aspect of personal and individual responsibility is absent, and this wealth becomes something less than true property.

> Property is relative to *human need*. That which is accumulated beyond an amount necessary to suffice for the human needs of its owner and his family is no longer property, but surplus wealth.
>
> Although to own a home and a car is to own property, and although possession of these consumer goods may have important effects upon the owner and his community, Locke and his successors thought of property as productive—yielding goods or services for exchange with others in the community—concentrated wealth means concentrated power—power to dominate other men, power to protect privilege, power to stifle the American Dream.

Maximum limits on income and wealth were an implicit part of the philosophy of all the prominent statesmen of early America except Alexander Hamilton.

Maximum income and wealth would remove many of the incentives to monopolistic practices. Why conspire to corner markets, fix prices, and so forth, if you cannot keep the loot? As for labor, the minimum income would enable the outlawing of strikes, which are rapidly becoming intolerably exploitative of the general public. Unions would not be needed as a means of confronting the power of concentrated wealth, since wealth would no longer be concentrated. Indeed, the workers would have a share of it and thus would not be at the mercy of an employer. In addition, some limit on corporate size would be needed, as well as a requirement that all corporate profits be distributed as dividends to stockholders.

With no large concentrations in wealth and income, savings would be smaller and would truly represent abstinence from consumption rather than surplus remaining after satiation. There would be less expansionary pressure from large amounts of surplus funds seeking ever new ways to grow exponentially and leading to either physical growth, inflation, or both.

The minimum income could be financed out of general revenues, which, in addition to a progressive income tax within the income limits, would also include revenues from the depletion quota auction (to be discussed below) and 100 percent marginal tax rates on wealth and income above the limits. Upon reaching the maximum, most people would devote their further energies to noneconomic pursuits, so that confiscatory revenues would be small. But the opportunities thus forgone by the wealthy would be available to the not so wealthy, who would still be paying taxes on their increased earnings. The effect on incentive would be negative at the top but positive at lower levels, leading to a broader participation in running the economy. If the maximum and minimum were to move so close together that real differences in effort could not be rewarded and incentives were insufficient to call forth the talent and effort needed to sustain the system, then we should have to widen the limits again or simply be content with the lower level of wealth that could be maintained within the narrower distributive limits. Since we would no longer be anxious to grow, the whole question of incentives would be less pressing. There might also be an increase in public service by those who have hit the maximum. As Jonathan Swift argued (1958, p. 1003):

> In all well-instituted commonwealths, care has been taken to limit men's possessions; which is done for many reasons, and, among the rest, for one which, perhaps, is not often considered; that when bounds are set to men's desires after they have acquired as much as the laws will permit them, their private interest is at an end, and they have nothing to do but to take care of the public.

Transferable Birth Licenses

This idea was first put forward in 1964 by Kenneth Boulding (1964, pp. 135–136). Hardly anyone has taken it seriously, as Boulding knew would be the case. Nevertheless, it remains the best plan yet offered, if the goal is to attain aggregate stability with a minimum sacrifice of individual freedom and variability. It combines macrostability with microvariability. Since 1964 we have experienced a great increase in public awareness of the population explosion and an energy crisis, and we are now experiencing the failures of the great "technological fixes" (green revolution, nuclear power, and space). This has led at least one respected demographer to take Boulding's plan seriously, and more will probably follow (Heer, 1975).

So many people react so negatively to the birth license plan that I should emphasize that the other two institutions (distributive limits and depletion quotas) do not depend on it. The other two proposals could be accepted and the reader can substitute his own favorite population control plan if he is allergic to this one.

The plan is simply to issue equally to every person (or perhaps only to every woman, since the female is the limitative factor in reproduction, and since maternity is more demonstrable than paternity) an amount of reproduction licenses that corresponds to replacement fertility. Thus each woman would receive 2.1 licenses. The licenses would be divisible in units of one-tenth, which Boulding playfully called the "deci-child." Possession of ten deci-child units confers the legal right to one birth. The licenses are freely transferable by sale or gift, so those who want more than two children and can afford to buy the extra licenses, or can acquire them by gift, are free to do so. The original distribution of the licenses is on the basis of strict equality, but exchange is permitted, leading to a reallocation in conformity with differing preferences and abilities to pay. Thus distributive equity is achieved in the original distribution, and allocative efficiency is achieved in the market redistribution.

A slight amendment to the plan might be to grant 1.0 certificates to each individual (or 2.0 to each woman) and have these refer not to births but to "survivals." If a female dies before having a child, then her certificate becomes a part of her estate and is willed to someone else, for example, her parents, who either use it to have another child or sell it to someone else. The advantage of this modification is that it offsets existing class differentials in infant and child mortality. Without the modification, a poor family desiring two children could end up with two infant deaths and no certificates. The best plan, of course, is to eliminate class differences in mortality, but in the meantime this modification may make the plan initially easier to accept. Indeed, even in the absence of class mortality differentials the modification has the advantage of building in a "guarantee."

Let us dispose of two common objections to the plan. First, it is argued that it is unjust because the rich have an advantage. Of course, the rich *always* have an advantage, but is their advantage increased or decreased by this plan? Clearly it is decreased. The effect of the plan on income distribution is equalizing because (1) the new marketable asset is distributed equally; and (2) as the rich have more children, their family per capita incomes are lowered; as the poor have fewer children their family per capita incomes increase. From the point of view of the children, there is something to be said for increasing the probability that they will be born richer rather than poorer. Whatever injustice there is in the plan stems from the prior existence of rich and poor, not from Boulding's idea, which actually reduces the degree of injustice. Furthermore, income and wealth distribution are to be controlled by a separate institution, discussed above, so that in the overall system this objection is more fully and directly met.

A more reasonable objection concerns the problem of enforcement. What to do with law-breaking parents and their illegal children? What do we do with illegal children today? Often they are put up for adoption. Adoption could be encouraged by paying the adopting parents the market value, plus subsidy if need be, for their license, thus retiring a license from circulation to compensate for the child born without a license. Like any other lawbreakers, the offending parents would be subject to punishment. The punishment need not be drastic or unusual. Of course, if everyone breaks a law, no law can be enforced. The plan presupposes the acceptance by a large majority of the public of the morality and necessity of the law. It also presupposes widespread knowledge of contraceptive practices and perhaps legalized abortion as well. But these presuppositions would apply to any institution of population control except the most coercive.

Choice may be influenced in two ways: by acting on or "rigging" the *objective* conditions of choice (prices and incomes in a broad sense), or by manipulating the *subjective* conditions of choice (preferences). Boulding's plan imposes straightforward objective constraints and does not presumptuously attempt to manipulate people's preferences. Preference changes due to individual example and moral conversion are in no way ruled out. If preferences should change so that, on the average, the population desired replacement fertility, the price of a certificate would approach zero and the objective constraint would automatically vanish. The current decline in the birth rate has perhaps already led to such a state. Maybe this would be a good time to institute the plan, so that it would already be in place and functioning, should preferences change toward more children in the future. The moral basis of the plan is that everyone is treated equally, yet there is no insistence upon conformity of preferences, the latter being the great drawback of "voluntary" plans that rely on official moral suasion, Madison Avenue techniques, and even Skinnerian behavior control. Which is the greater affront to the individual—to be forbidden what he wants for objective reasons that he and everyone else ought to be able to understand, or to get what he "wants" but to be badgered and manipulated into "wanting" only what is collectively possible? Some people, God bless them, will never be brainwashed, and their individual nonconformity wrecks the moral basis (equal treatment) of "voluntary" programs.

Kingsley Davis (1973, p. 28) points out that population control is not a technological problem.

The solution is easy as long as one pays no attention to what must be given up. For instance a nation seeking ZPG could shut off immigration and permit each couple a maximum of two children, with possible state license for a third. Accidental pregnancies beyond the limit would be interrupted by abortion. If a third child were born without a license, or a fourth, the mother would be sterilized and the child given to a sterile couple. But anyone enticed into making such a suggestion risks being ostracized as a political or moral leper, a danger to society. He is accused of wanting to take people's freedom away from them and institute a Draconian dictatorship over private lives. Obviously then reproductive freedom still takes priority over population control. This makes a solution of the population problem impossible because, by definition, population control and reproductive freedom are incompatible.

The key to population control is simply to be willing to pay the cost. The cost of the plan here advocated seems to me less than the cost of Davis's hypothetical suggestion because it allows greater diversity—families need not be so homogeneous in size, and individual preferences are respected to a greater degree. Moreover, should it become necessary or desirable to have negative population growth (as I believe it will), the marketable license plan has a great advantage over those plans that put the limit on a flat child-per-family basis. This latter limit could be changed only by an integral number, and to go from two children to one child per family in order to reduce population is quite a drastic change. In the Boulding scheme of marketable licenses issued in deci-child units or one-tenth of a certificate, it would be possible gradually to reduce population by lowering the issue to 1.9 certificates per woman, to 1.8, and so on, the remaining 0.1 or 0.2 certificates being acquired by purchase or gift.

Part of our difficulty in accepting the transferable license plan is that it is so direct. It frankly recognizes that reproduction must henceforth be considered a scarce right and logically faces the issue of how best to distribute that right and whether and how to permit voluntary reallocation. But there is an amazing preference for indirect measures—find new roles for women, change the tax laws, restrict public housing to small families, encourage celibacy and late marriage, be more tolerant of homosexuality, convince people to spend their money on consumer durables rather than having children, make it popular to have children only between the ages of twenty and thirty, and so forth.

Whence this enormous preference for indirectness? It results partly from our unwillingness to really face the issue. Limiting reproduction is still a taboo subject that must be approached in contorted and roundabout ways rather than directly. Furthermore, roundaboutness and indirectness are the bread and butter of empirical social scientists, who get grants and make their reputations by measuring the responsiveness of the birth rate to all sorts of remote "policy variables." The direct approach makes estimation of all these social parameters governing tenuous chains of cause and effect quite unnecessary. If the right to reproduce were directly limited by the marketable license plan, then the indirect measures would become means of adjusting to the direct constraint. For example, with reduced childbearing, women would naturally find other activities. The advantage of the direct approach is that individuals would be free to make their own personal specific adjustments to the general objective constraint, rather than having a whole set of specific constraints imposed on them in the expectation that it would force them indirectly to decide to do what objectively must be done. The direct approach is more efficient and no more coercive. But the direct approach requires clarity of purpose and frank objectives, which are politically inconvenient when commitment to the objective is halfhearted to begin with.

There is an understandable reluctance to couple money and reproduction—somehow it seems to profane life. Yet life is physically coupled to increasingly scarce resources, and resources are coupled to money. If population growth and economic growth continue, then even free resources, such as breathable air, will become either coupled to money and subject to price or allocated by a harsher and less efficient means. Once we accept the fact that the price system is the most efficient mechanism for rationing the right to scarce life-sustaining and life-enhancing resources, then perhaps rather than "money profaning life" we will find that "life sanctifies money." We will then take the distribution of money and its wise use as serious matters. It is not the exchange relationship that debases life (indeed, the entire biosphere runs on a network of material and energy exchanges), it is the underlying inequity in wealth and income beyond any functional or ethical justification that loads the terms of free exchange against the poor. The same inequality also debases the "gift relationship," since it assigns the poor to the status of a perpetual dependent and the rich to the status of a weary and grumbling patron. Thus gift as well as exchange relationships require limits to the degree of inequality if they are not to subvert their legitimate ends. The sharing of resources in general is the job of the distributist institution. Allocation of particular resources and scarce rights is done by the market within the distribution limits imposed.

In view of the fact that so many liberals, not to mention the United Nations, have declared it to be a human right to have whatever number of children the parents desire, it is worthwhile to end this discussion with a statement from one of the greatest champions of liberty who ever lived, John Stuart Mill (1952, p. 3191):

> The fact itself, of causing the existence of a human being, is one of the most responsible actions in the range of human life. To undertake this responsibility—to bestow a life which may be either a curse or a blessing—unless the being on whom it is to be bestowed will have at least the ordinary chances of a desirable existence, is a crime against that being. And in a country either over-peopled, or threatened with being so, to produce children, beyond a very small number, with the effect of reducing the reward of labor by their competition, is a serious offence against all who live by the remuneration of their labor. The laws which, in many countries on the Continent, forbid marriage unless the parties can show that they have the means of supporting a family, do not exceed the legitimate powers of the State: and whether such laws be expedient or not (a question mainly dependent on local circumstances and feelings), they are not objectionable as violations of liberty. Such laws are interferences of the State to prevent a mischievous act—an act injurious to others, which ought to be a subject of reprobation and social stigma, even where it is not deemed expedient to superadd legal punishment. Yet the current ideas of liberty, which bend so easily to real infringements of the freedom of the individual in things which concern only himself, would repel the attempt to put any restraint upon his inclinations when the consequence of their indulgence is a life or lives of wretchedness and depravity to the offspring, with manifold evils to those sufficiently within reach to be in any way affected by their actions.

Depletion Quotas

The strategic point at which to impose control on the throughput flow seems to me to be the rate of depletion of resources. If we limit aggregate depletion, then, by the law of conservation of matter and energy, we will also indirectly limit aggregate pollution. If we limit throughput flow, then we also indirectly limit the size of the stocks maintained by that flow. Entropy is at its minimum at the input (depletion) end of the throughput pipeline and at its maximum at the output (pollution) end. Therefore, it is physically easier to monitor and control depletion than pollution—there are fewer mines, wells, and ports than there are smokestacks, garbage dumps, and drainpipes, not to mention such diffuse emission sources as runoff of insecticides and fertilizers from fields into rivers and lakes and auto exhausts. Land area devoted to mining is only 0.3 percent of total land area (National Commission on Materials Policy, 1973).

Given that there is more leverage in intervening at the input end, should we intervene by way of taxes or quotas? Quotas, if they are auctioned by the government rather than allocated on nonmarket criteria, have an important net advantage over taxes in that they definitely limit aggregate throughput, which is the quantity to be controlled. Taxes exert only an indirect and very uncertain limit. It is quite true that given a demand curve, a price plus a tax determines a quantity. But demand curves shift, and they are subject to great errors in estimation even if stable. Demand curves for resources could shift up as a result of population increase, change in tastes, increase in income, and so forth. Suppose the government seeks to limit throughput by taxing it. It then spends the tax. If government expenditures on each category of commodity were equal to the revenues received from taxing that same category, then the limit on throughput would be largely canceled out, with the exact degree of canceling depending on the elasticity of demand. If the government taxes resource-intensive items and spends on time-intensive items, there will be a one-shot reduction in aggregate physical throughput but not a limit to its future growth. A credit expansion by the banking sector, an increase in velocity of circulation of money, or deficit spending by the government for other purposes could easily offset even the one-shot reduction induced by taxes. Taxes can reduce the amount of depletion and pollution (throughput) per unit of GNP down to some irreducible minimum, but taxes provide no limit to the increase in the number of units of GNP (unless the government runs a growing surplus) and therefore no limit to aggregate throughput. The fact that a tax levied on a single resource could, by inducing substitution, usually reduce the throughput of that resource very substantially should not mislead us into thinking that a general tax on all or most resources will reduce aggregate throughput (fallacy of composition). Recall that there is no substitute for low-entropy matter-energy. Finally, it is *quantity* that affects the ecosystem, not price, and therefore it is ecologically safer to let errors and unexpected shifts in demand result in price fluctuations rather than in quantity fluctuations. Hence quotas.

An example will illustrate the reason for putting the control (whether tax or quota) on resources rather than on commodities. Suppose the government taxes automobiles heavily and that people take to riding bicycles instead of cars. They will save money as well as resources (Hannon, 1975). But what will the money saved now be spent on? If it is spent on airline tickets, resource consumption would increase above what it was when the money was spent on cars. If the money is spent on theater tickets, then perhaps resource consumption would decline. However, this is not certain, because the theater performance may entail the air transport of actors, stage sets, and so on, and thus indirectly be as resource-consumptive as automobile expenditures. If people paid the high tax on cars and continued buying the same number of cars, then they would have to cut other items of consumption. The items cut may or may not be more resource-intensive than the items for which the government spends the revenue. If the revenue is spent on B-1 bombers, there would surely be a net increase in resource consumption. The conclusion is that the tax or quota should be levied on the resource itself rather than on the commodity.

Pollution taxes would provide a much weaker inducement to resource-saving technological progress than would depletion quotas, since, in the former scheme, resource prices do not necessarily have to rise and may even fall. The inducement of pollution taxes is to "pollution avoidance," and thus to recycling. But increased competition from recycling industries, instead of reducing depletion, might spur the extractive industries to even greater competitive efforts. Intensified search and the development of technologies with still larger jaws could speed up the rate of depletion and thereby lower short-run resource prices. Thus new extraction might once again become competitive with recycling, leading to less recycling and more depletion and pollution—exactly what we wish to avoid. This perverse effect could not happen under a depletion quota system.

The usual recommendation of pollution taxes would seem, if the above is correct, to intervene at the wrong end with the wrong policy tool. Intervention by pollution taxes also tends to be microcontrol, rather than macro. There are, however, limits to the ability of depletion quotas to influence the qualitative nature and spatial location of pollution, and at this fine-tuning level pollution taxes would be a useful supplement, as would a bureau of technology assessment. Depletion quotas would induce resource-saving technological change, and the set of resource-saving technologies would probably overlap to a great degree with the set of socially benign technologies. But the coincidence is not complete, and there is still a need, though a diminished one, for technology assessment.

How would a depletion quota system function? The market for each resource would become two-tiered. To begin with, the government, as a monopolist, would auction the limited quota rights to many buyers. Resource buyers, having purchased their quota rights, would then have to confront many resource sellers in a competitive resource market. The competitive price in the resource market would tend to equal marginal cost. More efficient producers would earn differential rents, but the pure scarcity rent resulting from the quotas would have been captured in the depletion quota auction market by the government monopoly. The total price of the resource (quota price plus price to owner) would be raised as a result of the quotas. All products using these resources would become more expensive. Higher resource prices would compel more efficient and frugal use of resources by both producers and consumers. But the windfall rent from higher resource prices would be captured by the government and become public income—a partial realization of Henry George's ideal of a single tax on rent (George, 1951).

The major advantage is that higher resource prices would bring increased efficiency, while the quotas would directly limit depletion, thereby increasing conservation and indirectly limiting pollution. Pollution would be limited in two ways. First, since pollution is simply the other end of the throughput from depletion, limiting the input to the pipeline would naturally limit the output. Second, higher prices would induce more recycling, thereby further limiting materials pollution and depletion up to the limit set by the increased energy throughput required by recycling. The revenue from the depletion quota auction could help finance the minimum-income component of the distributist institution, offsetting the regressive effect of the higher resource prices on income distribution. Attempts to help the poor by underpricing resources are totally misguided, because the greatest benefit of subsidized prices for energy, for example, goes to those who consume the most energy—the rich, not the poor. This is hardly progressive.

Higher prices on basic resources are absolutely necessary. Any plan that refuses to face up to this necessity is worthless. Back in

1925 economist John Ise made the point in these words (Ise, 1925, p. 284):

> Preposterous as it may seem at first blush, it is probably true that, even if all the timber in the United States, or all the oil or gas or anthracite, were owned by an absolute monopoly, entirely free of public control, prices to consumers would be fixed lower than the long-run interests of the public would justify. Pragmatically this means that all efforts on the part of the government to keep down the prices of lumber, oil, gas, or anthracite are contrary to the public interest; that the government should be trying to keep prices up rather than down.

Ise went on to suggest a general principle of resource pricing: that nonrenewable resources be priced at the cost of the nearest renewable substitute. Therefore, virgin timber should cost at least as much per board foot as replanted timber; petroleum should be priced at its Btu equivalent of sugar or wood alcohol, assuming they are the closest renewable alternatives. In the absence of any renewable substitutes, the price would merely reflect the purely ethical judgment of how fast the resources should be used up—that is, the importance of the wants of future people relative to the wants of present people. Renewable resources are assumed to be exploited on a sustained-yield basis and to be priced accordingly.

The Ise principles could also be used in setting the aggregate quota amounts to auction. For renewables, the quota should be set at an amount equivalent to some reasonable calculation of maximum sustainable yield. For nonrenewables with renewable substitutes, the quota should be set so that the resulting price of the nonrenewable resource is at least as high as the price of its nearest renewable substitute. For nonrenewables with no close renewable substitute, the quota would reflect a purely ethical judgment concerning the relative importance of present versus future wants. Should these resources be used up by us or by our descendants? The price system cannot decide this, because future generations cannot bid in present resource markets. The decision is ethical. We have found it too easy to assume that future generations will be better off due to inevitable "progress" and therefore not to worry about the unrepresented claims of the future on exhaustible resources.

In addition to the Ise principles, which deal only with depletion costs that fall on the future, the quotas must be low enough to prevent excessive pollution and ecological costs that fall on the present as well as on the future. Pragmatically, quotas would probably

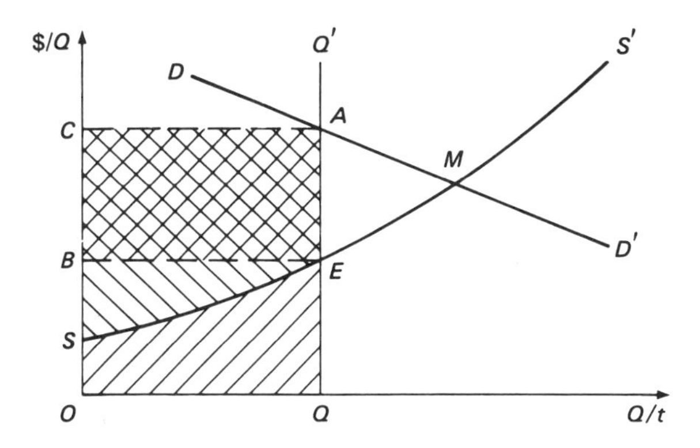

Figure 19.1
The depletion quota scheme. From *Steady-State Economics* by Herman E. Daly. W. H. Freeman and Company. Copyright 1977.

at first be set near existing extraction rates. The first task would be to stabilize, to get off the growth path. Later, we could try to reduce quotas to a more sustainable level, if present flows proved too high. Abundant resources causing little environmental disruption would be governed by generous quotas and therefore relatively low prices, with a consequently strong incentive to technologies that make relatively intensive use of the abundant resource.

Depletion quotas would capture the increasing scarcity rents but would not require the expropriation of resource owners. Quotas are clearly against the interests of resource owners, but not unjustly so, since rent is by definition unearned income from a price in excess of the minimum supply price. The elimination of this unearned increment would no doubt reduce the incentive to exploration and new discovery. Geological exploration has many aspects of a natural monopoly and probably should, in any case, be carried on by a public corporation. As the largest resource owner by far, the government should not have to lease public lands to private companies who have more geological information than the government about the land. If private exploration is thought desirable, it could be encouraged by a government bounty paid for mineral discoveries. The current resource owners would suffer a one-time capital loss when depletion limits are imposed and, in fairness, should be compensated.

For many readers a graphical exposition of the depletion quota scheme will be helpful, as shown in figure 19.1. *DD′* is the market demand curve for the resource in question. *SS′* is the supply curve of the industry. A depletion quota, in the aggregate amount *Q*, is imposed, shown by the vertical line *QQ′*. The total price paid per unit of the resource (price paid to resource owner plus price paid to government for the corresponding quota right) is *OC*. Of the total price *OC*, the amount *OB* is the price paid to the resource owner for one unit of the resource, and *BC* is the price paid to the government for a quota right to purchase one unit of the resource. Of the total amount paid, *OQAC*, the amount *OSEQ* is cost, reflecting the necessary supply price. The remainder, *SEAC*, is surplus, or rent.

Rent is defined as payment in excess of necessary supply price. Of the total rent area, the amount *BES* is differential rent, or surplus that arises from the difference in the supply price of the marginal amount produced, which is *QE*, and all previous amounts produced. Price is determined at the margin, and is equal to *QE*, the marginal cost of production. Since the cost of production of all inframarginal units is less than *QE*, and since all units sell at the same price, equal to *QE*, a profit, or differential rent, is earned on all inframarginal units produced. The profit on the first unit is *BS* and declines slightly for each additional unit until it is zero for the last unit at *Q*. Thus *BES* is the sum of the diminishing series of inframarginal per-unit profits. It is called differential rent because its amount depends on the schedule of cost differences between the first and last units. The remainder of the surplus, the amount *CAEB*, is pure scarcity rent. It does not arise from cost differentials but simply from the excess of the market price above the marginal cost of production, by the amount *AE*. In effect, *AE* represents a kind of price per unit of resources in the ground that prior to the quota auction had implicitly been priced at zero. At the market equilibrium *M*, the entire surplus would be differential rent, and scarcity rent would be zero. Hence scarcity rent, as the name implies, emerges when the resource is made scarce relative to the quantity corresponding to market equilibrium, which, of course, is what happens when quotas are imposed.

The scarcity rent *CAEB is* captured by the government quota auction. The differential rent *BES* remains in the hands of the resource owners. The reason for this particular division of the surplus is that the resource market is assumed to be competitive (many sellers and buyers), while the quota auction market is monopolistic (many buyers, one seller). The government has monopoly power; the resource owners and buyers have none. The price in the resource market is set by competition at an amount equal to marginal cost, *QE*. The

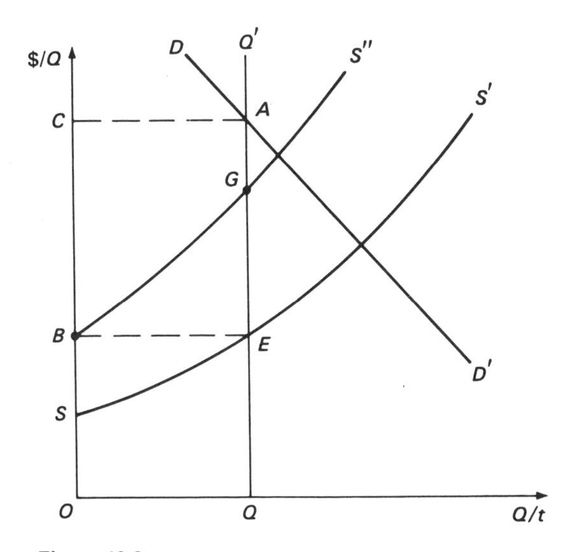

Figure 19.2
The depletion quota scheme over time. From *Steady-State Economics* by Herman E. Daly. W. H. Freeman and Company. Copyright 1977.

government, by charging what the market will bear with no fear of being undercut by competitors, is able to extract the remainder of the full demand price, or the amount *AE*. If the resource market were also monopolized, then the division of scarcity rent between the government and private monopolies would be indeterminate. Even in that case, however, the government would have an advantage in that the quota right has to be purchased first. Thus even if competition is less than perfect in the resource market, we would still expect the government to capture all monopoly profits (scarcity rents), because it constitutes the first tier of the market and controls the entry of buyers into the second tier.

Over time the supply curve for nonrenewable resources would shift upward as more accessible resources become depleted and previously submarginal mines and wells have to be used. In figure 19.2 the higher supply curve is represented by *BS″*, which may be thought of as the "unused" segment of the original supply curve, *ES′*, shifted horizontally to the left until it touches the vertical axis. Assuming an unchanged demand curve and quota, it is clear from figure 19.2 that rising cost of production (now shown by the larger area, *OBGQ)* will eventually eliminate the pure scarcity rent, leaving only differential rent. Quotas will slow down the upward shift of the supply curve relative to what it would have been with faster depletion, but of course they cannot arrest the inevitable process. Probably the quota would have to be reduced as the supply curve shifted up in order to pass along the higher cost signals to users and to maintain some scarcity rent for public revenue.

For renewable resources, where the quota is set at maximum sustainable yield, there would be no upward shift of the supply curve. However, the demand curve for renewables would shift up as nonrenewable resource usage became more restricted and expensive and efforts were made to substitute renewables for nonrenewables. The quota on renewables would then protect those resources from being exploited beyond capacity in order to satisfy the rising demand while at the same time rationing access to the limited amount and diverting the windfall profits into the public treasury. In sum, the depletion quota auction is an instrument for helping us to make the transition from a nonrenewable to a renewable resource base in a gradual, efficient, and equitable manner.

The depletion quota scheme allows a reconciliation between the two conflicting goals of efficiency and equity. Efficiency requires high resource prices. However, equity is not served by high prices, because they have a regressive effect on income distribution in the same way that a sales tax does, and also because the windfall rents arising from the higher prices accrue to resource owners, not to the poor. The latter effect can be reversed by capturing the scarcity rent through the depletion quota auction and using it to finance a minimum income, and/or to replace the most regressive taxes.

Two further efficiency increases could be expected. First, taxing rent causes no allocative distortions and is the most efficient way to raise government revenue. To the extent that a rent tax (or its equivalent in this case) replaces other taxes, then static allocative efficiency should be improved. Second, as conservatives and radicals alike have noted, the minimum income could substitute for a considerable number of bureaucratic welfare programs. Of course, the major increase in efficiency would result directly from higher resource prices, which would give incentives to develop resource-saving techniques of production and patterns of consumption. Equity is not served by low prices, which, in effect, give a larger subsidy to the rich than to the poor, since the rich consume more resources. Equity is served by higher incomes to the poor and by a maximum limit in the incomes of the rich.

A Coordinated Program

Let us now consider all three institutions as a unified program.

The allocation among firms of the limited aggregate of resources extracted during a given time period would be accomplished entirely by the market. The distribution of income within the maximum and minimum boundaries imposed would also be left to the market. The initial distribution of reproductive licenses is done outside the market on the basis of strict equity—one person, one license—but reallocation via market exchange is permitted in the interest of efficiency. The combination of the three institutions presents a nice reconciliation of efficiency and equity and provides the ecologically necessary macrocontrol of growth with the least sacrifice in terms of microlevel freedom and variability. The market is relied upon to allocate resources and distribute incomes within imposed ecological and ethical boundaries. The market is not allowed to set its own boundaries, but it is free within those boundaries. Setting boundaries is necessary. No one has ever claimed that market equilibria would automatically coincide with ecological equilibria or with a reasonably just distribution of wealth and income. Nor has anyone ever claimed that market equilibria would attain demographic balance. The very notions of "equilibrium" in economics and ecology are antithetical. In growth economics equilibrium refers not to physical magnitudes at all but to a balance of desires between savers and investors. As long as saving is greater than depreciation, then net investment must be positive. This implies a *growing* flow of physical inputs from and outputs to nature, that is, a biophysical disequilibrium. Physical conditions of environmental equilibrium must be imposed on the market in aggregate quantitative physical terms. Subject to these quantitative constraints, the market and price system can, with the institutional changes just discussed, achieve an optimal allocation of resources and an optimal adjustment to its imposed physical system boundaries. The point is important because the belief is widespread among economists that internalization of externalities, or the incorporation of all environmental costs into market prices, is a sufficient environmental policy and that once this is accomplished the market will be able to set its own proper boundaries automatically. This is not so. Nor, as we have already seen, is it possible to incorporate all ecological costs in rigged money prices.

The internalization of externalities is a good strategy for fine-tuning the allocation of resources by making relative prices better measures of relative marginal social costs. But it does not enable the market to set its own absolute physical boundaries with the larger ecosystem. To give an analogy: Proper allocation arranges the weight in a boat optimally, so as to maximize the load that can be carried. But there is still an absolute limit to how much weight a boat can carry, even optimally arranged. The price system can spread the weight evenly, but unless it is supplemented by an external absolute limit, it will just keep on spreading the increasing weight evenly until the evenly loaded boat sinks. No doubt the boat would sink evenly, *ceteris paribus*, but that is less comforting to the average citizen than to the neoclassical economist.

Two distinct questions must be asked about these proposed institutions for achieving a steady state. First, would they work if people accepted the goal of a steady state and perhaps voted the institutions into effect? Second, would people ever accept either the steady-state idea or these particular institutions? I have tried to show that the answer to the first question is probably "yes." Let the critic find any remaining flaws; better yet, let him suggest improvements. The answer to the second question is clearly "no" in the short run. But several considerations make acceptance more plausible in the not-too-long run.

The minimum-income side of the distributist institution already has some political support in the United States; the maximum limits will at first be thought un-American. Yet, surely, beyond some figure any additions to personal income would represent greed rather than need, or even merit. Most people would be willing to believe that in most cases an income in excess of, let us say, $100,000 per year has no real functional justification, especially when the highly paid jobs are usually already the most interesting and pleasant.

In spite of their somewhat radical implications, the proposals presented in this chapter are, as we have seen, based on impeccably respectable conservative institutions: private property and the free market.

By fixing the rate of depletion we force technology to focus more on the flow sources of solar energy and renewable resources. The solar flux cannot be increased in the present at the expense of the future. Thus let technology devote itself to learning how to live off our solar income rather than our terrestrial capital. Such advances will benefit all generations, not just the present. Indeed, the main goal of the depletion quota plan is to turn technological change away from increasing dependence on the terrestrial stock and toward the more abundant flow of solar energy and renewable resources. As the stock becomes relatively more expensive, it will be used less in direct consumption and more for investment in "work gates" that increase our ability to tap the solar flow. Instead of taking long-run technical evolution as a parameter to which the short-run variables of price and quantity continually adjust, the idea is to take short-run quantities (and hence prices) as a social parameter to be set, so as to induce a direction of technological evolution more in harmony with mankind's long-run interests.

This new direction of technological change is likely also to be in mankind's short-run interests, if we accept the view that man's evolution in a solar-based and stable economy has programmed him for that kind of life rather than for the stresses of a growing industrial economy. The future steady state could be a good deal more comfortable than past ones and much more human than the overgrown, overcentralized, overextended, and overbearing economy into which growth has pushed us.

The depletion quota plan should appeal to both technological optimists and pessimists. The pessimist should be pleased by the conservation effect of the quotas; the optimist should be pleased by the price inducement to resource-saving technology. The optimist tells us not to worry about running out of resources because technology embodied in reproducible capital is a nearly perfect substitute for resources. As we run out of anything, prices will rise and substitute methods will be found. If we believe this, then how could we object to quotas, which simply increase the scarcity and prices of resources a bit ahead of schedule and more gradually? This plan simply requires the optimist to live up to his faith in technology.

Like the maximum limits on income and wealth, the depletion quotas could also have a trust-busting effect if accompanied by a limit—for example, no single entity can own more than x percent of the quota rights for a given resource or more than y percent of the resource owned by the industry of which it is a member. We could set x and y so as to allow legitimate economies of scale, while curtailing monopoly power.

The actual mechanics of quota auction markets for three or four hundred basic resources would present no great problems. The whole process could be computerized, since the function of an auctioneer is purely mechanical. It could be vastly simpler, faster, more decentralized, and less subject to fraud and manipulation than today's stock market. In addition, qualitative and locational variation among resources within each category, though ignored at the auction level, will be taken into account in price differentials paid to resource owners.

The depletion quota and birth quota systems bear an obvious analogy. The difference is that the birth quotas are privately held and equally distributed initially, and then redistributed among individuals through the market; the depletion quotas are collectively held initially and then distributed to individuals by way of an auction market. The revenue derived from birth quotas is private income; the revenue from depletion quotas is public income.

The scheme could, and probably must, be designed to include imported resources. The same depletion quota right could be required for importation of resources, and thus the market would determine the proportions in which our standard of living is sustained by depletion of national and foreign resources. Imported final goods would now be cheaper relative to national goods, assuming foreigners do not limit their depletion. Our export goods would now be more expensive relative to the domestic goods of foreign countries. Our terms of trade would improve, but we would tend to a balance of payments deficit. However, with a freely fluctuating exchange rate, a rise in the price of foreign currencies relative to the dollar would restore equilibrium. The balance of payments can take care of itself. If foreigners are willing to sell us goods priced below their true full costs of production, we should not complain.

It might be objected that limiting our imports of resources will work a hardship on the many underdeveloped countries that export raw materials. This is not clear, because such a policy will also force them to transform their own resources domestically rather than through international trade. Foreign suppliers of raw materials will be treated no differently than domestic suppliers. Finished goods would not be subject to quotas. In any case, it is clear that in the long run we are not doing the underdeveloped countries any favor by using up their resource endowment. Sooner or later (sooner, in the case of OPEC), they will begin to drive a hard bargain for their nonrenewable resources, and we had better not be too dependent on them. Probably they will limit their raw material exports, thus making unnecessary any limits that we might place on our raw material imports. Eventually, population control and environmental protection policies might become preconditions for membership in a new free-trade bloc or common market. Free trade would be the rule among all countries that limited their own populations and rates of domestic depletion, while controls could be put on trade with other countries whenever desirable.

All three of the institutions we have discussed are capable of gradual application during the transition to a steady state. The birth quota does not have to be immediately set at negative or zero growth, or even at replacement, but could begin at any currently prevailing level and gradually approach replacement or lower fertility. Initially the certificate price would be zero, and it would rise gradually as the number of certificates issued to each person was cut from, for instance, 1.1, to 1.0, to 0.9, or to whatever level is desired. The depletion quotas could likewise be set at present levels or even at levels corresponding to a slower rate of increase than in the recent past. They could be applied first to those materials in shortest supply and to those whose wastes are hardest to absorb. Initial prices on quota rights would be low but then would rise gradually as growth pressed against the fixed quotas or as quotas were reduced in the interest of conservation. In either case the increased scarcity rent would become revenue to the government. The distribution limits might begin near the present extremes and slowly close to a more desirable range. The three institutions are amenable to any degree of gradualism we may wish. However, the distribution limits must be tightened faster than the depletion limits if the burden on the poor is to be lightened. All three control points are price system parameters, and altering them does not interfere with the static allocative efficiency of the market.

But it is also the case that these institutions could be totally ineffective. Depletion quotas could be endlessly raised on the grounds of national defense, balance of payments, and so forth. Real estate and construction interests, not to mention the baby food and toy lobbies and the military, might convince Congress to keep the supply of birth licenses well above replacement level. People at the maximum income and wealth limit may succeed in continually raising that limit by spending a great deal of their money on TV ads extolling the Unlimited Acquisition of Everything as the very foundation of the American Way of Life. Everything would be the same and all justified in the sacred name of growth. Nothing will work unless we break our idolatrous commitment to material growth.

A definite US policy of population control at home would give us a much stronger base for preaching to the underdeveloped countries about their population problem. So would the reduction in US resource consumption resulting from depletion quotas. Without such a base to preach from we will continue to waste our breath, as we did at the 1974 Population Conference in Bucharest.

Thus we are brought back to the all-important moral premises. A physical steady state, if it is to be worth living in, absolutely requires moral growth. Future progress simply must be made in terms of the things that really count rather than the things that are merely countable. Institutional changes are necessary but insufficient. Moral growth is also necessary but insufficient. Both together are necessary and sufficient, but the institutional changes are relatively minor compared to the required change in values.

On Moral Growth

Let us assume for a moment that the necessity of the steady state and the above outline of its appropriate technologies and social institutions are accepted. Logic and necessity are not sufficient to bring about social reform. The philosopher Leibnitz observed, "If geometry conflicted with our passions and interests as much as do ethics, we would contest it and violate it as much as we do ethics now, in spite of all the demonstrations of Euclid and Archimedes, which would be labeled paralogisms and dreams" (quoted in Sauvy, 1970, p. 270). Leibnitz is surely correct. However logical and necessary the above outline of the steady state, it is, on the assumption of static morality, nothing but a dream. The physically steady economy absolutely requires moral growth beyond the present level.

Economists and other social scientists of positivistic bias seem to consider appeals to morality as cheating, as an admission of intellectual defeat, like bending the pieces of a jigsaw puzzle. In economics there is a long and solid tradition of regarding moral resources as static and too scarce to be relied upon. In the words of the great British economist Alfred Marshall, "progress chiefly depends on the extent to which the *strongest* and not merely the *highest* forces of human nature can be utilized for the increase of social good" (quoted in Robertson, 1956, p. 148).

Presumably self-interest is stronger and more abundant than brotherhood. Presumably "progress" and "social good" can be defined independently of the driving motive of society.

Another British economist, D. H. Robertson, once asked the illuminating question: What is it that economists economize? His answer was "love, the scarcest and most precious of all resources" (Robertson, 1956, p. 154). Paul Samuelson quotes Robertson approvingly in the latest edition of *Economics,* his influential textbook. Nor are economists alone in ruling out reliance on moral resources. The reader will recall that in his "Tragedy of the Commons" biologist Garrett Hardin identifies a class of problems with no technical solution. He rules out moral solutions as self-eliminating on a somewhat far-fetched evolutionary analogy, and advocates a political solution: mutual coercion mutually agreed upon. This is fine, but where is the mutual agreement to come from if not from shared values, from a convincing morality? Political scientist Beryl Crowe (1969), in revisiting the tragedy of the commons, argues that the set of no-technical-solution problems coincides with the set of no-political-solution problems and that Hardin's "mutual coercion mutually agreed upon" is politically impossible. Between them they present a convincing case that "commons problems" will not be solved technically nor politically, assuming static morality. Mutual coercion does not substitute for, but presupposes, moral growth.

Going back to Robertson's repulsive but correct idea that economists economize love, one may ask, "How?" Mainly by maximizing growth. Let there be more for everyone year after year so that we need never face up to sharing a fixed total. Unequal distribution can be justified as necessary for saving, incentive, and hence growth. This must continue, otherwise the problem of sharing a fixed total will place too heavy a strain on our precious resource of love, which is so scarce that it must never be used. I am reminded of Lord Thomas Balough's statement that one purpose of economic theory is to make those who *are* comfortable *feel* comfortable.

To paraphrase the above, we are told "Don't worry about today's inequities, but anxiously fix your attention on tomorrow's larger total income." Compare that with the Sermon on the Mount: "Do not be anxious about tomorrow, for tomorrow will be anxious for itself. Let the day's own evil be sufficient for the day." The morality of the steady state is that of the Sermon on the Mount. Growthmania requires the negation of that morality. If we give our first attention to the evils of the day, we will have moral growth though not so much economic growth. If we anxiously give our first attention to tomorrow's larger income, we will have economic growth but little or no moral growth. Since economic growth is reaching physical limits anyway, we may now find the Sermon on the Mount more appealing and easier to accept.

The same idea is stated in Alexander Solzhenitsyn's *Cancer Ward,* in the chapter entitled "Idols of the Market Place," in which the position of "ethical socialism" is advocated. The main theme is "ethics first and economics afterwards"—a theme that finds as little acceptance in the Soviet Union as it does in the United States, perhaps even less. The following words are from the character Shulubin (1968, p. 513):

> Happiness is a mirage—as for the so-called "happiness of future generations" it is even more of a mirage. Who knows anything about it? Who has spoken with these future generations? Who knows what idols they will worship? Ideas of what happiness is have changed too much through the ages. No one should have the effrontery to try to plan it in advance. When we have enough loaves of white bread to crush them under our heels, when we have enough milk to choke us, we still won't be in the least happy. But if we share the things we don't have enough of, we can be happy today! If we care only about "happiness" and about reproducing our species, we shall merely crowd the earth senselessly and create a terrifying society.

There are other sources of moral support for the steady state besides the Sermon on the Mount. From the Old Testament we have two creation myths, the Priestly and the Yahwistic, one that gives value to creation only with reference to man, and one that gives value to creation independently of man. In Western thought the first tradition has dominated, but the other is there waiting to receive its proper emphasis. Also, Aldo Leopold's "land ethic" is extremely appealing and would serve admirably as the moral foundation of the steady state. Finally Karl Marx's materialism and objection to the alienation of man from nature can be enlisted as a moral foundation of the steady state. Marx recognized that nature is the "inorganic body of man" and not just a pile of neutral stuff to be dominated (Marx, 1963, p. 127).

In writing this chapter, I've considered the steady state only at a national level. Clearly the world as a whole must eventually adjust to a steady state. Perhaps ultimately this recognition will promote unity among nations—or, conversely, the desire for unity may promote the recognition. However, when nations cannot even agree to limit the stock of "bads" through disarmament, it is hard to be optimistic about their limiting the stocks of "goods." There is no alternative except to try, but national efforts need not wait for international agreement.

Finally, one rather subtle yet very powerful moral force can be enlisted in support of the steady-state paradigm. That is wholeness. If the truth is the whole, as Hegel claimed, then our current splintered knowledge is so far from truth that it is hardly worth learning. I believe this is why many of our best university students do not work very hard at their studies. Why continue mining the deep, narrow, disciplinary shafts sunk into man's totality by the intellectual fragment makers? Why deepen the tombs in which we have buried the wholeness of knowledge? Why increase the separation of people by filling separate heads with separate fragments of knowledge? The malaise reflected in these questions is very grave, and is, in my view, a major reason for the new surge of interest in ecology. Ecology is whole. It brings together the broken, analyzed, alienated, fragmented pieces of man's image of the world. Ecology is also a fad, but when the fad passes, the movement toward wholeness must continue. Unless the physical, the social, and the moral dimensions of our knowledge are integrated in a unified paradigm offering a vision of wholeness, no solutions to our problems are likely. John Stuart Mill's idea of the stationary state seems to me to offer such a paradigm.

But there remains a deeper question. Is it realistic in our secular, "pluralistic" society to expect any kind of moral consensus? Where is this moral consensus to come from? Not from a spineless relativism or from the hallucinatory psychic epiphenomena that seem to haunt complex mechanisms. Let us state it directly in the strongest terms. Ultimately, the possibility of moral consensus presupposes a dogmatic belief in objective value. If values are subjective, or thought to be merely cultural artifacts, then there is nothing objective to which appeal can be made or around which a consensus might be formed. Consensus based upon what everyone recognizes to be a convenient cultural myth (like belief in Santa Claus) would not bear much stress. Only real objective values can command consensus in a sophisticated self-analytical society. We have no guarantee that objective value can be clarified, nor that, once clarified, it would be accorded the consensus it merits. But without faith in the existence of an objective hierarchy of value and in our ability at least vaguely to perceive it, we must resign ourselves to being driven by technological determinism into an unchosen, and perhaps unbearable, future. On what other grounds is technical determinism to be resisted? Just as physical research must be based on a dogmatic faith that nature is orderly, so research into policy questions must presuppose the reality of an ordered hierarchy of value. If *better* or *worse* are meaningless terms, then all policy is nonsense.

In C. S. Lewis's words (1959, p. 46), "A dogmatic belief in objective value is necessary to the very idea of a rule which is not tyranny or an obedience which is not slavery." The same insight underlies Edmund Burke's famous dictum that "society cannot exist unless a controlling power upon will and appetite be placed somewhere, and the less of it there is within, the more there must be without." Control from within can only result from obedience to objective value. If interior restraints on will and appetite diminish, then exterior restraints, coercive police powers, or Malthusian positive checks must increase. In Burke's words, "Men of intemperate minds cannot be free. Their passions forge their fetters."

A major reason for pessimism about the course of human affairs is that the very words "*dogmatic* belief in *objective* value" automatically shut the minds of most modern intellectuals. Why is this so? Probably because *dogmatic* has come to be almost synonymous with *egotistic,* and because the term *objective value* has connotations of absolutism and intolerance. The confusions underlying these two misinterpretations have been well stated by others whose words I will borrow.

G. K. Chesterton informs us (n.d., pp. x–xi),

> To be dogmatic and to be egotistic are not only not the same thing, they are opposite things. Suppose, for instance that a vague skeptic eventually joins the Catholic church. In that act he has at the same moment become less egotistic and more dogmatic. The dogmatist is by the nature of the case not egotistical, because he believes that there is some solid, obvious and objective truth outside him which he has perceived and which he invites all men to perceive. And the egotist is in the majority of cases not dogmatic, because he has no need to isolate one of his notions as being related to the truth; all his notions are equally interesting because they are related to him. The true egotist is as much interested in his own errors as in his own truth; the dogmatist is interested only in the truth, and only in the truth because it is true. At the most the dogmatist believes that he is in the truth; but the egotist believes that the truth, if there is such a thing, is in him.

A related clarification was made by E. F. Schumacher (1977, p. 58):

> The result of the lopsided development of the last three hundred years is that Western man has become rich in means and poor in ends. The hierarchy of his knowledge has been decapitated; his will is paralyzed because he has lost any grounds on which to base a hierarchy of values. What are his highest values?
>
> A man's highest values are reached when he claims that something is good in itself, requiring no justification in terms of any higher good. Modern society prides itself on its pluralism, which means that a large number of things are admissable as "good in themselves," as ends rather than as means to an end. They are all of equal rank, all to be accorded *first priority.* If something that requires no justification may be called an "absolute," the modern world, which *claims* that everything is relative, does, in fact, worship a very large number of "absolutes.". . . Not only power and wealth are treated as goods in themselves—provided they are mine and not someone else's—but also knowledge for its own sake, speed of movement, size of market, rapidity of change, quantity of education, number of hospitals, etc., etc. In truth, none of these sacred cows is a genuine end; they are all means parading as ends.

Science and technology, with their analytic-empirical mode of thinking, have led many into a kind of scientism that seeks to debunk all knowledge that does not have an analytic-empirical basis. Knowledge about ends—about objective value and right purpose—derives from an "illicit" source and is considered "forbidden knowledge" by the priests of the scientistic inquisition. Unless this error is recognized, unless we come around to a "dogmatic belief in objective value," or what Boris Pasternak called "the irresistible power of unarmed truth," then it makes no sense to concern ourselves with economics. Why strain our gnats of marginal inefficiency in the allocation of means to serve ends while swallowing camels of total incoherence in the ordering of those ends? Indeed, if our ends are perversely ordered, then it is better that we should be *inefficient* in allocating means to their service.

It is one thing to insist on the logical necessity of a dogmatic belief in objective value as a basis for resisting technical determinism, but it is something else to have clear and certain knowledge of what objective value is. We must be open-minded regarding differing understandings of the nature of objective value and corresponding principles of right action. But we must make an effort to state the general principles that should guide our decisions and apply them to economic questions.

Probably the rule of right action most accepted in practice is Jeremy Bentham's greatest good for the greatest number. Economists have avoided the difficult problem of defining *good* by substituting the word *goods,* in the sense of commodities. The principle thus became the greatest per capita product for the greatest number. More products per capita and more people to enjoy those products lead, in this view, to the greater social good. Our commitment to growth is no doubt based in considerable degree on this principle, which implies that right action is that which leads to more goods for more people.

But there are two problems with the greatest per capita product for the greatest number. First, as others have pointed out, the dictum contains one too many "greatests." It is not possible to maximize more than one variable. It is clear that numbers of people could be increased by lowering per capita product, and per capita product could be increased by lowering numbers, since resources taken from one goal can be devoted to the other. Second, it makes a big difference whether "greatest number" refers to those simultaneously alive or to the greatest number ever to live *over time.*

To resolve the first of these difficulties, we must maximize one variable only and treat some chosen level of the other as a constraint on the maximization. For one of the "greatests" we must substitute *sufficient.* There are two possible substitutions: the greatest per capita product for a sufficient number, or a sufficient per capita product for the greatest number. Which is the better principle? I suggest that we adopt the latter, and that "greatest number" be understood as greatest number over time, which takes care of the second problem. The revised principle thus becomes *sufficient per capita product for the greatest number over time.*

It is hard to find any objection to maximizing the number of people who will ever live at a material level sufficient for a good life. However, this certainly does *not* mean maximizing the number alive at any one time. On the contrary, it means the avoidance of any destruction of the earth's capacity to support life, a destruction that results from overloading the life support system by having too many people—especially high-consuming people—alive at once. The opportunity cost of those extra lives in the present is fewer people alive in all subsequent time periods, and consequently a reduction in total lives ever to be lived at the sufficient level. Increasing per capita product beyond the sufficient level (extravagant luxury) may also overburden life support systems and have the same long-run life-reducing effect as excess population.

Maximizing number while satisficing per capita product does not imply that quantity of life is a higher value than quality. It does assume that beyond some level of sufficiency further increase in per capita goods does not increase quality of life and, in fact, may well diminish it. But sufficiency is the first consideration. To put it more concretely, the basic needs of all present people take priority over future numbers, but the existence of more future people takes priority over the trivial wants of the present. The impact of this revised utilitarian rule is to maximize life, or, what is the same thing, to economize the long-run capacity of the earth to support life at a sufficient level of individual wealth. The sufficient level may be thought of as a range of limited inequality rather than a single specific per capita income applicable to everyone. Some inequality is necessary for fairness.

This modified utilitarian principle certainly offers no magic philosopher's stone for making difficult choices easy. But it does seem superior to the old Benthamite rule in that it draws our attention to the concept of sufficiency, and it extends our time horizon. It forces us to face the question of purpose: sufficient *for what?* needed for what? It will be very difficult to define sufficiency and build the concept into economic theory and practice. But I think it will prove far more difficult to continue to operate on the principle that there is no such thing as enough.

Note

1. To *satisfice,* as used here, means to seek enough rather than the most. The concept of "enough" is difficult to define but even more difficult to deny.

References

Boulding, Kenneth E. 1964. *The Meaning of the Twentieth Century.* New York: Harper & Row.

Boulding, Kenneth E. 1966. "The Economics of the Coming Spaceship Earth." In Henry Jarrett, ed., *Environmental Quality in a Growing Economy.* Baltimore: Johns Hopkins University Press.

Chesterton, G. K. N.d. Introduction to *Poems by John Ruskin.* London: George Routledge and Sons.

Crowe, Beryl. 1969. "The Tragedy of the Commons Revisited." *Science,* 28 November.

Daly, Herman E. 1968. "On Economics as a Life Science." *Journal of Political Economy,* May/June.

Daly, Herman E. 1974. "The Economics of the Steady State." *American Economic Review,* May.

Davis, Kingsley. 1973. "Zero Population Growth." *Daedalus,* Fall.

Energy Policy Project of the Ford Foundation. 1974. *A Time to Choose.* Cambridge, Mass.: Ballinger.

Fisher, Irving. 1906. *The Nature of Capital and Income.* London: Macmillan.

George, Henry. 1951. *Progress and Poverty.* New York: Robert Schalkenbach Foundation. (Originally published in 1879.)

Georgescu-Roegen, Nicholas. 1966. *Analytical Economics.* Cambridge: Harvard University Press.

Goldberg, Michael. 1976. "Less Is More." Unpublished manuscript.

Hannon, Bruce. 1975. "Energy, Growth, and Altruism." Urbana: University of Illinois, Center for Advanced Computation (mimeographed).

Heer, David M. 1975. "Marketable Licenses for Babies: Boulding's Proposal Revisited." *Social Biology,* Spring.

Ise, John. 1925. "The Theory of Value as Applied to Natural Resources." *American Economic Review,* June.

Lewis, C. S. 1959. "The Abolition of Man." From *The Abolition of Man.* New York: Macmillan. (Reprinted in this volume.)

Luten, Daniel S. "Teleoeconomics: The Microdynamic, Macrostatic Economy." Department of Geography, University of California, Berkeley (mimeographed).

McClaughry, John. 1974. "The Future of Private Property and Its Distribution." *Ripon Quarterly,* Fall.

Marx, Karl. 1963. *Karl Marx's Early Writings.* Trans. and ed. T. B. Bottomore. New York: McGraw-Hill.

Mill, John Stuart. 1952. *On Liberty.* Chicago: Encyclopedia Brittanica Great Books. (Originally published in 1859.)

Mill, John Stuart. 1881. "Of Property." In *Principles of Political Economy.* Book 2. New York: Appleton-Century-Crofts.

National Academy of Sciences. 1975. *Mineral Resources and the Environment.* Washington, D.C.: US Government Printing Office.

National Commission on Materials Policy. 1973. *Material Needs and the Environment Today and Tomorrow.* Washington, D.C.: US Government Printing Office.

Okun, Arthur. 1975. *Equality and Efficiency: The Big Tradeoff.* Washington, D.C.: Brookings Institution.

Ordway, Samuel H., Jr. 1953. *Resources and the American Dream.* New York: Ronald Press.

President's Commission on Population Growth and the American Future. 1972. *Final Report to President Nixon, Congress, and the American People.* Washington, D.C.: US Government Printing Office, March.

Robertson, D. H. 1956. *Economic Commentaries.* London: Staples Press.

Sauvy, A. 1970. *The General Theory of Population.* New York: Basic Books.

Schumacher, E. F. 1977. *A Guide for the Perplexed.* New York: Harper & Row.

Solzhenitsyn, Aleksandr I. 1968. *The Cancer Ward.* New York: The Dial Press.

Swift, Jonathan. 1958. "Thoughts on Various Subjects." In G. B. Woods et al., eds., *The Literature of England.* Glenview, Ill.: Scott, Foresman.

Weinberg, Alvin M., and H. E. Goeller. 1976. "The Age of Substitutability." *Science,* 20 February.

20 Postscript: Some Common Misunderstandings and Further Issues Concerning a Steady-State Economy

Herman E. Daly

In a survey article on recent contributions of economists to the study of scarcity and growth, Professor Richard B. Norgaard of the University of California at Berkeley commented: "Daly and others argue that some sort of a steady state system relying largely on flow resources would both reduce many environmental and social problems and be viable over the long run. An invigorating, productive debate has not developed largely because economists have ignored or put down the challenge."[1]

Unfortunately, Professor Norgaard is correct both in his statement that the challenge has been largely ignored and in lamenting the absence of a productive debate. Such a debate is badly needed, since it is only through the critical interplay of many minds that errors can be uncovered and further implications discerned. Although such a debate has not yet emerged, there have been some scattered criticisms, objections, and comments that merit consideration. A productive debate will be facilitated by clearing the agenda of a number of nonissues.

Problems of Terminology

Previously, following J. S. Mill, I used the term *stationary state* in the same basic sense as the classical economists used it, to refer to an economy in which population and capital stock had ceased growing. This led to some confusion, because the neoclassical economists had redefined the term to refer to an economy in which tastes and technology were unchanging but in which population and capital stock

Written especially for the preceding edition of this book.

could be growing. The classical sense referred to an actual physical state toward which the economy was presumably tending. The neoclassical meaning referred to a hypothetical concept or an ideal case, such as a frictionless machine or an ideal gas, which is a useful step in analysis but is not meant as a description of any real world state.

To avoid confusion, I adopted the term *steady state* from the physical and biological sciences. This seemed a happy choice, because I was arguing from biophysical first principles anyway and because the term *steady state* meant to physical scientists very nearly what the term *stationary state* had meant to the classical economists before the neoclassicals redefined it. The main difference and continuing problem is that in physical science a steady-state system implies both quantitatively and qualitatively identical replacement, whereas a SSE, as here defined, implies quantitatively identical replacement only, with quality free to change.

Unfortunately, that is not the end of the story. Some modern economists specializing in growth models have adopted the term *steady-state growth* to refer to models in which population and capital are growing absolutely but in which certain ratios between absolutely growing magnitudes remain constant. This case could more reasonably be referred to as *proportional growth*. The noun *state* literally means the standing or stability of a thing (from the Latin *stare*, to stand). The adjectives *stationary* and *steady* simply amplify this idea of standing as opposed to running, of constancy as opposed to increase or decrease. The term *steady-state growth* is therefore etymologically inept as well as contradictory to common usage in physics and biology.

The upshot is that, for orthodox modern economists, the terms *stationary state* and *steady state* now refer to special cases of growth rather than to the contrary of growth. One is almost led to suspect a conspiracy among growth economists to find every term in the English language that is in any way descriptive of nongrowth and redefine it to refer to some special case of growth. Thus robbed of any words with which to express an idea contrary to growth, all economists would have to become growth economists by preemption of vocabulary! The one word the growth economists have reserved for expressing nongrowth is, of course, *stagnation*, with its strong connotations of foulness, dullness, and putrefaction.

Is the SSE a Counsel of Despair Based on a Willingness to Accept Present Evils and a Desire to Stop Economic Evolution?

Professor William Nordhaus of Yale University, a recent member of the President's Council of Economic Advisors, noted that "the political and social impediments to metering or internalizing the undesirable consequences of our activities are becoming unmanageably high." He concluded that "it should not be surprising, then, that many responsible analysts simply throw up their hands in despair and opt for a steady-state society, accepting the familiar ills of today in preference to the unforeseen consequences of continued economic evolution."[2]

Since these "responsible analysts" remain unnamed, no one can check a reference to test the possibility that Professor Nordhaus is setting up a straw man. As one who has identified himself with the advocacy of a SSE, I hope I may be forgiven the assumption that Professor Nordhaus had me in mind, among others. If so, I certainly appreciate being referred to as a "responsible analyst."

Be that as it may, however, I really am grateful to Professor Nordhaus for two things—first, for recognizing the existence of the problem to which the SSE is a response; and second, for making explicit three misunderstandings that result when critics make up rather than look up the position of those nameless advocates of the steady state whom they are criticizing. Let us consider each mistaken allegation in turn.

Does the SSE Offer a Counsel of Despair?

On the contrary, from its first full expression at the hands of John Stuart Mill, it has been a counsel of hope. As Mill put it,

> I cannot . . . regard the stationary state of capital and wealth with the unaffected aversion so generally manifested towards it by political economists of the old school. I am inclined to believe that it would be, on the whole, a very considerable improvement on our present condition.
>
> . . . It is scarcely necessary to remark that a stationary condition of capital and population implies no stationary state of human improvement. There would be as much scope as ever for all kinds of mental culture, and moral and social progress; as much room for improving the Art of Living and much more likelihood of its being improved, when minds cease to be engrossed by the art of getting on.[3]

The truly despairing position, in my opinion, is that we are prisoners of economic growth and that when "impediments to metering or internalizing the undesirable consequences of our activities become unmanageably high," we are not allowed to follow the simple expedient of reducing the level of our activities until these undesirable consequences become manageable again.

Does the SSE Accept the Familiar Ills of Today?

Not at all. The position is that further growth is not the answer to present poverty, injustice, and alienation in developed countries. Rather, we must face up to the necessity of imposing limits to inequality, limits to population growth, and limits to the throughput of matter-energy. Smaller-scale, more decentralized, less arcane and dangerous technologies are called for. A just, sustainable, and participatory economy requires a steady state rather than a growth mode of operation. The steady state is not an end in itself, but a means, a constraint imposed by the ends of justice, sustainability, and participation, as well as by the approximate steady-state nature of the total ecosystem of which the economy is a part.

The familiar evil of mass poverty in the Third World is by no means accepted. The SSE seeks to maintain levels "sufficient for a good life and sustainable for a long future." Rich countries have reached levels more than sufficient for a good life, though not on a sustainable basis. Poor countries have usually not reached a sufficient per capita level, but neither have they moved out quite so far on the limb of unsustainability. Present poverty makes further economic growth necessary in poor countries. But further population growth in poor countries, and further demographic and economic growth in rich countries, will make the attainment of a sufficient level in poor countries impossible. Furthermore, if the poor countries adopt the same pattern of large-scale capital- and resource-intensive growth characteristic of the rich countries, then they too will climb further out on the limb of unsustainability, adding the weight of their resource demands to the already excessive demands of the rich countries. Growth is not the cure for the familiar ills of today; in fact, it is the cause of many of them.

Does the SSE Imply a Halt to Economic Evolution?

The SSE imposes the condition that economic evolution (like the evolution of the planet earth) work itself out under the constraint of limited stocks and flows of matter and energy. Has the constant mass of the earth and the constant flux of solar energy halted terrestrial evolution? Certainly not. This limitation comes closer to being a precondition for evolution than an obstruction. Neither will the cessation of quantitative growth in the physical dimensions of the economy halt the qualitative evolution of wants, technologies, and institutions. Rather, such limits will channel economic evolution away from its current self-destructive path. Not all the consequences of continued economic evolution along the physical growth path are "unforeseen." Some consequences are quite foreseeable, and it is these discernible risks of catastrophe, not the unforeseeable novelty inherent in evolution, that the SSE seeks to avoid.

Misunderstandings of the "Impossibility Theorem"

The "impossibility theorem" simply states that a US-style high-resource-consumption standard for a world of 4 billion people is impossible. Even if by some miracle it could be attained, it would be very short-lived. Even less is it possible to support an ever-growing standard of per capita consumption for an ever-growing population. Crises of depletion, pollution, and ecological breakdown would be the immediate consequences of generalizing US resource consumption standards to the whole world. Development plans that take this generalization of the US economy as their explicit or implicit goal, as most development plans do, are simply unrealistic.

I am aware of two attempts by economists to refute this "impossibility theorem." One, by Professor Lester Thurow, is an apparent refutation on logical grounds, but in fact it is a misunderstanding. The other, by Professor Wassily Leontief and his colleagues, is an empirical refutation or, rather, can easily be misinterpreted as such. Let us consider each in turn.

Thurow comments:

> In the context of zero economic growth and other countries, a fallacious "impossibility" argument is often made to demonstrate the need for zero economic growth. The argument starts with a question. How many tons of

this or that non-renewable natural resource would the world need if everyone in the world now had the consumption standards enjoyed by those in the U.S.? The answer is designed to be a mind-boggling number in comparison with the current supplies of such resources. *The problem with both the question and the answer is that it assumes that the rest of the world is going to achieve the consumption standards of the average American without at the same time achieving the productivity standards of the average American* [emphasis added]. This is, of course, algebraically impossible. The world can only consume what it can produce. When the rest of the world has consumption standards equal to those of the U.S., it will be producing at the same rate and providing as much of an increment to the world-wide supplies of goods and services as it does to the demands for goods and services.[4]

No one denies that consumption must be matched by production. But production of goods and services requires natural resources, and more production of goods and services does not provide more natural resources—it uses them up. It is the US economy as a whole (resource productivity as well as resource use rates; production as well as consumption) that is generalized in the impossibility argument. The fact that consumption must be matched by production is an obvious truism having nothing to do with the adequacy of resources to the task of scaling up the US economy to the whole world.

In fairness to Professor Thurow one should recognize that perhaps someone, somewhere, did actually commit the fallacy of generalizing US consumption without generalizing US production levels. Unfortunately, Professor Thurow gives no reference to any specific case of this unlikely fallacy, which he says is "often made." And even if someone did commit that fallacy, it hardly invalidates the correct version of the impossibility argument, which would seem to have a greater claim on the attention of scholars than does a fallacious substitute.

The 1977 United Nations study *The Future of the World Economy* (by Wassily Leontief and others) seems, on superficial reading, to offer an empirically based refutation of the impossibility argument.[5] Although the tone of the report is reassuring and optimistic, what it in fact says is more a confirmation than a refutation of the impossibility argument. The report says that if population grows only at the UN low-rate projection—and if the developing countries' GNP can grow at 6.9 percent, while the developed countries lower their growth from the historic 4.5 percent to 3.6 percent—then, if we look ahead no further than the year 2000 (a mere twenty years, less than one generation), we can conclude that the average per capita income gap between rich and poor countries will be reduced from 12 to 1 down to 7 to 1, and that world resources will likely be adequate to the scenario.

What happens in the year 2025?[6] What happens if we reduce the gap from 12 to 1 all the way to 1 to 1? Are resources adequate to that task? The report does not ask questions of this sort. The impossibility theorem says, in effect, "this old car cannot make it from New York to San Francisco." The UN report says, "this resilient previously owned automobile can very likely get you all the way to Cleveland." The report notes (p. 5):

> In spite of the new more rational and economic ways of using mineral resources the world is expected to consume during the last 30 years of the twentieth century from 3 to 4 times the total volume of minerals that has been consumed throughout the whole previous history of civilization. Are the finite reserves of minerals in the earth's crust adequate to sustain this demand?

The report answers its own question in the affirmative for the next twenty years, but is silent about the future beyond that. History is somehow discontinuous, beginning anew on the year 2001. Even the adequacy of resources to the year 2000 is a matter of dispute, but it will not be pursued here.[7]

There is an exceedingly interesting asymmetry in the report's treatment of the growth rates of developed countries (DCs) and underdeveloped countries (UDCs). The UDCs' high growth rate is a policy goal to be brought about by planned action which is assumed to be successful. The lower growth rate of the DCs represents not a planned, conscious lowering of the GNP growth rate, but rather a failure to maintain past rates of growth. This is a curious mixture of planned conscious achievement and unplanned failure. The report does not advocate slower growth in DCs in order to allow UDCs to catch up. Rather, it just predicts that DC growth rates will, in fact, fall in spite of efforts to keep them at historic levels or even to raise them. No one is asked to change goals. Growth remains the goal in the DCs, and any reduction in the gap simply results from the fortuitous failure of DCs to grow as fast as UDCs.

It would seem that a careful justification of this crucial assumption is in order. All the report tells us is that "it was felt that an assumption of gradually declining growth rates in the developed countries would be more realistic than would a simple extrapolation of their past performance" (p. 3).

There may be good reasons for this assumption, and I do not want to make too much of the fact that the report does not mention any. But it is at least conceivable that the DCs could be successful in maintaining or increasing their historic rates of growth. They are certainly trying to. Would the success of the DCs in achieving higher growth make life easier or more difficult for the UDCs? Some economists argue that the best thing the DCs can do for the UDCs is to grow more rapidly. Others argue that rapid growth in the DCs preempts resources and makes growth more difficult for the UDCs, and of course also widens the gap. The UN report is silent on this critical issue—an issue that would become more obtrusive if DCs were assumed to be able to grow at historic rates.

A Corollary to the Impossibility Theorem

The demographic transition thesis has led many to advocate concentrating on economic growth in poor countries and trusting that population growth will take care of itself. "Development is the best contraceptive" was the slogan often quoted at the UN Population Conference in Bucharest. There are many objections to be raised against such a policy. It relies too mechanistically on an inverse correlation between development and fertility. There is probably an equally strong inverse correlation between development and illiteracy, yet no one invents a "literacy transition thesis" and argues that "development is the best teacher of reading and writing" and that when the stage of development reaches the point at which it is in people's interest to control their illiteracy they will do so, and not before. Of course, literacy programs and population control programs are not sufficient conditions for development, but they help. The logical consequence of "development is the best contraceptive, teacher, political reformer, public health program, and so on" would be somehow to work for "development" directly without investing in any of the correlates of development, which would all presumably come about automatically. But what is "development" shorn of all its correlates? What is left to invest in?

But the impossibility theorem puts the demographic transition policy in an even worse position. The whole idea of such a policy is to buy a reduction in the rate of increase of human bodies by an increase in the rate of production of artifacts—in other words, to trade off an increase in consumption per capita in exchange for a reduction in "capitas." The total load on the ecosystem is the product of the number of people times per capita consumption. The demographic transition policy simply lowers one factor by increasing the other, with no guarantee that the product of the two factors will decrease. In fact, it is likely even to increase if we believe the oft-cited figure that one American child equals 50 Indian children in terms of lifetime demands on the ecosystem. Is the demographic transition's tradeoff between production and reproduction 50 to 1? Will Indian per capita consumption have to reach US levels before Indian reproduction falls to US levels? Will that require an increase in Indian per capita consumption by a factor of 50? If not by 50, then by how much? Are world resources adequate to pay such a high price for population stability via the demographic transition policy? Certainly not, if the impossibility theorem is correct, and no proponent of the demographic transition policy has bothered to refute the impossibility theorem. The warranted conclusion is that direct population control measures cannot be avoided by appeals to economic growth and the "demographic transition." Indeed, if rich countries must lower their per capita consumption in the future, would we expect the demographic transition mechanistically to work in reverse, causing a rise in the birth rate? Probably not, but since all truly mechanical models work in reverse, maybe the demographic transition theorists would expect falling consumption to be accompanied by a rising birth rate. If so, all the more need for a direct policy of population control.

A Steady-State Economy versus a Failed Growth Economy

A situation of nongrowth can come about in two ways: as the failure of a growth economy to grow, or as the success of conscious policies aiming at a SSE. No one denies that when a growth economy fails to grow, the result is unemployment and suffering. The main reason for advocating a SSE is precisely to avoid the suffering of a failed growth economy, because we know that, sooner or later, the growth economy will not be able to continue growing—or rather that the marginal social cost of growth will be greater than the marginal social benefits, so that growth will cost more than it is worth, even if it is physically possible.

This simple distinction is often ignored, as when the editors of *Fortune* wrote, "the country has just gone through a real life tryout of zero growth [the period 1973–1975, which is remembered not as an episode of zero growth but as the worst recession since the 1930s]."[8] The distinction between a SSE and a failed growth economy is erased by referring to both cases as "zero growth." The failures of a growth economy really should not be used as arguments against a SSE! The fact that airplanes fall to the ground if they try to remain stationary in the air merely reflects the fact that airplanes are designed for forward motion. It does not constitute a proof that helicopters cannot remain stationary. A growth economy and a SSE are as different as an airplane and a helicopter. Needless to add, one would not attempt to attain a SSE by simply reversing current policies aimed at growth. Economies, like airplanes, do not fly in reverse, or neutral either.

Price-Determined versus Price-Determining Decisions

There is a widespread belief among economists that market prices determine everything. It was recently claimed by three economists that

> Only if we eliminate the market incentives for innovation and investment, or reduce the scope of market forces through further attenuation of private property in resources, must we face a real long-term "resource crisis." The only nonrenewable and nonsubstitutable resource is the set of institutions known as a market order, which eliminates crises with respect to physical resources.[9]

In other words, as long as we base all our decisions on free market prices, resource constraints disappear as a long-run concern. We do not, in this view, need any ecologically or ethically determined limits on the total flow of resources.

Prices are important but are not all-powerful. Market prices are relevant only to temporally and ecologically parochial decisions, whose major consequences lie wholly within the human economy of commodity exchange and within the present generation. Market prices are excellent means for efficiently allocating a given resource flow from nature among alternative uses in the service of a given population of already existing people with a given distribution of wealth and income. Market prices should not be allowed to decide the rates of flow of matter-energy across the economy-ecosystem boundary or to decide the distribution of resources among different people (or among different generations, which, of course, are different people). The first must be an ecological decision, the second an ethical decision. These decisions of course will and should influence market prices, but the whole point is that these ecological and ethical decisions are *price-determining*, not *price-determined*. Many economists simply fail to grasp this point.

Economic Growth Required to Maintain Full Employment?

It is clear from the previous discussion that a SSE cannot be identified with a failed growth economy, and thus it does not imply unemployment. In fact, there are several reasons for believing that full employment will be easier to attain in a SSE than in our failing growth economies. One condition of a SSE is limits to inequality in the form of a minimum and a maximum income. The minimum income would substitute for the unemployment-causing minimum wage in providing a guaranteed subsistence. The maximum limit on income and wealth, and the more equal distribution generally, would reduce the aggregate saving rate, which would bolster aggregate demand and employment. Furthermore, the policy of limiting the matter-energy throughput would raise the price of energy and resources relative to the price of labor. This would lead to the substitution of labor for energy in production processes and consumption patterns, thus reversing the historical trend of replacing labor with machines and inanimate energy, whose relative prices have been declining. Another policy of the SSE, zero population growth, would also ease unemployment by lowering the number of job seekers, though only after a lag of about twenty years.

I do not claim that these considerations provide a sufficient answer to the question of how to provide full employment in a SSE. But I think they do demonstrate that this question is easier to answer than the analogous one faced by a growth economy—namely, how can full employment be maintained in an economy that becomes ever more capital- and energy-intensive in its technology while at the same time facing ever greater scarcity of the nonrenewable resources upon which its technology is based?

Occasionally the employment argument for growth becomes truly absurd, as in the case of the Concorde airplane. We are told that 40,000 British workers' jobs depend on the success of Concorde, and whoever opposes that technically overdeveloped white elephant must be a hard-hearted elitist with no feeling for working people. A moment's reflection will show that if the billions squandered on Concorde were spent on any of hundreds of less capital-intensive projects, employment would be *increased*. Also, working people would be much more likely to benefit from the services of their own product. With Concorde fares 20 percent above first-class fares, not many working people will be riding Concorde. They will remain on the ground to have their ears assaulted by the flatulent sonic booms of their jet-set betters, to have their skins absorb the extra cancer-inducing ultraviolet radiation resulting from ozone depletion, and to have their stomachs develop ulcers from worrying about how long their livelihoods can possibly derive from such an absurd product.

Is the SSE Capitalistic or Socialistic?

The one thing that capitalism and socialism have agreed upon is the importance of economic growth. Both have accepted the criterion that whichever system can grow faster must be better, and each strives to win the growth race. The notion of a SSE is rejected by both. Both systems suffer from growthmania, and the SSE presents as much a challenge to Big Socialism as to Big Capitalism. But it also offers a point of reconciliation for the future. When old adversaries discover that they have made the same error, their brotherhood in humility should facilitate reconciliation.

The growth versus steady state debate really cuts across the old left-right rift, and we should resist any attempt to identify either growth or steady state with either left or right, for two reasons. First, it will impose a logical distortion on the issue. Second, it will obscure the emergence of a third way, which might form a future synthesis of socialism and capitalism into a SSE and eventually into a fully just and sustainable society. Neither capitalism nor socialism can make much of a claim to being just, sustainable, and participatory. Let us not insist on pouring new wine into old wineskins.

The difficulty of deciding whether the SSE is capitalistic can be seen from the following. Since the SSE would rely mainly on private property and decentralized market decision making, we might consider it ipso facto capitalist. But according to Karl Marx's definition of capitalism, the SSE would not be capitalistic, because with maximum and minimum income and wealth limits there would be no monopoly class ownership of the means of production and no correlative class of proletarians who must sell their labor power to the capitalist on his terms in order to survive. Nor with maximum wealth limits would there be the unrestrained drive to accumulate, which Marx said was "Moses and the Prophets" for the industrial capitalist and eventually would lead to collapse of the system. Whether the SSE is capitalistic or not depends on how one defines capitalism. I suggest that it is more profitable to work out the concept of steady state, of a just and sustainable society, as a third way—neither capitalistic nor socialistic but a way to which traditional capitalists and socialists are invited to contribute and within which they might find an embracing synthesis.

Space Colonization versus the Steady State

One of the more interesting arguments against a steady state takes the form of the following syllogism:[10]

Premise I: Continuing growth is necessary for freedom. Only an iron-fisted worldwide dictatorship could control growth.

Premise II: Continuing growth cannot occur on earth because earth really is finite.

Conclusion: Space colonization is necessary to preserve freedom.

There is agreement between space colonizers and steady-state advocates regarding Premise II. The disagreement centers on Premise I. What evidence is there for such a statement? Have dictatorships come into being because nations were striving to become steady states or because they were dedicated to growth? Would a substitute for Premise I to the effect that only a dictatorship could provide the compulsion and discipline necessary to make space colonization work be any less plausible?

But there is yet a more basic problem with the argument. Even if the total number of space colonies can grow indefinitely, each individual colony must be managed in a steady-state mode. A single space colony cannot tolerate a population explosion either of human bodies or of artifacts. The aggregate of all individual habitats may grow and grow, but *each* habitat must be managed as a steady state with births plus immigration equal to deaths plus emigration, with an analogous materials balance for artifacts. The very discipline that is alleged to imply dictatorship on earth is encountered again on each and every space habitat. If a steady state requires a dictatorship on spaceship earth, then why not also on spaceship L-5X351? Inasmuch as each space colony must be run as a steady state, does it not make sense, even if one favors space colonization, to learn to live in the steady-state mode on the large, resilient, forgiving, and relatively self-operating spaceship earth before attempting to manage an entire artificial ecosystem within a fragile, rotating torus protected from the cold vacuum of space by a few inches of aluminum and glass? I doubt that such an environment, vulnerable to sabotage in thousands of ways, would be able to tolerate much dissent. In all probability a military style of discipline and security, referred to no doubt as "rationality," would be necessary, and civil liberties would be deemed too dangerous to tolerate.

The space colonizers consider freedom to migrate as one of the most basic freedoms to be protected. But Garrett Hardin has raised a pertinent question. Suppose we have a new space habitat waiting for 50,000 inhabitants. Who gets to—or has to—go? Either way, people will have to be chosen. How? Perhaps by HEW guidelines requiring a fair share of blacks, whites, Chicanos, Indians, Catholics, Jews—not to mention Cajuns, Creoles, Mormons, and Unitarians? Or if that seems a recipe for tribal warfare, should we strive for "ethnic purity" and religious homogeneity on each habitat? What then happens to religious freedom and ethnic pluralism?

The alleged impossibility of a steady state on earth provides a poor intellectual launching pad for space colonies. A better (but still less than compelling) argument for space colonization might be made along the lines suggested by Kenneth Boulding ("Spaceship Earth Revisited") as a way to reintroduce isolation and variety into the evolutionary process. But even if space colonization were desirable, its feasibility presupposes the ability to live under steady-state rules of management.

Does the SSE Imply Ecological Salvation and Eternal Life for the Species?[11]

The SSE cannot last forever because of the entropy law. In the very strict and inclusive sense, a steady-state economy is impossible. Indeed, any steady-state process is impossible. At some point in the past it had to have a beginning, and at some point in the future it will have to have an end. A steady-state economy cannot last forever, but neither can a growing economy, nor a declining economy.

Consider a candle. The flame is lit and grows to mature equilibrium size. It then burns in a steady state until the candle burns down; finally it flickers and dies. We must recognize that at some time in the past the candle had to be lit, and that it must go out sometime in the future. Therefore, if we draw temporal boundaries around the process so as to include the beginning and the end, we cannot call the process a steady state. But if we draw temporal boundaries after lighting and before going out, we can describe the greater part of the flame's life as a steady state, without implying that it will last forever or that it had no beginning.

The stocks of artifacts and people in the economy can exist in a quasi-steady state for as long as the resource "candle" lasts. We can turn our resource candle into a Roman candle and burn it rapidly and extravagantly, or we can seek to maintain a steady flame for a long time, or we can put out the flame before the candle has burned down. The steady-state view advocates the middle course. That any choice among the three alternatives represents a value judgment is quite evident.

The candle analogy, though useful, fails in two respects. First, we have two candles—the sun, and the stock of terrestrial minerals. The sun cannot be burned up like a Roman candle. It burns with a steady flame as far as human time scales are concerned. But the terrestrial resource candle can be, and is being, used as a Roman candle. The second failing of the analogy is that, unlike the combustible material fed from the candle to the flame, the terrestrial resources fed to our economy do not remain constant in quality or accessibility as they are depleted. We first exploit the best and most accessible resources known to us. As depletion forces us to exploit progressively poorer-grade resources, the gross throughput of matter and energy will have to increase in order to yield the same net throughput of the minerals required to maintain stocks constant. Also, a larger fraction of the constant capital stock would have to be devoted to ever more capital-intensive means of winning mineral resources. At some point the extra minerals are not worth the extra sacrifice required to win them, and the remaining minerals cease to be "resources." The SSE seeks to guide the economy toward maximum feasible reliance on solar energy and renewable resources and away from the current unsustainable practice of living largely on accumulated geological capital.

In sum, the SSE does not proclaim "ecological salvation and eternal life for the species." Far from promising any future escape from the status of creaturehood and mortality, the SSE is based on the assumption that creation will have an end—that it is finite temporally as well as spatially. But creation is affirmed as good while it lasts, and its longevity is not a matter of indifference. Just as individuals strive to keep their bodies in a healthy steady state even while knowing that time and entropy are slowly but continuously defeating them, so we must strive to keep our collective exosomatic body, the economy, in a quasi-steady state as a strategy of good stewardship. Only God can raise any part of his creation out of time and into eternity. As mere stewards of creation, all we can do is to avoid wasting the limited capacity of creation to support present and future life. Indeed, if we thought that that capacity were infinite, we would not be concerned about wasting some of it. Conversely, those who are not concerned about the waste of life support capacity are often those who believe that human beings can achieve an infinite life span for their species, thanks to the Faustian powers of technology. While recognizing the important, but limited, role that technology can play in increasing life support capacity, and in "relieving man's estate," the SSE perspective rejects as idolatrous the widespread belief that technology can in any fundamental way raise humanity from the status of creature to that of creator.

It is curious that the traditional Christian view takes it for granted that creation will have an end and places its faith in immortality at the personal level and in the realm of the spirit. The modern world, having lost its faith in personal immortality and in the spirit as well, seeks to substitute a belief in species immortality which is in thorough contradiction to the most basic tenets of physical science, on which its idolized technology depends. A believer in personal immortality can at least claim to have some direct inner experience of his own immortal spiritual nature. But no one can claim to have any direct intimations of species immortality. And any reasoned argument for the possibility of species immortality must fly in the teeth of the entropy law and therefore forfeit any claim to scientific standing.

Far from having any truck with "eternal life for the species," the SSE is simply a strategy of good stewardship for taking care of God's creation for however long he wills it to last. In taking care of that creation, special, but not exclusive, attention must be given to humanity, including not only the present but also future generations and in a sense past generations as well. Just as our bequests to the future are meant for many future generations, not just the next one, so our inheritance from the past was meant for many generations, not just ours. To treat it as exclusively ours is to break faith with the past as well as with the future. Nor should our concern be exclusively with human beings, past, present, and future. If a man is worth many sparrows, then a sparrow cannot be worthless. Stewardship requires an extension of brotherhood first to all presently existing people but also to future and past people, and to nonhuman life, all in some appropriate degrees.

Of course, these issues of the just and sustainable society far transcend the notion of a SSE, which is advocated not as a solution to the problems of justice and sustainability, but as a framework of economic life that at least allows these problems to be taken seriously.

Notes

1. "Scarcity and Growth: How Does It Look Today?" *American Journal of Agricultural Economics* (December 1975), 811.

2. "Metering Economic Growth," in Kenneth D. Wilson, ed., *Prospects of Growth* (New York: Praeger, 1977), p. 208.

3. J. S. Mill, *Principles of Political Economy,* vol. 2 (London: John W. Parker and Son, 1857), pp. 320–326.

4. Lester Thurow, "The Implications of Zero Economic Growth," in *U.S. Prospects for Growth: Prospects, Problems, and Patterns,* vol. 5, *The Steady-State Economy,* Joint Economic Committee of Congress, Washington, D.C.: US Government Printing Office, December 2, 1976, p. 46.

5. Wassily W. Leontief and others, *The Future of the World Economy* (New York: Oxford University Press, 1977).

6. It is worth remembering that the year 2025 is well within the expected lifetime of the average reader of this book.

7. One of the coauthors of this study, Professor Anne Carter, at a World Council of Churches Conference in Zurich (June 1978), vigorously disassociated herself from the conclusion that growth as usual is feasible, which was drawn from the report by many, including, it would appear, Leontief himself. In Professor Carter's view the model has very little of value to say on the subject of environmental and resource constraints on growth.

8. "Well, How Do You Like Zero Growth," *Fortune* (November 1976), 116.

9. G. Anders, W. Gramm, and S. Maurice, *Does Resource Conservation Pay?* International Institute for Economic Research, Original Paper 14, July 1978, p. 42.

10. This is essentially the position taken by Princeton physicist Gerard K. O'Neill in his opening statement at the Edison Electric Institute's Symposium on Prospects for Growth (Washington, D.C., June 1978), and later confirmed in private correspondence.

11. This section deals with some of the criticisms of the SSE raised by N. Georgescu-Roegen's "The Entropy Law and the Economic Problem" (reprinted in this volume).

Index